Global Communication Electric

»Global History«
Volume 15

Edited by Sebastian Conrad, Andreas Eckert and Ulrike Freitag

M. Michaela Hampf is Professor for North American History at the John-F. Kennedy Institute for North American Studies at Freie Universität Berlin. *Simone Müller-Pohl* is Assistant Professor for North American History at the University of Freiburg.

M. Michaela Hampf, Simone Müller-Pohl [eds.]

Global Communication Electric

Business, News and Politics in the World of Telegraphy

Campus Verlag
Frankfurt/New York

Distribution throughout the world except Germany, Austria and Switzerland by

The University of Chicago Press
1427 East 60th Street
Chicago, IL 60637

Bibliographic Information published by the Deutsche Nationalbibliothek.
The Deutsche Nationalbibliothek lists this publication in the Deutsche Nationalbibliografie; detailed bibliographic data are available in the Internet at http://dnb.d-nb.de
ISBN 978-3-593-39953-9

Cover: »The Progress of the Century«, New York 1876 (Currier & Ives) © picture alliance / akg-images
Typesetting: Fotosatz L. Huhn

Printed on acid free paper.
Printed in the United States of America

This book is also available as an E-Book.

For further information:
www.campus.de
www.press.uchicago.edu

Content

Global Communication Electric: Business, News and Politics in the World of Telegraphy

M. Michaela Hampf and Simone Müller-Pohl

On April 16, 1879 Emma Pender, wife of cable magnate and head of the Eastern and Associated Companies, John Pender, was having tea with her lady friends at her London house, when she received a package from her daughter in Fiji by the Melbourne steamer. Her daughter's letter accompanying the package was dated January 15, a good three months before the package interrupted Emma Pender's tea engagement. Her daughter asked Pender to send her a telegram to indicate how long the package had taken to arrive. While Emma Pender swiftly set her servant to this task, the package stirred a tea time discussion among the ladies present who all declared "the size of [their] world cruel." Pender later recalled in a letter to her daughter that it fell to the hostess to remind her excited guests of "older days" without the telegraph and "no regular mail at all."[1] For Victorian upper class women, telegraphy and the global communications system swiftly became integral parts of their everyday lives. Indeed, Pender's interaction with her daughter was embedded in a world-wide system of regular mail and parcel service, railway and steamship transportation and messenger boys as well as telegraphy. The size of the world might have seemed cruel to ladies enjoying high tea in Victorian London, but the global communciations system had enabled them to idealize an electric world in union in the first place.

Telegraphs are an emblem of modernity as well as catalysts of our present global condition. The establishment of an extensive and world-spanning network of landline and submarine cable connections in the mid-nineteenth century fostered the emergence of structures and patterns of interaction on a global scale. World politics, a global economy and a global media system only became possible with the creation of *global communication electric*. Moreover, the telegraphs caused the most dramatic globalization effects among all new technologies of the nineteenth century, as telegraph lines were easier to lay

1 Emma Pender, "Letter to her daughter Marion," March 30, 1879, Emma Pender Papers, Cable and Wireless Archive.

than railway tracks and transmitted news with a higher speed and greater capacity than steamships.[2] Significant changes in long-distance communication occurred particularly after the successful completion of the transatlantic telegraph cable in 1866 and the subsequent expansion of land and submarine lines to cross the Euro-American continents and establish connections from Europe and North America to India, Australia, Japan and Brazil in the 1870s. Message and messenger became more independent of each other as the former could now travel faster than the latter. Communication over large distances became telecommunication – the focus of modern media culture changed from a movement of goods to a transmission of information.[3] This had enormous effects: news not only travelled ever faster, it also became ever more current – albeit not necessarily more newsworthy. Despite exorbitant transmission rates, increasing numbers of people used the telegraphs for political, economic and, to a lesser extent, private communication.[4] Starting out with a few hundred telegrams per year passing through this global network, numbers rose to six million telegrams a year by the turn of the twentieth century.[5] By 1900, much of the world had 'logged on' to global communication *electric.*

"The universe is a procession, with measured and beautiful motion," wrote Walt Whitman in "I Sing the Body Electric," an 1855 poem that explored how the human body mediates between the spiritual and the material world. Whitman's body electric combines poetic discoveries and aesthetic manifestos with the physical body, portraying the soul as inseparable from its corporeal manifestation.[6] So too contemporaries saw the world as a body with nerves of submarine and telegraph cables that spanned its circumference. Like Whitman's body electric, telegraph cables seemed to offer a new means to understand the soul of the world and to offer a connection that promised universal peace and harmony akin to a body in perfect working

2 Osterhammel and Petersson, *Geschichte der Globalisierung*, p. 54.

3 Hartmann, *Globale Medienkultur*, pp. 9–10; Similarly Wenzlhuemer, "The Dematerialization of Telecommunication: Communication Centres and Peripheries in Europe and the World, 1850–1920," or Ursula Lehmkuhl, "Producing and Consuming Knowledge: Comparative Perspectives on the Development and Usage of the Telegraph in Nineteenth-Century Germany and America," p. 171.

4 On the use of Weltcommunication see Müller-Pohl, 'By Atlantic Telegraph': A Study on Weltcommunication in the 19th Century.

5 Bright, *Submarine Telegraphs*, p. 167.

6 Whitman and Cowley. *Walt Whitman's Leaves of Grass*; Blum, *How to write about Walt Whitman*, p. 103.

order. Contemporaries heralded telegraphy as a tool of *Weltcommunication*, to use a term coined by the German philosopher of technology Ernst Kapp, which practically "annihilate[d] any terrestrial dimension," meaning distance and time.[7] Nowadays, scholars often interpret telegraphy as the chief globalizer of communication technologies in the mid-nineteenth century.[8] As such it represented the necessary precondition for modern interaction across space and time and ultimately processes of globalization. Some studies, such as Tom Standage's 1998 monograph, *The Victorian Internet*, even argue for the commensurability of the nineteenth-century global telegraph network and today's use of the internet.[9] A large majority of the existing literature on global communication and telegraphy further accentuates the telegraphs' importance for imperial control and Euro-American nationalist power politics. Scholars portray telegraphs as "tools of empire" that aided the formation and consolidation of nation states and empires or helped to generate narratives of national technological progress and development.[10]

This book critically reconsiders these grand narratives of the annihilation of time and space, the Victorian internet, imperial control and nationalist power politics. The contributions emphasize the importance of other aspects of telegraphy, such as transboundary processes of scientific and business exchanges, intergovernmental modes of governance and alternative notions of identity formation beyond and outside of the (primarily Euro-American) nation state. Neither news nor communication stopped at national boarders and in particular the ocean cable companies laid cables within the openness of maritime regions rather than within the restrictive boundaries of imperial territories. The book hence aims at exploring how global business, news and politics worked not only by means of, but also in the world of telegraphy.

The methodology of global history provides a new avenue to explore telegraphy as a historical force of globalization. Informed by postcolonial theory and subaltern studies, global history expands beyond histories of globalization. Global history challenges Eurocentric narratives of modern-

7 Kapp, *Grundlinien einer Philosophie der Technik.*

8 Squier, The Influence of Submarine Cables upon Military and Naval Supremacy, p. 600.

9 Standage, *The Victorian Internet.*

10 Headrick, *The Tools of Empire*; similarly Headrick, *The Tentacles of Progress* or Headrick, *The Invisible Weapon*; The following examples from India and Brazil also illustrate the potency of treatises on technological progress for national history projects Berthold, *History of the Telephone and Telegraph in Brazil, 1851–1921*; Shridharani, *Story of the Indian Telegraphs*; Kunert, *Telegraphen-Seekabel*; Haigh and Wilshaw, *Cableships and Submarine Cables*; Hennig, Die historische Entwicklung der deutschen Seekabelunternehmungen.

ization or Westernization alongside the methodological nationalism that thus far allowed scholars to assign cable connections a distinct nationality or imperial agenda.[11] Global historians of telecommunication, meanwhile, use concepts of agency as well as reconceptualizations of space to approach telegraphy from a perspective that is not bound to the nation state.[12] For example, cable companies, contractors and agents often propagated a cable cosmopolitanism that challenged and expanded state boundaries and narratives of national modernization.[13] These actors' spaces of action aligned with the maritime space of their submarine cables, their frame of identity with their profession and its international scientific networks and their working schemes with the logic of emerging global capitalism far more than nationalist interpretations of their history have shown. The global history approach also reminds us to analyze networks of connections against the backdrop of their disconnections; to see users in relation to non-users.[14] Finally, the global communication network also supported processes of othering by reconfiguring mental maps of the globe.[15] As Eric Hobsbawm points out, while global news makers at the time depended upon a 'shrinkage of the globe' through the instantaneity of news coverage, their reports, such as the 'discovery' of David Livingstone, created the notion of the 'dark continent' or 'far-away' places that lay outside of the Euro-American system.[16]

Global communication electric looks at the emergence of a global media system between 1860 and 1930 from the global history perspective in order to broaden and challenge popular conceptions of telegraphy as *Tools of Empire*, the *Victorian Internet* or as a means to 'annihilate space and time.' The book contains four sections, *Inter|Nationalisms*, *Agents|Actors*, *Use|News* and *Space|Time*, each of which highlights one particular aspect of *global com-*

11 On the approach of global history see: Geyer and Bright, World History in a Global Age; Osterhammel and Petersson, *Globalization: A Short History*; Sachsenmaier, *Global Perspectives on Global History*; Conrad, (ed.), *Globalgeschichte*, for a practical example Bayly, *The Birth of the Modern World*; Osterhammel, *Die Verwandlung der Welt*; On methodological nationalism Wimmer and Glick-Schiller, Methodological Nationalism and Beyond.

12 See for example Jorma Ahvenainen's focus on regions: Ahvenainen, *The Far Eastern Telegraphs*, Ahvenainen, *The European Cable Companies in South America: Before the First World War*.

13 On cable cosmopolitanism see Müller-Pohl, Working the Nation State.

14 In this context see Latour's distinction between the 'literatue' and the 'illiterate.' Latour, Die Logistik der Immutable Mobiles.

15 See the concept of mental maps in Christoph Conrad, Vorbemerkung.

16 Hobsbawm, *The Age of Capital*, p. 60.

munication electric.[17] The authors highlight collaborative modes of operating the system as well as cosmopolitan, internationalist and even socialist ideas of telegraphic cooperation. The volume's contributors widen the scope of people involved with the telegraph from the conventional groups of national and international administrations, telegraph companies and telegraph users to integrate the telegraphers, journalists or visitors to world fairs. The chapters redefine telegraphic space and concomitant notions of connectivity and entanglement by delineating different and sometimes competing telegraphic geographies and questioning established notions of instantaneity. In its thirteen chapters, *global communication electric* explores the varied uses of telegraphy, real or imagined, and so narrates the history of global communication from multiple perspectives.[18] Returning to the notion of the telegraph as globalizer, our analyses expand this picture to construct the image of a medium of connection as well as friction, of political, social and economic entanglement as well as disentanglement and of crossing as well as creating distance in space and time.

Inter|Nationalisms

Telegraphy emerged shortly before the zenith of European nationalism and undoubtedly played an important role in the creation and administration of national territoriality. At the same time, however, like the wireless, the telephone, or the internet later, telegraphy held the promise of overcoming national borders and boundaries and contributing to a more peaceful world order. From the 1850s to the 1870s, after almost every new cable landing, contemporaries expressed their wishes that the connection would bring 'peace and goodwill' to the 'civilized' nations.[19] The almost contradictory desires to create exclusive national territories but also world peace provoke questions about the role of the national in the global media system, the emergence of international cooperation and standardization and the relationship between regional and global media markets and audiences. This volume shows that

17 Hills, *The Struggle for Control of Global Communication.*

18 Edgerton, From Innovation to Use.

19 See for example Queen Victoria's welcome telegram of 1858. Queen Victoria to President James Buchanan, telegram, August 18, 1858, cited in Telegraph Construction and Maintenance Company, *The Telcon Story 1850–1950* (London, 1950), p. 49.

the national and the global were not mutually exclusive categories, but part and parcel of a complex story of simultaneous and interlinked processes of globalization and nationalization.

A multitude of studies deals with the interrelation of telecommunication systems and globalization processes, such as the development of global markets, global politics or a global public sphere/ *Weltöffentlichkeit*.[20] For a long time, the telegraphs' 'nationality' played an important role in historiography: as an assumed *a priori*, nationality and imperial competition provided the analytical framework. Interpretations rested upon the assumption that national and imperial interests correlated with corporate interests and that this strong interrelation would help imperial powers, in particular the British Empire, to control their far-flung territory. This scholarship tied in with the idea that imperialism was something made in the European capitals and then implemented 'out there.'[21] Telegraph networks thus represented essential foundations for nations' economic and political benefits and military security.[22] Since the publication of Harold Innis' *Empire and Communications* in 1950, one of the dominant themes was the interrelation between communication and the 'rise and fall' of empires. Thereafter, the transnational nature of submarine telegraphs was predominantly characterized as the nations' *Struggle for Control of Global Communication*.[23] In particular historians of new imperialism have analyzed the field of telegraphy as one of the main sites for "great power rivalries." They argue that the principal motive for telegraphic expansion was to secure independent communication channels between mother country and colony.[24] Aside from controlling communication against other imperial rivals, scholarship portrays the technology's great importance within a narrative of European colonial conquests in Asia and

20 Rudolf Stichweh suggests that there is one *global public* as the internal system of the arising *Weltpolitik*. Jürgen Habermas on the other hand uses the term *Öffentlichkeit* to denote a distinct bourgeois society. We think that both approaches, Stichweh's idea of a world system as well as Habermas' concept of public sphere fall short. Taken into consideration the high degree of disconnections and fragments as well as the high degree of difference instead of homogeneity, it makes more sense to speak in the plural of public spheres. Stichweh, The Genesis and Development of a Global Public Sphere; Habermas, *The Structural Transformation of the Public Sphere*.

21 Rosenberg, Introduction, p. 18.

22 Headrick, Strategic and Military Aspects of Submarine Telegraph Cables.

23 Hills, *The Struggle for Control of Global Communication*; similarly Hugill, The Geopolitical Implications of Communication under the Seas.

24 Griset and Headrick, Submarine Telegraph Cables, p. 543.

Africa.[25] The presence of almost all major submarine telegraph companies, such as the Eastern and Associated Telegraph Companies, in addition to Reuter's news agency in London seemed to support an argument for Britain's hegemony over world communication. Both control over news transmission as well as news production, according to Robert Boyce, eventually secured "Britain's ascendency as a world power."[26]

While we do not want to refute the imperial thesis entirely, we wish to complement it and reconsider the role of empires and nation state within and for the global media system. New imperial history has shown that imperial systems displayed many racial, gendered, and economic forms and constituted multiple territorializing regimes.[27] Without any doubt, the imperial framework played an important role in structuring cable routes or garnering resources such as gutta percha for cable manufacture. Before the existence the Pacific cables in 1902 and 1904 respectively, the submarine cables never served as a means for exploring new markets or connections, but were instruments to accelerate existing economic, political and cultural connections. The entrepreneurs refused to lay cables where there was "not even a sandbank on which to catch fish" in the words of James Anderson, managing director of the Eastern and Associated Telegraph Companies.[28] Global communication thus often followed colonial pathways and benefited from a global imperial coloniality constructing, as Lars Bluma exemplifies in his chapter on the telegraphs and world fairs, a system of global communication that was deeply entrenched in the logic of imperial power relations as well as Eurocentric notions of civilization.[29]

In a similar fashion, communication devices furthered imperial and national strategies. Communication devices featured prominently in Japan's schemes for technology development and represented one core pillar of Meiji Japan's policy of economic transformation and industrialization. After the

25 Headrick, *The Tools of Empire*, p. 11.

26 Boyce, Submarine Cables as a Factor in Britain's Ascendancy as a World Power, 1850–1914. The oldest proponent of this imperial thesis is Kennedy, Imperial Cable Communications and Strategy, 1870–1914, p. 729.

27 Rosenberg, Introduction, p. 18; on new imperial history see Wilson, *A New Imperial History;* Cain and Harrison, *Imperialism: Critical Concepts in Historical Studies*; Hall, *Cultures of Empire.*

28 James Anderson 1887, cited in J. H. Heaton, "Penny-a-Word Telegrams throughout the Empire," *Proceedings of the Royal Colonial Institute* 40 (1908–1909), p. 4.

29 On global coloniality see Conrad and O'Hagan, *German Colonialism: A Short History* as well as Ballantyne and Burton, Empires and the Reach of the Global.

end of the Tokugawa Shogunate and Japan's isolationist policy (*sakoku*) in 1868, Japanese leaders quickly realized that in order to avoid colonialism, their country not only had to become militarily powerful, but also technologically sophisticated. Thus starting in the 1870s and "[w]aving the flag of techno-nationalism," the Japanese government invested heavily in sectors deemed essential for economic development, such as mining, railroads, electric power and communication.[30] Yet ironically, techno-nationalism depended upon transnational cooperation. As Daqing Yang explores in his chapter, until 1940, the Danish-based Great Northern Telegraph Company featured prominently within Japan's imperial logic through the build-up of telegraphic connections as the company's economic interests went hand in hand with those of the Japanese government.[31]

Despite this obvious interdependence of capitalist and imperialist interests, the cables and cable companies' implied nationality in previous studies is problematic in several ways. As Dwayne Winseck and Léonard Laborie illustrate in their chapters, methodological nationalism disguises not only the importance of global capitalism, but also the large degree of inter-imperial and international collaboration and cooperation. Territorial and capitalist imperialism formed two distinct aspects of the relationship between communciations and empire between 1860 and 1930.[32] The spaces of capital and communication were not necessarily coterminous with the space of empire. Moreover, these communication spaces were governed internationally and based upon principles of cooperation: the International Telegraph Union was established in 1865 to regulate trans-national landline traffic.[33] As Laborie shows, debates and negotiations on the international never erased the importance of national company monopolies or national borders, but rather provided new fora to negotiate these relationships.[34]

Finally, the medium of telegraphy as actual and imagined technology-in-use always also allowed for alternative constructions of community, such as an imagined global communist community in Martin Doll's chapter

30 Anchordoguy, Nippon Telegraph and Telephone Company (NTT), pp. 508–12.

31 Similarly also in his other works: Yang, *Technology of Empire* as well as Yang, Submarine Cables and the Two Japanese Empires.

32 Harvey, *The New Imperialism*, p. 26.

33 Laborie, *L'Europe Mise en Réseaux;* Ahvenainen, "The International Telegraph Union: The Cable Companies and the Governments"; International Telecommunications Union, (ed.), *International Telecommunications Union: Celebrating 130 Years, 1865–1965.*

34 On the transnational nation further see Conrad and Osterhammel, *Das Kaiserreich transnational.*

or the emergence of Indian nationalism as discussed in the chapters of Michael Mann and Amelia Bonea. In India, the telegraphs were essential for not only British colonial control over the subcontinent, but also fostering an Indian newspaper culture at the end of the nineteenth century, which became increasingly nationalistic. At the beginning of the twentieth century, this all-Indian public sphere heavily attacked the British colonial regime as well as the collaborative politics of the Indian National Congress. Turning this protest into a mass phenomenon, Mohandas K. Gandhi drew heavily upon the telegraphs and newspapers. Technology was used to uphold as well as to counter imperial space and thus enabled the empire to 'talk back.'[35]

Agents|Actors

While telegraphy provides the central focus for this book, the chapters examine telegraphy as part of a broader global media system. Emma Pender and her daughter used multiple media to communicate; indeed, the global media system emerged from an ensemble of technologies, including telephony, the typewriter, print and photography in addition to railway, steamship and automobile transport systems. This assemblage of technologies acted in concerto and created the world-spanning network which fostered processes of integration and fragmentation. While the chapters mainly focus on the telegraphs as prime means of *Weltcommunication*, they portray telegraphs as the central puzzle piece within a larger media array of global communication.

Furthermore, different people put telegraphy to different uses in different times and places. Contemporaries noted during their debates on submarine cables and world peace in the 1850s that the telegraphs were merely a technical and mechanical extension of human interaction. They remained "a musical instrument, on which operators may play any tune they choose."[36] This book thus moves away from examining technology in terms of invention or innovation or the technological or innovative 'firsts' marked by the mythical dates of 1858/1866 (the Atlantic cable), 1870 (the cable to India), 1878

35 On how the empire talked back see Headrick, A double-edged Sword: Communications and Imperial Control in British India; Ewing, 'A most Power Insturment for a Despot': The Telegraph as Trans-national Instrument of Imperial Control and Policital Mobilization in the Middle East or Bayly, *Empire and Information*.

36 Pastor Rev. J. Bray, "On the Atlantic Cable," *New York Morning Express*, August 17, 1858.

(duplex on Atlantic cable) or 1902 (the Pacific cable & transatlantic wireless transmission). Rather, drawing on David Edgerton's approach, the chapters focus on the telegraphs, wired and wireless, and other communication devices of the time as technology-in-use. Edgerton's distinction between innovation and use integrates a history of technology into more general history, provoking questions about class, race and gender in the use of *global communication electric*.[37] Such an angle further marks the shift in attributing agency from the technology itself to the cable agents.[38] Engineers, telegraphers, journalists or media entrepreneurs feature prominently throughout the pages of this book as the driving forces behind the globalizing forces of telegraphy.

Global historians argue not only for a variety of globalization processes, but also a multiplicity of modernities driven by contemporaries themselves.[39] Globalization is not an impersonal force over which ordinary people have no power, but it is "driven by identifiable actors working through institutions they own and/or control."[40] Contemporaries' imagination and practices shaped and formed the content of global communication electric: they imagined who would communicate with whom to what purpose and used their assumptions to draw conclusions on telegraph routes, tariffs and practices. Submarine telegraphy's exorbitant tariffs, for instance, necessitated short telegrams, a feature that soon came to characterize the submarine cable system. On the transatlantic telegraph market, tariffs quickly dropped from initially £1 per word in 1866 (breaking down the £20/20 words tariff to a one word tariff) to 1 shilling per word in 1888, where they remained stable – and financially out of reach for most – until 1923.[41] Consequently, users followed the basic rule that "[t]he wordier a message and the greater the distance, the higher the charge."[42] The pressure to be cost-efficient led to ever shorter telegrams and eventually to the development of the 'telegram style,' which left

37 Edgerton, From Innovation to Use, p. 12.

38 In this context also see Heilbronner's essay on technological determinism Heilbronner, Do Machines Make History?

39 Eisenstadt, *Multiple Modernities*.

40 Sklair, *The Transnational Capitalist Class*, p. 1.

41 Bright, *Submarine Telegraphs*, pp. 143–4; Winseck and Pike, *Communication and Empire*, p. 146. According to Bernhard Finn there was another tariff reduction in 1904, for which, however, I can find no primary source. Coates and Finn, *A Retrospective Technology Assessment*, p. 87.

42 Wilke, The Telegraph and Transatlantic Communication Relations, p. 119.

out anything that was redundant or not essential for the message, making its content, in the words of Emma Pender, "suggestive to the imagination."[43]

Yet contemporaries' interactions with telegraphy were entangled in and dependent on larger macrostructures such as the nation state or the empire. One core pillar for John Pender's success in building up his cable network of the Eastern and Associated Telegraph Companies, for example, were his close ties to "Herbert," meaning Sir Robert George Wyndham Herbert, Under-Secretary of the British Colonial Office and a key cog in the machinery of British imperialism.[44] Following individual actors thus provides the link between micro- and macro levels of analysis. Still, nation states could be actors, too, and remained important players in global communications.[45] In her chapter, Wendy Gagen argues that the submarine cable operators of the Eastern and Associated Telegraph Companies were simultaneously entangled in imperial and other, in this case company, structures. The 'manly telegrapher' represents a hybrid figure between the English gentleman and the company man. Although the nation remains a vital part of global communications, it was one among many players within a system of worldwide entanglements and interactions.

Use|News

At the time, this new kind of *Weltcommunication* fostered ideologies of world citizenship and cosmopolitanism influenced by European enlightenment.[46] According to Frank Hartmann, this notion morphed into the current historiographical perception of a larger global entity organized in a global media system.[47] Many scholars see the development of a global submarine cable network as the first step in the history of telecommunications from a *Victorian Internet*, via a "global public sphere" to the world as a *global village* in

43 Pender, "Letter to Daughter Marion," February 10, 1878, Cable and Wireless Archive.

44 Pender, "Letter to her son in Law William," September 14 1877, Cable and Wireless Archive.

45 Isabella Löhr makes a similar argument for the edited volume Löhr and Wenzlhuemer, *The Nation State and Beyond.*

46 Winder, Imagining World Citizenship in the Networked Newspaper; further see, Conrad, Enlightenment in Global History.

47 Hartmann, *Globale Medienkultur*, pp. 11, 79.

the information age.[48] But how inclusive was the concept of *Weltcommunication* really? The assumption of a unidirectional progression of global communication densification is, at least for the nineteenth century, erroneous. As the exorbitant tariffs of the telegraph system blatantly show, the promise of an all-inclusive *Weltcommunication* was as empty as the analogy of a Victorian internet is misleading. Given that a British farm laborer earned on average 12–14 shillings a week and a skilled artisan between 28 shillings and £2 a week in the late nineteenth century, the 1 shilling/ word tariff still characterized telegraphy as a last resort means of communication for many.[49] Submarine and also terrestrial telegraphs never enabled social or mass communication, but remained a specialty service for an exclusive clientele until wireless and the telephone replaced telegraphy as the dominant means of global communication in the 1920s.[50]

Apart from economic communication on the price of gold, cotton and the like, one of the most obvious uses of *global communication electric* was the transmission of news.[51] Early on, newsmakers, such as Julius Reuter, recognized the benefits of telegraphy's speed and instantaneity for making news faster, albeit not necessarily more 'newsworthy.' Before the Atlantic cable of 1866, news gatherers on both sides employed optical telegraphs, hired spotters to locate ships en route or employed rowboats to obtain news from across the ocean as fast as possible.[52] In the mid-nineteenth century, news agencies, such as Havas (Agence France-Presse after 1944), Reuters or Wolff's Telegraphisches Bureau (WTB) were set up and soon took over 'global' news distribution.[53] In a global history of news, journalists, newspapers and in particular news agencies play an important role.[54] They were news wholesalers and through the commodification of news, they operated globally "in the

48 Standage, *The Victorian Internet.* Stichweh, The Genesis and Development of a Global Public Sphere, pp. 26–9. McLuhan, *The Global Village.*

49 Harrison, *Late Victorian Britain*, p. 30.

50 John, *Network Nation*, p. 2.

51 On economic communication see Preda, Socio-Technical Agency in Financial Markets.

52 Ibid., 79; Blondheim, *News over the Wires.*

53 On the news agencies see Palmer, *Des petits journaux aux grandes agencies;* Storey, *Reuters' Century;* Basse, *Wolffs Telegraphisches Bureau;* Lefebure, *Havas: Les arcanes du pouvoir;* Read, *The Power of News;* Schwarzlose, *The Rush to Institution, from 1865 to 1920*; A comprehensive history of the major news agencies is Boyd-Barrett, *The International News Agencies*; Boyd-Barrett and Rantanen, *The Globalization of News* and Rantanen, Foreign Dependence and Domestic Monopoly; Daya K. Thussu, *Internationalizing Media Studies.*

54 Boyd-Barrett and Rantanen, The Globalization of News, p. 6.

production and distribution of 'consciousness'."[55] As we see in the chapters of Michael Mann, Amelia Bonea and Volker Barth, the definition of 'what is news' directly related to the workings of the global media system. The definition of news was strongly intertwined with processes of globalization tied to modern capitalism: news, seen as a reformulation of information as a commodity, was constituted in its generic form by the technology-in-use, journalistic debates about 'objectivity' and 'truth,' as well as the emergence of mass-media markets.[56]

From the very beginning, news providers of any kind, journalists, editors or correspondents, enjoyed special rates from the cable companies. At times, these 'press telegrams'. cost as little as half price and in the beginning of transatlantic telegraphy, American newspapers could sign up for a dispatch of twenty lines for the morning papers for $900 a day. Accredited agents of newspapers or news agencies could obtain such special rates, if only for political and general news, which had to be intended for publication and written in plain, unabbreviated words.[57] For news providers, the world-spanning system opened up unprecedented opportunities. Its speed allowed news to be transmitted ever faster and thus become ever more current. Already in the summer of the 1866 and with a durable working of the first Atlantic cable, *The Era* marveled over how people could "hear to-day what was done in New York yesterday, and probably, ere long, [they should] have, in the evening's papers in the Strand, the gossip of the morning in Broadway."[58] By 1885, it was at all not uncommon to have "a message sent and the answer received within the space of ten minutes."[59] However, as Volker Barth points out, the communication network also engendered new problems and concerns as business opportunities of making news ever speedier conflicted with evolving journalistic ideals of accuracy and objectivity.

Finally, the globalization of news not only related directly to the policy and cost of transmission or emerging standards and practices of (English-language) journalism, but also furthered processes of news inequality. In the case of India, the country's dependency on foreign news provided by Reuters created certain asymmetries in communication which persisted long after the demise of colonialism. The global history perspective could fur-

55 Ibid., p. 5.
56 Ibid., pp. 1–2.
57 See articles LXV–LXVII of the ITU conventions for press telegrams.
58 *The Era*, Topics of the Week, August 5, 1866.
59 General News, *Birmingham Daily Post*, December 29, 1885.

ther enrich debates on this 'media imperialism,' which argues that Western media organizations have come to dominate global exchanges of news. After all, global history takes into account the multi-directionality of information flows while still not disregarding established power constituencies and their creation. From 1860 to 1930, tariffs, routes and working routines enabled and restrained journalists and newsmakers in the creation of 'global' publics and the commodification of news by both including and excluding users from the benefits and pitfalls of global instantaneity.

Space|Time

While analyzing yesterday's ocean cables or today's fiber optic cables and satellites, *global communication electric* always centers upon questions of the system's structure and construction as communicational space. The geographies of these spaces may differ widely as they are not only bound to concepts of race, class and gender, but also, as Gordon Winder and Jonas Harvard remind us, to actual topographic restraints.[60] Little remained of global connectivity when wires kept freezing over in the cold of Northern Europe or when there was simply no direct cable connection across the Pacific in 1902. As Emma Pender, buoyant commentator of her husband's work, pointed out in one of her letters to her daughter, many a telegraph route was "confessedly urgently desired & wanted."[61] The map of global communication electric changed rather little after the 1880s. At a time when processes of global integration accelerated ever more, the system which provided the basis for these processes became increasingly stagnant. After 1879, when all major trading routes had their cables and all continents could be reached from Europe, no new cable routes of importance were exploited, but only old ones duplicated. Thirteen transatlantic cables existed by 1900, yet none across the Pacific.[62] Any communication between Canada and Australia, for example, still had to pass through London. Communicational space must not necessarily overlap

60 Mathew Zook and Martin Dodge make a similar argument concerning the relevance of geography for today's internet: Zook, Connected is a Matter of Geography, Dodge and Kitchin, *Mapping Cyberspace.*

61 Pender, "Letter to her daughter," January 19, 1878, Cable and Wireless Archive.

62 On the context of Atlantic and/versus Pacific history see Mapp, Atlantic History from Imperial, Continental, and Pacific Perspectives.

with the network's physical geography and rather than diminishing the role of geography or space, the electric telegraph also increased the differences between the locations able to make good use of the possibilities of the new technology and those which could not.

In the mid-nineteenth century, the invention of telegraphy ignited a crisis of space. Although the increasing density of a global maritime transport network as well as the emergence of steamships had accelerated communication in the decades before the telegraphs, the telegraphs sped up communication by a wholly different order of magnitude.[63] The experience of distance in relation to the time needed to cross that distance changed dramatically. Prior to these times, corresponding over long distances was trying. Ordinary mail required a fortnight between Europe and North America. In the case of South America, letters took about a month, for the Far East, one to two months, for Australia and New Zealand, about 70 days' transmission time.[64] The telegraphs changed not only the speed of communication, but also created new possibilities for control over large distances. Prior to the telegraph, if a merchant sent out his ship to India or the Middle East, there was no way of controlling its course or freight after departure and he could only wait for its return. Now, as Charles Bright, telegraph expert of the time, pointed out, the middle man had almost become obsolete in some departments of international commerce.[65] The advent of wireless at the beginning of the twentieth century expedited this process by discovering the niche of the sea as its biggest potential market.[66] By enabling for ship-to-ship communication and a point-to-many communicational stream, wireless made geographical obstacles, such as oceans or markers, such as the cables' landing places, even more irrelevant.

One of the most vivid nineteenth-century discourses accompanying the laying of long-distance submarine telegraphs concentrated on the topos of radically altered perceptions of space and time.[67] These expressed themselves in statements concerning the "annihilation of time and distance" or the idea

63 Kaukiainen, Shrinking the World: Improvements in the Speed of Information Transmission, c. 1820–1870.

64 Ahvenainen, The Role of Telegraphs in the 19th Century Revolution of Communication, p. 21.

65 Bright, *Submarine Telegraphs*, p. 172.

66 Evans, 'The Path to Freedom'? Transocean and German Wireless Telegraphy, 1914–1922, p. 227.

67 Kern, *The Culture of Time and Space, 1880 – 1918.*

of "instantaneous intercourse."[68] Contemporaries were puzzled by the fact that electromagnetic telegraphy seemingly had no speed and "that electricity was present at both ends of the wire at the same instant – instantaneous, immediate," as Florian Sprenger puts it.[69] The perception of space was also in crisis. Building upon discourses of progress and the railways, contemporaries divided their time period into a pre-telegraphic era, when long distances impeded communication, and a telegraphic era when distance seemed to be shrinking through the dramatic reduction in the time taken to cross it (time-space convergence).[70] Soon, contemporary writers such as Karl Knies, Ernst Kapp or Charles Bright solidified this notion of a "practical shrinkage of the globe."[71] In media and cultural theory, these ideas found expression in McLuhan's implosion of place with the electronic age, when the world contracted into a 'global village,'[72] Paul Virilio's and David Harvey's idea of a "time-space compression" or Anthony Giddens' "time-space-distanciation."[73]

This reduction of ocean spaces, such as the Mediterranean or the Atlantic as well as other large terrestrial distances to a "whispering gallery" by means of telegraphy further evoked the sense that global communication networks had constructed *new* spaces.[74] In the case of the United States southwestern borderlands, as Torsten Kathke shows in his chapter, the creation of 'new' spaces through telegraphic communication correlated with processes of identity formation and significantly affected the construction of regional and national identity within and for the territory of Arizona. These new telegraphic spaces more often than not intersected with and reenforced con-

68 On the annihilation of time and space see for example *The Belfast News-Letter*, "The Belfast News-Letter," July 28, 1866.

69 Roland Wenzlhuemer speaks in this context of a dematerialization of information. Wenzlhuemer, *Connecting the Nineteenth-Century World*, p. 30.

70 Schivelbusch, *Geschichte der Eisenbahnreise*.

71 Bright, *Submarine Telegraphs*, pp. 169, 202; Knies, *Der Telegraph als Verkehrsmittel*, pp. 242–3.

72 McLuhan, *The Gutenberg Galaxy*.

73 Virilio and Beitchman, *The Aesthetics of Disappearance*; Harvey, Between Space and Time: Reflections on the Geographical Imagination. Giddens' thesis is based upon the idea that via modern means of communication and transport, social relations stretch across distance. He distinguishes between a tribal society when all communication is virtually face to face and which therefore has a low degree of time-space distanciation and capitalist society which is characterized by the fact that much of its communication and interaction takes place across time and space. Kaspersen, *Anthony Giddens: An Introduction*, p. 54.

74 *Trewman's Exeter Flying Post*, "Wednesday, August 1, 1866: The Atlantic Telegraph," August 1, 1866.

cepts of class, gender and, most of all, race. The mastery of the material world by Euro-America from the late eighteenth-century onwards was spurred on in the mid-nineteenth century by a wave of industrialization and technological development, which "increased Europe's [and North America's] superiority exponentially in virtually all fields of science and technology".[75] Of all technologies of the nineteenth century, the telegraph in particular embodied to many Euro-Americans the "ultimate symbol of man's power over nature" by appearing to harness lightning and put it to work for man's pleasure.[76] Telegraphy also helped Euro-Americans to set themselves apart from 'the rest'. As Jürgen Osterhammel put it, in the eighteenth century, Europe often compared itself with Asia; in the nineteenth century, it considered itself incomparable.[77] The example of Arizona provides a particularly illuminating case study of how the telegraph 'helped' the region to 'become white' by nourishing the dominance of an Anglo-American elite over a largely Anglo-Mexican population.

The telegraphs as emblems of the industrial and communicational revolution of the mid-nineteenth century asked and allowed contemporaries to re-evaluate and re-construct major concepts such as space and time, but also territoriality, agency and belonging. As technology-in-use, they furthered processes of global integration as well as fragmentation, clearly marking those inside and outside the network along the lines of race, class and gender.[78] The electric telegraphs played a key role in multiplying and diversifying the experience of modernity and as such their analysis is of utmost importance for understanding the various layers and meanings of globalization.

The Chapters

Each of the four sections of the book highlights a particular theme of *global communication electric*: *Inter|Nationalisms*, *Agents|Actors*, *Use|News* and *Space|Time*. In the opening chapter, Dwayne Winseck examines the global media system in relation to empire and state. He argues for deemphasizing

75 Adas, *Machines as the Measure of Men*, p. 141; similarly Adas, *Dominance by Design*.

76 Morus, The Nervous System of Great Britain: Space, Time and the Electric Telegraph in the Victorian Age.

77 Osterhammel, *Europe, the 'West' and the Civilizing Mission*.

78 Hampf, Radio as Dispositif: The History of the Yet-to-be Written User Manuals.

the struggle for control model in favor of investigating the cosmopolitan character of the cable industry and the cartels that cut across national lines and defined the global media system. Similarly to Léonard Laborie, he sees the period between 1860 and 1930 as defined by an incipient globalization of control framework where markets were as important as politics and both domains were intertwined. In his chapter on *the ITU and the Governance of the First Globalization of Telecommunications*, Laborie follows this critique of the struggle for control model. Focusing on the International Telegraph Union, he shows how cooperation between the telegraphic nations and empires not only preceded but survived disagreement among them. As the final contribution in the *Inter|Nationalisms* section, Martin Doll considers communication technologies from socialist and communist point of views and points to political interpretations of telegraphy which scholarship has thus far overlooked. As he illustrates, the wiring of the world not only created a global capitalist economy but – from a Marxist perspective – was also understood as a powerful element of its dialectical reversal by offering the promise of the universal connectibility of the international working class.

In the second section on *Agents|Actors*, Daqing Yang explores the relationship between the state and the company as agents of technological progress in Japan. Highlighting Japan's new ambitions in Asia and its bureaucratic policies as well as the Danish Great Northern Company's own business calculations, he shifts the focus from a simplistic narrative of Japan's continuous struggle for autonomy. Lars Bluma in turn looks at the utopian linkage of telegraphy and the world fairs between 1851 and 1880 as complementary expressions of Western utopian visions of an imaginary future. To contemporaries, both represented modernity and they linked fairs and telegraphs to the vision that technology was the prime mover for social modernization. In the end both produced connected as well as unconnected, civilized and barbarian people. Gender too played a significant role in constructing telegraphic identities. Wendy Gagen's chapter explores how gender influenced processes of identity formation among the Eastern and Associated Telegraph Cable Companies' employees. While a distinct company culture and identity as the 'manly telegrapher' was important for the companies' success, it was simultaneously embedded in an ideal of the heroic imperial adventurer and offered a particular expression of 'Englishness' lived at the various cable stations around the world. The manly telegrapher thus represented a hybrid figure whose geographical location may have been transnational, while his identity remained very much tied to the national.

The third section on *Use|News* starts with Michael Mann's chapter on *Telegraphy and the Emergence of an All-India Public Sphere.* He focuses on the social and communicative role of the telegraphs and illustrates how newspapers deeply transformed and shaped the emerging all-India public sphere. He challenges the technological success story of Western ascendancy by examining the colonial context of global communication. Amelia Bonea continues Mann's analysis by highlighting the complex nature of news in the colonial environment of India as a continutally contested field of power relations. Focusing on Reuters in India, she illustrates how negotiations between Reuters, the colonial government of India and Anglo-Indian editors revolved around both access to technologies of communication and the privilege to distribute news to the Indian population. News was a function of both available technology and socio-economic and political factors. With his chapter *Making the Wire Speak. Transnational Techniques of Journalism,* Volker Barth examines how telegraphy changed the role of journalists and their techniques of news production, transmission and verification. In dealing with new, revolutionary means of communication, news businesses and media enterprises had to adapt their daily routines to the requirements of an increasingly well-established world-wide network.

The final section on *Space|Time* starts with Gordon Winder's chapter on *Telecommunications Technology and News of Disaster.* Winder's analysis of the *LA Times* from 1917 to 1939 shows how telecommunication technology shaped disaster coverage in two ways. First, in covering distant earthquake disasters, the newspaper used the latest telecommunications technology to produce the news, and in the process celebrated its superior access to the world's latest stories. Second, the main story often revolved around the newspaper's own ability to report on earthquakes in faraway places. This chapter thus analyses media fascination with telegraph and radio transmission of the news of earthquakes, and the geographical and social implications of this fascination for how such disasters were understood. Focusing on the global telegraph networks' margins, Jonas Harvard's chapter *Northern Experiences of Global Telegraphy* presents an alternative to the grand narratives of steady progress by examining the realities of the electric telegraph at the local level. Harvard examines how two Swedish cities, Helsingborg and Piteå, sparsely populated areas with adverse geographical conditions, interacted with the telegraph. While the possibility of bridging distances through technology might initially have seemed a godsend, Harvard shows that the realities of global communication often faltered on the realities of geography, weather and local circumstances. In his chapter *Power Lines,* Torsten Kathke explores how telegraphic communication could create spatial identities. The virtual networks of telegraphy connecting Arizona with Washington D.C. helped to replace one elite with another, as an established Anglo-Mexican elite, rooted in the region and its Hispanic cultural traditions, slowly acquiesced to a national-minded Anglo one. Finally, Florian Sprenger reminds us how space was not necessarily annihilated by telegraphy but that contemporaries described and experienced electricity in contradictory ways as both instantaneous and having speed, as both immediate and mediated. As he posits, this distinction between slow speed and no speed is crucial for understanding electrical telegraphy as a medium of global communication: accepting the notion of an annihilation of time and space would risk losing sight of the phantasmatic dimension of technical media, their actual modi operandi and their micro-temporal politics.

Over the course of writing and editing this book, we have received support, advice and guidance from several individuals and organizations to whom we are strongly indebted. This book emerged from papers presented at the conference Global Communication Electric at the Museum of Communication in Berlin in 2010 for which we received generous financial support from the gender equality office of the John F. Kennedy Institute for North American Studies as well as the German Research Foundation (DFG). We would like to thank the museum staff, in particular Anita Knobloch-Böhme, Lioba Nägele, Oliver Götze, and Thomas Jander, for their generous support of the conference. The museum provided the perfect venue for us to think through concepts of global communication electric and we relished the rare opportunity for scholars to be able to wander back and forth between our intellectual engagement with the theories concerned and the actual objects of our studies. Our discussions laid the basis for this volume and we thank all participants, our chairs and commentators, Roland Wenzlhuemer, Graeme Gooday, Gordon Winder, Andreas Etges and Jürgen Wilke as well as our keynote speaker, Friedrich Tietjen, for sharing their thoughts and comments with us. We would also like to express our appreciation to the editors of the series Global History, Sebastian Conrad, Andreas Eckert and Ulrike Freitag, for their constructive criticism and support. We are also very grateful to Heidi Tworek without whose invaluable feedback this book would not have taken the shape it has now. Our research assistants, Ulrike Stedtnitz and Hans Anders, have contributed immensely to this volume with their keen eye for detail and their patience with the formatting. Finally, we would like to thank our authors for undertaking this book project with us, for re-thinking concepts of global communication under the auspices of global history and for making this book project a valuable and enjoyable experience.

Bibliography

Adas, Michael. *Machines as the Measure of Men: Science, Technology, and Ideologies of Western Dominance.* Ithaka, NY 1989.

—, *Dominance by Design: Technological Imperatives and America's Civilizing Mission.* Cambridge, MA, 2006.

Ahvenainen, Jorma. *The Far Eastern Telegraphs: The History of Telegraphic Communications between the Far East, Europe, and America before the First World War.* Helsinki 1981.

—, "The Role of Telegraphs in the 19th Century Revolution of Communication." In *Kommunikationsrevolutionen: Die neuen Medien des 16. und 19. Jahrhunderts.* Edited by Michael North, Köln 1995, pp. 73–80.

—, *The European Cable Companies in South America: Before the First World War.* Helsinki 2004.

—, "The International Telegraph Union: The Cable Companies and the Governments." In *Communications under the Seas: The Evolving Cable Network and its Implications.* Edited by Bernard S. Finn and Daqing Yang, Cambridge MA, 2009, pp. 61–80.

Anchordoguy, Marie. "Nippon Telegraph and Telephone Company (NTT) and the Building of a Telecommunications Industry in Japan." *The Business History Review* 75, no. 3 (2001), pp. 507–541.

Basse, Dieter. *Wolff's Telegraphisches Bureau: 1849 bis 1933; Agenturpublizistik zwischen Politik und Wirtschaft.* München 1991.

Bayly, Christopher A. *Empire and Information: Intelligence Gathering and Social Communication in India, 1780–1870.* New Delhi 2007.

—, *The Birth of the Modern World: 1780–1914*, Malden, MA, 2009.

Berthold, Victor M. *History of the Telephone and Telegraph in Brazil, 1851–1921*, New York 1922.

Blondheim, Menahem. *News over the Wires: The Telegraph and the Flow of Public Information in America, 1844 – 1897.* Cambridge, MA, 1994.

Boyce, Robert. "Submarine Cables as a Factor in Britain's Ascendancy as a World Power, 1850–1914." In *Kommunikationsrevolutionen: Die neuen Medien des 16. und 19. Jahrhunderts.* Edited by Michael North, Köln 1995, pp. 81–99.

Boyd-Barrett, Oliver. *The International News Agencies.* London, Beverly Hills, CA, 1980.

Boyd-Barrett, Oliver, and Theri Rantanen. "The Globalization of News." In *The Globalization of News*. Edited by Oliver Boyd-Barrett and Theri Rantanen, 1–15. London 1998, pp. 1–15.

—, ed. *The Globalization of News*. London 1998.

Bright, Charles. *Submarine Telegraphs: Their History, Construction and Working*. London 1898.

Cain, P. J., and Mark Harrison. *Imperialism: Critical Concepts in Historical Studies*. London, New York 2001.

Coates, Vary T., and Bernard S. Finn. *A Retrospective Technology Assessment: Submarine Telegraphy – the Transatlantic Cable of 1866*. San Francisco, CA 1979.

Codell, Julie F. (ed.). *Imperial co-Histories: National Identities and the British and Colonial Press*. Madison, NC, 2003.

Coe, Lewis. *The Telegraph: A History of Morse's Invention and its Predecessors in the United States*. Jefferson, NC, 1993.

Conrad, Christoph. "Vorbemerkung." *Geschichte und Gesellschaft* 28 (2002), pp. 339–342.

Conrad, Sebastian, and Jürgen Osterhammel. *Das Kaiserreich transnational: Deutschland in der Welt 1871–1914*. Göttingen 2006.

Conrad, Sebastian. "Doppelte Marginalisierung: Plädoyer für eine transnationale Perspektive auf die Deutsche Geschichte." *Geschichte und Gesellschaft* 28 (2002), pp. 145–169.

—, "Enlightenment in Global History: A Historiographical Critique." *American Historical Research* (October 2012), pp. 999–1027.

—, ed. *Globalgeschichte: Theorien, Ansätze, Themen*. Frankfurt a. M. 2007.

Conrad, Sebastian, and Sorcha O'Hagan. *German Colonialism: A Short History*. Cambridge, New York 2011.

Dodge, Martin, and Rob Kitchin. *Mapping Cyberspace*. London 2001.

Döring, Jörg and Tristan Thielmann, (eds.) *Mediengeographie: Theorie – Analyse – Diskussion*. Bielefeld 2009.

Edgerton, David. "From Innovation to Use: Ten Eclectic Theses on the Historiography of Technology." *History and Technology* 16 (1999), pp. 1–26.

Eisenstadt, Shmuel N. *Multiple Modernities*. New Brunswick, NJ, 2002.

Evans, Heidi. "'The Path to Freedom'? Transocean and German Wireless Telegraphy, 1914–1922." *Historical Social Research. Historische Sozialforschung* 35, no. 1 (2010), pp. 209–233.

Ewing, Thomas E. "'A most Power Insturment for a Despot': The Telegraph as Trans-national Instrument of Imperial Control and Policital Mobilization in the Middle East." In *The Nation State and Beyond: Governing Globalization Processes in the Nineteenth and Early Twentieth Centuries*. Edited by Isabella Löhr and Roland Wenzlhuemer, New York 2012, pp. 83–100.

Finn, Bernard S. and Daqing Yang, (eds.) *Communications under the Seas: The Evolving Cable Network and its Implications*. Cambridge, MA, 2009.

Finzsch, Norbert and Ursula Lehmkuhl (eds.). *Atlantic Communications: The Media in American and German History from the Seventeenth to the Twentieth Century.* Oxford 2004.

Geyer, Michael, and Charles Bright. "World History in a Global Age." *The American Historical Review* 100, no. 4 (1995), pp. 1034–1060.

Gilmore, Paul. "The Telegraph in Black and White." *ELH* 69, no. 3 (2002), pp. 805–833.

Griset, Pascal, and Daniel R. Headrick. "Submarine Telegraph Cables: Business and Politics, 1838–1939." *Business History Review* 75, no. 3 (2001), pp. 543–578.

Habermas, Jürgen. *The Structural Transformation of the Public Sphere: An Inquiry into a Category of Bourgeois Society.* Cambridge 2010.

Hall, Catherine. *Cultures of Empire: Colonizers in Britain and the Empire in Nineteenth and Twentieth Centuries; a Reader.* Manchester 2000.

Haigh, Kenneth R., and Edward Wilshaw. *Cableships and Submarine Cables.* London 1968.

Hampf, Michaela. "Radio as Dispositif: The History of the Yet-to-Be-Written User Manuals." In *Atlantic Communications: The Media in American and German History from the Seventeenth to the Twentieth Century.* Edited by Norbert Finzsch and Ursula Lehmkuhl, Oxford, New York 2004, pp. 309–319.

Harrison, John F. C. *Late Victorian Britain, 1875–1901.* London 2000.

Hartmann, Frank. *Globale Medienkultur: Technik, Geschichte, Theorien.* Wien 2006.

Harvey, David. "Between Space and Time: Reflections on the Geographical Imagination." *Annals of the Association of American Geographers* 80, no. 3 (1990), pp. 418–434.

—, *The New Imperialism.* Oxford 2005.

Headrick, Daniel R. *The Tools of Empire: Technology and European Imperialism in the Nineteenth Century.* New York 1981.

— *The Tentacles of Progress: Technology Transfer in the Age of Imperialism, 1850–1940.* New York, Oxford 1988.

—, *The Invisible Weapon: Telecommunications and International Politics, 1851–1945.* New York 1991.

—, "Strategic and Military Aspects of Submarine Telegraph Cables." In *Communications under the Seas: The Evolving Cable Network and its Implications.* Edited by Bernard S. Finn and Daqing Yang, pp. 185–208. Cambridge MA, 2009.

—, "A double-edged Sword: Communications and Imperial Control in British India." *Historical Social Research. Historische Sozialforschung* 35, no. 1 (2010), pp. 51–65.

Heilbroner, Robert. "Do Machines Make History?" *Technology and Culture* 8, no. 3 (1967), pp. 335–345.

Hennig, Richard. "Die historische Entwicklung der deutschen Seekabelunternehmungen." *Beiträge zu Geschichte der Technik und Industrie. Jahrbuch des Vereines Deutscher Ingenieure* 1 (1909), pp. 241–250.

Hills, Jill. *The Struggle for Control of Global Communication: The Formative Century.* Urbana, ILL 2002.

Hobsbawm, Eric J. *The Age of Capital: 1848 – 1875.* New York 1996.

Hugill, Peter J. "The Geopolitical Implications of Communication under the Seas." In: *Communications under the Seas: The Evolving Cable Network and its Implications.* Edited by Bernard S. Finn and Daqing Yang, 257–78. Cambridge MA, 2009.

International Telecommunications Union (ed.). *International Telecommunications Union: Celebrating 130 Years, 1865–1965,* 1995.

John, Richard R. *Network Nation: Inventing American Telecommunications.* Cambridge MA, 2010.

Kapp, Ernst. *Grundlinien einer Philosophie der Technik: Zur Entstehungsgeschichte der Cultur aus neuen Gesichtspunkten.* Braunschweig 1877.

Kaspersen, Lars B. *Anthony Giddens: An Introduction to a Social Theorist.* Oxford 2000.

Kennedy, P. M. "Imperial Cable Communications and Strategy, 1870–1914." *The English Historical Review* 86, no. 341 (1971), pp. 728–752.

Kern, Stephen. *The Culture of Time and Space, 1880–1918.* Cambridge, MA, 2003.

Knies, Karl. *Der Telegraph als Verkehrsmittel: Mit Erörterungen über den Nachrichtenverkehr überhaupt.* Tübingen 1857.

Kunert, A. *Telegraphen-Seekabel.* Geschichte der deutschen Fernmeldekabel 2. Köln–Mühlheim 1962.

Laborie, Léonard. *L'Europe mise en réseaux: La France et la coopération internationale dans les postes et les télécommunications (années 1850–années 1950).* Brussels 2010.

Latour, Bruno. "Die Logistik der Immutable Mobiles." In *Mediengeographie: Theorie – Analyse – Diskussion.* Edited by Jörg Döring and Tristan Thielmann, pp. 111–144. Medienumbrüche 26. Bielefeld 2009.

Lefebure, Antoine. *Havas: Les arcanes du pouvoir.* Paris 1992.

Lehmkuhl, Ursula. "Producing and Consuming Knowledge: Comparative Perspectives on the Development and Usage of the Telegraph in Nineteenth-Century Germany and America." In *Atlantic Communications: The Media in American and German History from the Seventeenth to the Twentieth Century.* Edited by Norbert Finzsch and Ursula Lehmkuhl, Oxford 2004, pp. 171–179.

Löhr, Isabella and Roland Wenzlhuemer (eds.). *The Nation State and Beyond: Governing Globalization Processes in the Nineteenth and Early Twentieth Century.* Heidelberg, 2012.

McLuhan, Marshall. *The Global Village: Transformations in World Life and Media in the 21st Century.* New York 1992.

—, *The Gutenberg Galaxy: The Making of Typographic Man.* Toronto 2008.

Morus, Iwan R. "The Nervous System of Great Britain: Space, Time and the Electric Telegraph in the Victorian Age." *The British Journal for the History of Science* 33, no. 4 (2000), pp. 455–475.

Müller, Simone. "The Transatlantic Telegraphs and the Class of 1866: The Formative Years of Transnational Networks in Telegraphic Space, 1858–1884/89." *Historical Social Research. Historische Sozialforschung* 35, no. 4 (2010), pp. 237–259.

Müller-Pohl, Simone. "'By Atlantic Telegraph': A Study on Weltcommunication in the 19th Century." *Medien & Zeit*, no. 4 (2010), pp. 40–54.

—, "Working the Nation State: Submarine Cable Agents and Cable Transnationalism, 1858–1914." In *The Nation State and Beyond: Governing Globalization Processes in the Nineteenth and Early Twentieth Centuries*. Edited by Isabella Löhr and Roland Wenzlhuemer, New York 2012, pp. 101–123.

North, Michael, (ed.) *Kommunikationsrevolutionen: Die neuen Medien des 16. und 19. Jahrhunderts*. Köln 1995.

Osterhammel, Jürgen. *Europe, the "West" and the Civilizing Mission*. The 2005 Annual Lecture. London 2006.

—, *Die Verwandlung der Welt: Eine Geschichte des 19. Jahrhunderts*. München 2009.

Osterhammel, Jürgen, and Niels P. Petersson. *Globalization: A Short History*. Princeton, NJ, 2005.

—, *Geschichte der Globalisierung: Dimensionen, Prozesse, Epochen*. München 2007.

Palmer, Michael. *Des petits journaux aux grandes agences: Naissance du journalisme moderne, 1863–1914*. Paris 1983.

Paul W. Mapp. "Atlantic History from Imperial, Continental, and Pacific Perspectives." *The William and Mary Quarterly* 63, no. 4 (2006), pp. 713–724.

Pender, Emma. Papers. Cable and Wireless Archive.

Potter, Simon J. *News and the British World: The Emergence of an Imperial Press System, 1876 – 1922*. Oxford 2003.

Preda, Alex. "Socio-Technical Agency in Financial Markets: The Case of the Stock Ticker." *Social Studies of Science* 36, no. 5 (2006), pp. 753–782.

Rantanen, Terhi. "Foreign Dependence and Domestic Monopoly: The European News Cartel and U.S. Associated Presses, 1861–1932." *Media History* 12, no. 1 (2006), pp. 19–35.

Read, Donald. *The Power of News: The History of Reuters*. Oxford 1999.

Rosenberg, Emily S. (ed.). *A World Connecting: 1870–1945*. Cambridge, MA, 2012.

—, "Introduction." In *A World Connecting: 1870–1945*. Edited by Emily S. Rosenberg, Cambridge, MA, 2012, pp. 3–28.

Sachsenmaier, Dominic. *Global Perspectives on Global History: Theories and Approaches in a Connected World*. Cambridge, New York 2011.

Schivelbusch, Wolfgang. *Geschichte der Eisenbahnreise: Zur Industrialisierung von Raum und Zeit im 19. Jahrhundert*. Wien/München 1977.

Schwarzlose, Richard A. *The Nation's Newsbrokers. The Rush to Institution, from 1865 to 1920*. Evanston, ILL 1990.

Shridharani, Krishnalal. *Story of the Indian Telegraphs: A Century of Progress*. New Delhi 1953.

Sklair, Leslie. *The Transnational Capitalist Class*. Oxford 2003.

Squier, George. "The Influence of Submarine Cables upon Military and Naval Supremacy." *Proceedings of the U.S. Naval Institute* XXVI, no. 4 (1901), pp. 599–622.

Standage, Tom. *The Victorian Internet: The Remarkable Story of the Telegraph and the Nineteenth Century's Online Pioneers.* London 2003.

Stichweh, Rudolf. "The Genesis and Development of a Global Public Sphere: Revised Version of a Paper first published in Development 46, 2003, 26–29." (2006), pp. 1–11.

Thussu, Daya K. *Internationalizing Media Studies.* London 2009.

Tony Ballantyne, and Antoinette Burton. "Empires and the Reach of the Global." In *A World Connecting: 1870 – 1945.* Edited by Emily S. Rosenberg, Cambridge, 2012, pp. 285–434.

Virilio, Paul, and Philip Beitchman. *The Aesthetics of Disappearance.* Cambridge, London 2009.

Wenzlhuemer, Roland. *Connecting the Nineteenth-Century World: The Telegraph and Globalization.* Cambridge, New York 2013.

Wilke, Jürgen. "The Telegraph and Transatlantic Communication Relations." In *Atlantic Communications: The Media in American and German History from the Seventeenth to the Twentieth century.* Edited by Norbert Finzsch and Ursula Lehmkuhl, Oxford 2004, pp. 107–34.

Wimmer, Andreas, and Nina Glick-Schiller. "Methodological Nationalism and Beyond: Nation-State Building, Migration and the Social Sciences." *Global Networks* 2, no. 4 (2002), pp. 301–334.

Wenzlhuemer, Roland. "The Dematerialization of Telecommunication: Communication Centres and Peripheries in Europe and the World, 1850–1920." *Journal of Global History* 2 (2007), pp. 345–372.

Wilson, Kathleen. *A New Imperial History: Culture, Identity, and Modernity in Britain and the Empire, 1660–1840.* Cambridge UK, New York 2004.

Winder, Gordon M. "Imagining World Citizenship in the Networked Newspaper: La Nación Reports the Assassination at Sarajevo, 1914." *Historical Social Research. Historische Sozialforschung* 35, no. 1 (2010), pp. 140–166.

Winseck, Dwayne R., and Robert M. Pike. *Communication and Empire: Media, Markets, and Globalization, 1860 – 1930.* Durham 2007.

Yang, Daqing. *Technology of Empire: Telecommunications and Japanese Imperialism 1930–1945.* Cambridge, London 2003.

—, "Submarine Cables and the Two Japanese Empires." In *Communications under the Seas: The Evolving Cable Network and its Implications.* Edited by Bernard S. Finn and Daqing Yang, 227–56. Cambridge, 2009.

Yrjö Kaukiainen. "Shrinking the World: Improvements in the Speed of Information Transmission, c. 1820–1870." *European Review of Economic History* 5, no. 1 (2001), pp. 1–28.

Zook, Mathew. "Connected is a Matter of Geography." *netWorker* 5, no. 3 (2001), pp. 13–17.

Inter|Nationalisms

Globalizing Telecommunications and Media History: Beyond Methodological Nationalism and the Struggle for Control Model of Communication History

Dwayne Winseck

During the late nineteenth and early twentieth century, the foundations of the modern media were put into place. Undersea cables spanned the planet and interconnected with national and urban telegraph networks, news agencies, and the commercial press. It was a time that many see as the closest predecessor of our own.[1]

The story of the this early period is often told as one where the hegemony of *Pax Britannica* over *international* communications was consolidated in the late nineteenth century but then lost to the US after World War II – albeit with important moves occurring in this direction by the 1910s.[2] I call this the "struggle for control" model. A more recent view is offered by the "circuits of culture" approach that builds on Harold A. Innis' classic *Empire and Communication* and the "struggle for control" model to assess the role of communications media *within* the British and European empires.[3] This approach focuses not just on administration and control, but how "the empire talked back." In other words, the focus is on how overlapping and reciprocal circuits of communication furthered British imperial control while also underpinning the rise of nationalism, anti-imperialism and a politics of cultural identity inside the imperial edifice. For two generations of scholarship, another group of scholars has highlighted how Reuters, Havas, Wolff and, after 1893, the Associated Press forged a cartel amongst themselves to control world news production and markets. Recent versions of these "global

1 See Held et al., *Global Transformations*; Rourke and Williamson, *Globalization and History*; Jones, *Multinationals and Global Capitalism*.

2 See Schiller, *Mass Communication*; Headrick, *Invisible Weapon*; Hills, *Struggle for Control*.

3 Harold Innis, *Empire and Communication*. See notably: Bayly, Informing Empire and Nation; Kaul, *Media and the British Empire*; Potter, *News and the British World*; Potter, Webs, Networks and Systems; Headrick, Double-Edged Sword.

news" studies stress how the agencies were both influential and influenced by their interaction with powerful local and national press combinations.[4] This chapter adds to these perspectives by putting less emphasis on the "*struggle* for control" model of history that still looms large across the field in favor of putting more emphasis on the following characteristics:

1. the cosmopolitan character of the cable industry,
2. the cartels that cut across national lines and defined the global media system,
3. the role of fast communication and the global media system in the New York and South American financial crises of 1873 and 1874

I see the period of the telegraph and media history between, roughly, 1860 and 1930 as defined more by an incipient "globalization of control" framework, where markets were at least as important as politics, and both domains were fundamentally intertwined, rather than the Machiavellian world of the "struggle for control" model, where politics and war reigned supreme, while communications and media companies, technologies and services were the play-things of nation-states.

The Rise of the Global Media

The consolidation of the telegraph industry in Britain, Europe and North America by the mid-1860s gave rise to large corporate entities that had pockets deep enough to incur the huge fixed capital investment and risks involved in the early attempts to wire the world. The British government's take-over of all telegraph systems in Britain provided another boost by transferring a huge sum of 6 million pounds (40 million US dollars) to those, notably John Pender and Julius Reuter, who plowed their windfall and experience in the domestic telegraph and news markets into efforts to develop global markets for undersea cables, telegraphs and news.[5] Transnational cable ser-

4 Boyd-Barrett und Rantanen, *Globalization of News*; Rantanen, The Struggle for Control of Domestic News Markets; Putnis, How the International News Agency Business Model Failed; Desbordes, Western Empires and News Flows.

5 While this section draws heavily on chapters two and three in Dwayne Winseck and Robert M. Pike, *Communication and Empire*, I still cite primary sources so that others can find them easily and because it is these resources that I am most interested in using to tell

vices reached the Mediterranean Basin, the Middle-East and India between 1857 and 1865, across the North Atlantic in 1866 and to China and Japan by 1872, South America in 1874, and Australasia two years later. They finally encircled the continent of Africa in the mid-1880s. The Eastern Associated Telegraph Companies was the largest undersea cable firm by far and was led by the pre-eminent cable mogul of the time, John Pender, who was also a founding co-owner and director at the Anglo American Telegraph Company (AATC). James Anderson and John Brett played similar roles, while Cyrus Field and several other Western Union directors parlayed their status in the North American telegraph industry into key positions at the Anglo American Company and the International Ocean Telegraph Company. The two firms dominated the Northern Atlantic and Caribbean cable markets by the mid-1870s. Both firms also highlighted the multinational, and especially Anglo-American, cast of the cable business. Outside the Anglo-American cast, William Siemens, of Germany, further illustrated the multinational character of the cable industry, given his role as technology supplier to the industry and as a financier of the Direct U.S. Cable Company and Indo-European Telegraph Company in the early-1870s. Likewise, the Danish bankers and merchants, C. F. Tietgen and Henri Erichsen joined with British capital and technology firms to create two significant firms, one for Asia and another for South America: the Great Northern Telegraph Company (1870) and the Western and Brazilian Telegraph Company (1873), respectively.[6]

Different media sectors were also linked to one another by the fact that news agency and press owners financed some of the most ambitious projects to wire the world, and began to use wireless services at an early date. Julius Reuter, for instance, invested heavily in the French Atlantic Cable Company, the Indo-European Telegraph Company and a failed bid to extend telegraph lines and cables to the Far East via Russia. For a short time, he was also a director in the British-based Globe Telegraph and Trust Company, an agency allied with the backers of the Pender group of companies and the AATC that had audacious but ultimately unrealized plans to unify control over the world's cable system.[7] Reuter even invited the Globe to acquire his news

a more compact version of the story that Pike and I tell in *Communication and Empire*. Britain, *The Telegraph Act, 1869* in: *Public General Statutes*, pp. 32–33 Vict., Chap. 73, London 1869.

6 Ibid.

7 Globe Telegraph and Trust Company, *Memorandum of Articles of Association*, 11 July, 1873 as well as Minutes, 1 January, 1874.

agency in 1874, but was rebuffed on account of what its leading forces, John Pender and James Anderson, saw as the steep price he was asking. [8] Afterwards, Reuter focused on creating bureaus in major world cities as soon as they were connected to the cable system, as in Bombay (1870), Hong Kong (1872), Shanghai (1873) and Buenos Aires (1874).[9] James Bennett, owner of the largest circulating daily in the US, *The New York Herald*, co-founded the Commercial Cable Company (1884), while the notorious robber baron Jay Gould acquired the *New York World* and Western Union at the height of the Gilded Age in the 1880s.

The high demand for cable services initially attracted some competitors, but a mixture of high fixed capital costs, monopolistic concessions and cartels resulted in eight global cable firms and a dozen or so smaller regional companies towering over the industry by 1890: the EATC, Anglo American Company, Great Northern Company, Western Union, the French Cable Company, the German Cable Company, the Commercial Company and the All-America Company.[10] Private enterprise ruled the industry and even at the height of the new imperialism (1880–1910), less than 20 percent of cables were state-owned.[11]

As consolidation swept the cable business, the three main global news agencies – Reuters, Havas and Wolff – built on their earlier agreements in 1870 to form a cartel that divided the world's news markets amongst themselves, while bringing the Associated Press of New York into the fold in 1893.[12] Such arrangements were not unique but a standard feature of the global media business from the outset. By 1920, they were applied to the

8 Globe Telegraph and Trust (1872–1880) *Board Meeting Minute Book*, vol. 1, 25 April, 1873, City of London Libraries, Guildhall, MSS24, 212.

9 Read, *Power of News*, pp. 50–58.

10 I have condensed the names of the major cable companies to reflect their common ownership and conventional forms of reference in the literature. Thus, the French Cable Company was formally known as Compagnie Française des Câbles Télégraphiques, which itself was an almagamation, after 1892, of several companies. The German Cable Company, majority owned by the firm of Felten and Guilleaume, consisted of: Deutsche Atlantische Telegraphen Ges, Deutsche See Telegraphen Ges, Deutsche Sudamerikanische Telegraphen Ges and Deutsch-Niederlandische Telegraphen Ges. The All America Company was an amalgamation of the Mexican Telegraph Company and the Central and South America Telegraph Company and renamed as such in 1920.

11 See Britain, *Second Report of the Interdepartmental Committee on Cable Communications*, Cmd 1056, London 1902, Appendix J; United States, House of Representatives Committee of Interstate and Foreign Commerce, *Pacific Cable*, Washington 1902.

12 Rantanen, The Struggle for Control of Domestic News Markets, pp. 35–40.

new field of wireless communication. In sum, the private enterprises upon which the global media rested were early multinational corporations and their services provided the informational underpinnings for global capitalist modernity.

The "Struggle for Control" Model: A Critique of Conventional Knowledge

The dominant interpretation of international communication history rests on a handful of themes. The first is the strong tendency to assume that national interests and corporate interests tightly overlapped. The presence of the world's largest cable company – the EATC – and the powerful Reuters news agency in London is often used to demonstrate Britain's hegemony over world communications.[13] Control over the medium and the message did confer enormous commercial and strategic advantages, and this was a power that British officials ceded only grudgingly. Britain's free trade policy was designed to attract cables and capital in a bid to establish and maintain London as the hub of world communication. A second theme, however, is that by the turn-of-the-twentieth century, France, Germany, Japan and the US began to take strong efforts to expand their own international communication systems, engaging Britain in a "struggle for control of global communication."[14]

A closely-related, third point is that these efforts gained momentum as inter-imperial rivalry sharpened under the auspices of the "new imperialism" (1880–1910), a conquest that led to 90 percent of Africa falling under British and European control within the initial ten years of this period, while large swathes of Asia and the Caribbean met a similar fate. Within this context, communications media helped to consolidate powerful countries' control over their respective empires as they competed with one another for territory, wealth and influence. The geography of international communication became as a result, and fourth, largely coterminous with the geography of imperialism. As Kelley Lee (1996) observes, "the integration of […] Eu-

13 For more detailed expositions of this theme, see Headrick, *Invisible Weapon*; Hills, *Struggle for Control* and Innis, *Empire and Communication*.

14 See the title of Hills' book, The Struggle for Control of Global Communications.

ropean imperialism [...] was reinforced by telegraph (and later radio and telephone) networks *whose reach was historically defined by the boundaries of empire.*"[15] World War I magnified the importance of strategic communication as network surveillance; censorship and propaganda became systematic elements of total war. This is central to a fifth theme: namely the instrumentalist view of communications media as "tools of empire" and "weapons of politics."[16] Lastly, the rapid ascent of US commercial, political and military interests during World War I allowed it to displace Britain and Europe as the center of world communication, and more fully after World War II, which led to further dismantling of the British and European empires.

While the "struggle for control" model assumes that imperialism and military strife were largely synonymous with globalization, they were not. International communication, or the global media, developed first and foremost, between Europe and North America in the 1860s and subsequently in commercial cities along the coasts of South America and in China and Japan in the 1870s, largely following the flow of capital and commerce. Except in the case of India and the settler colonies of Australia, Canada and New Zealand, cable firms and news agencies entered Africa, the Caribbean and Asia only *after* receiving government subsidies. Yet, even then, these areas were some of the least connected, worst served places on the planet.[17]

The case of Africa illustrates the point best. Three affiliates of the Eastern Telegraph Company completed a network around the continent in the mid-1880s – two decades after the northern Atlantic was crossed. This was achieved after they obtained roughly 7.5 million US dollars from the British Treasury, or more than half the 14 million US dollars given to all British-based companies between 1880 and 1902. Perhaps most surprisingly, however, France, Germany and Portugal also used the Eastern Company's cables to meet their imperial, military and commercial needs in Africa until just after the turn of the century, giving another 2.5 million dollars in subsidies over and above the high rates they already paid for services.[18]

Of course, there were episodic bouts of conflict, for example, between Britain and France during the Fashoda incident (1899), Britain and Germany during the Boer War (1899–1902) and between France and Germany

15 Lee, *Global Telecommunications Regulation*, p. 60.

16 See Headrick, *The Invisible Weapon*, emphasis added, and Headrick, *The Tentacles of Progress.*

17 Britain, *Second Report of the Interdepartmental Committee*, p. 59 and Appendices J and G.

18 Ibid, fn 17.

over Morocco (1908).[19] But the collective reliance on, and large subsidies given to, the Eastern Company reflects a "collaborative model of empire," a version of which was set out in the Berlin Agreement of 1885.[20] The Berlin Agreement urged the imperial powers, as well as Turkey and the US who also signed, to develop their imperial projects cooperatively, but in high-minded words that twisted conquest into a liberal humanitarianism underpinned by the ideals of modernization, freedom and tolerance. Such notions did not last beyond 1914. Even before then, national interests began to be asserted more strongly at the turn of the century. This could be seen, for example, as the South American Cable Company, a firm previously owned by the Eastern Company, was acquired by the French Post and Telegraph Authority (1902), and the new German-based South American Telegraph Company in 1911, acquired or laid their own cables to South America, with spurs to the west coast of Africa – but only after both firms accepted a joint-purse agreement with the Eastern Company.[21]

The drawn-out efforts to persuade the Eastern Company to improve its service and the need for huge subsidies from multiple governments suggest that the company was hardly eager to put imperial interests above its own. Protracted efforts to have the company, and Reuters, offer better services to the press and people of the British Empire point in the same direction.[22] Most notable, perhaps, is Britain, Australia, Canada and New Zealand's decision to jointly-finance and manage the imperial Pacific Cable, which finally opened in 1902, after nearly two decades of unflagging pressure from imperial and global media reformers such as Sandford Fleming and Henniker Heaton. The initiative was only realized after overcoming the great reluctance of the private companies and parsimonious elements of the British state, notably the Treasury. Altogether, these cases cast doubt on the common claim that Reuters, and the Eastern Company, were "imperial institutions with close links to metropolitan and colonial governments."[23] Their interests often did coincide, but when those of private enterprise were at odds with national interests, the former often triumphed.

19 Headrick, *The Invisible Weapon*, pp. 84–88; Mattelart, *Mapping World Communication*, pp. 14–15; Kennedy, Imperial Cable Communications and Strategy.

20 As quoted in Ferguson, *Empire*, p. 236.

21 Ahvenainen, *The European Cable Companies in South America*, pp. 282–318.

22 Potter, *News and the British World* and id., Webs, Networks and Systems. Also, Winseck and Robert Pike, *Communication and Empire*, chaps. 5 and 6.

23 Kaul, *Media and the British Empire*, p. 6; see also Read, *Power of News*.

The Global Media and the Empire of Capital: Towards an Alternative View

The perspective proposed in this chapter moves the role of capital accumulation much further up the ladder of intellectual priorities than tends to be the case in the "struggle for control" model. As Giovanni Arrighi observes, capital has been a globalizing force from its inception.[24] The key, however, is not to develop a one-dimensional focus on either the globalizing force of capital *or* imperialism, but on the interplay between *both* processes. The approach proposed here does this by combining David Harvey's distinction between *territorial* and *capitalist* imperialism with a *structuralist view* of the global system.[25] As David Long and Brian Schmidt write, contemporaries saw these two forces as two sides of the same coin or the pivots upon which the woof and warp of the modern era turned.[26]

Harvey defines *territorial imperialism* as a "distinctly political project [...] based in the command of territory and [...] the strategies used by a state to assert its interests and achieve its goals in the world at large."[27] While territorial imperialism *was* important, it mainly targeted "weak states" in Africa, the Caribbean and Asia. Yet, to stop here is akin to "judging the size and character of icebergs solely from the parts above the water-line," as John Gallagher and Ronald Robinson put it in their seminal article, "The Imperialism of Free Trade."[28] The spaces of capital and communication were *not* coterminous with the space of territorial-based empires, but much larger. Harvey reflects this idea by defining *capitalist imperialism* as a system of power that

---

24 Arrighi, *The Long Twentieth Century: Money, Power and the Origins of our Time*, London 1994.

25 Harvey, *The New Imperialism*. This view draws on Lukes' distinction between three views of power: (1) an *instrumentalist view* in which power is based on distinctive interests and the ability of one group or individual to realize their interests in the face of opposition from others; (2) an *agenda-setting view* in which power is a function of being able to shape the agenda around which debates, decisions and conflict occurs; and (3) a *systems view* in which power is a function of the ability of actors to forge relationships with others, exercise control over resources (material and symbolic), set the 'rules of the game' and deform the acceptable range of action that are open to all actors a part of the *system*. See Lukes, *Power: A Radical View*. Hills, pp. 11–16, and Lee, pp. 32–48 offer a good explanation of Lukes and the structuralist view, although their reliance on secondary sources and the struggle for control view constrains their theoretical stance.

26 Long and Schmidt, *Imperialism and Internationalism*, p. 2.

27 Harvey, *The New Imperialism*, p. 26.

28 Gallagher and Robinson, The Imperialism of Free Trade.

aims to allow capital accumulation and "economic power to flow across and through continuous space," where dominant models of development are emulated and consent preferred to coercion.[29] He also suggests that while power is usually the preserve of single states under territorial imperialism, capitalist imperialism relies upon "the *collective accumulation of power* as the [...] basis of hegemony."[30] Further, he does not view corporate interests as subordinate to state interests nor does he portray nation-states as the simple handmaidens of capital.

Harvey's views are of recent vintage but they overlap remarkably well with the theory of "capitalist investment imperialism" put forth by Charles Conant in 1898.[31] Conant was a renowned journalist, first official international economic adviser to the US Government and sympathetic theorist of American Empire who saw advances in communication and transportation as underpinning the creation of a world market and opening new outlets for what he saw as a super-abundance of capital in Britain, Europe and the US. Conant also, however, emphasized that improvements in communication were a double-edged sword. Faster, more universal communication synchronized world markets but could also magnify and expand the reach of *financial* crises. In addition, weak states were unstable and needed to be "rationally managed" in ways similar to the international financial commissions and public debt administrations that had been imposed by British and European interests on the Balkans, Egypt, Ottoman Empire and Spain in the late nineteenth century.[32] While these measures might work in the post-imperial nation-states of South America, or the faltering Ottoman and Chinese empires, where modernizing trends were asserting some influence, other areas of the world were seen as requiring even more thorough-going reforms to modernize the entire administrative and social system of weak states. They required, in other words, a system of imperial control. Conant contrasted his version of capitalist investment imperialism from territorial imperialism, however, on the basis that it did not seek permanent control, but a transitional period of directed modernization during which 'weak countries' would be remade in the image of capitalist modernity.[33]

29 Harvey, *The New Imperialism*, p. 26.
30 Harvey, *The New Imperialism*, p. 37. Emphasis added in quote.
31 Conant, C., The Economic Basis of Imperialism, *North American Review*, pp. 326–46.
32 Feis, *Europe*.
33 See Rosenberg, *Financial Missionaries* and Sklar, *The Corporate Structuring of American Capitalism*.

Conant recast imperialism to bring the link between globalization, communication, capitalism, modernization and empire to the fore. In doing so, he shifted the emphasis from an overly-politicized, state-centric discourse to one that better resonated with contemporary views of liberal imperialism and post-modern Empire.[34] The empire of capital approach counters a realist view of international relations, where national interests and competition for resources, power and influence trump all else. While realists might be inclined to brand such an approach as a species of idealism, Conant's views were of a piece with other turn-of-the-twentieth century scholars who *rejected* idealism. One such scholar was University of Wisconsin professor Paul Reinsch. The French and Dutch international law scholars Louis Renault and Tobias Asser also saw imperialism and internationalism as two sides of the same coin and products of increasing *interdependence*, rather than the idealist "cloud castles" of Kantian cosmopolitanism.[35] According to Reinsch:

> Our age is realistic and practical. [...] We are building up cooperation in constantly widening circles, so that it transcends national bounds to become a universal joint effort. [...] Universal cooperation is the watchword which stands for positive action, for the development of concrete facts in human life corresponding to the actual needs of our economic and social order. [...] The great economic and financial system [...] is being centralized. The psychological unity of the world is being prepared by the service of news and printed discussions, by which in the space of one day or week the same events are reported to all the readers from Buenos Aires to Tokyo, from Cape Town to San Francisco.[36]

Global Communications Media and Crises in the Empire of Capital

New communication media were crucial to interdependence because they enlarged the mentality of the individual, expanded the space of social interaction, and quickened its pace.[37] More than this, they forged the organizational bonds of interdependence, as the creation of the International

34 For example, Michael Ignatieff, *Empire Lite*; Cooper, Why We still Need Empires; Hardt and Negri, *Empire*.

35 Koskiennemi, The Gentle Civilizer of Nations, pp. 11–97 and 276–280.

36 Paul Reinsch, *Public International Unions*, pp. 3–4.

37 See also Cooley, *Social Organization*, pp. 80–81.

Telegraph Union in 1865 exemplified. They were also viewed as globalizing markets and the operations of those who could afford their steep cost. As the Secretary of Lloyds of London told a British Parliamentary Inquiry in 1899, "telegraphs and cables increase the *information component of business transactions* and provide business with knowledge about markets before they enter them."[38] Both Carey and Hoag develop this point by showing that the increased availability of information reduced "knowledge gaps" between markets, allowing prices in local markets to align with those in national markets and those, in turn, to become more closely aligned with world markets.[39] Hoag uses the New York and London capital markets to illustrate this point, observing that "the telegraph shortened the information lag by about nine or ten days," bringing prices in both markets closer together as a result.[40] *The Times* (London) took this idea a step further in the wake of the New York Stock Exchange panic of 1873, stating that the Atlantic cables helped to stabilize world markets by "minimizing some of the most dangerous effects of the American panic."[41] By sending news "instantaneously" from New York to Europe and quickly sending information back that strong measures were being taken, they kept the panic "within far narrower limits than would otherwise have been the case:" in other words, the cables had helped to facilitate the quick dispatch of huge sums of gold from Britain and Europe to keep the New York markets afloat.[42] This, in turn, stopped events from spiraling into an even larger global crisis.

These events illustrated well Conant's later claims that modern communication media were double-edged swords which simultaneously synchronized markets whilst having the potential to amplify and spread crises.[43] Similarly, *The Times* stressed the benefits of market integration, but closed its commentary on the panic in New York markets with the darker warning that the cable "has a drawback at times in its *echoing power* – each money market is troubled by bad news of the other, and the bad news tends to accumulate."[44]

38 emphasis added, Britain, Pacific Cable Committee, *Report, Minutes and Proceedings, etc.*, London 1899, pp. 156.

39 Carey, *Communication as Culture*; Hoag, The Atlantic Telegraph Cable.

40 Hoag, The Atlantic Telegraph Cable, p. 350.

41 The Times, "The Atlantic Telegraph and the Panic," 6 Oct 1873, p. 7.

42 Ibid.

43 Charles Conant, "The Economic Basis of Imperialism," *The North American Review* CLXVII (1898), pp. 326–346.

44 The Times, "The Atlantic Telegraph and the Panic," p. 7.

This theme emerged again the following year when both *The Times* and *The Economist* devoted substantial coverage to the South American financial crisis and the question of whether Reuters, in particular, held some responsibility for making a bad situation worse through both the speed and poor quality of its reporting. The timing of the events was especially interesting given that *The Times* and *The Economist*'s coverage of the dispute between British and European financiers, on the one side, and Reuters, on the other, began just six months after the first direct cable service to South America opened in June 1874. The new cable was the product of a joint-purse agreement between the Brazilian Submarine Telegraph Company, another branch of the Eastern Company colossus, which laid a new cable across the southern Atlantic, and the Western and Brazilian Telegraph Company which had just completed a new network of coastal cables.[45] Both firms focused mainly on commercial cities along the Atlantic coast of South America and, especially, the River Plate, which was the hub where Brazil, Uruguay and Argentina's economies intersected and flowed into the broader global economy. Consistent with their usual practice, the Reuters and Havas agencies quickly entered the market and, perhaps less commonly, struck an agreement in early 1875 to jointly open news bureaus in Rio de Janeiro, Montevideo, Buenos Aires and a half dozen other important cities in South America.[46]

In the four months of coverage that ensued between December 19, 1874 and April 20, 1875, Reuters was accused of distributing "scandalous" reports "of the most strangely misleading character," with *The Economist*, in particular, strongly criticizing both the agency and members of the press "which published the false news."[47] Furthermore, *The Economist* called on Reuters to explain the "pure inventions" found in its reports and for the poor overall quality of its coverage. The agency's stories were also found wanting in comparison to what was seen as the more accurate accounts of the Buenos Aires' press.[48] As the coverage continued, it also became evident that there were two circuits of news at play in this context: the first, the speedy, "cable news" supplied by Reuters which had the ability, it seems, to move markets, and a second, slower circuit based on the coverage of events provided by the South American press and carried to London and other financial capitals by

45 The firm consisted of British and Danish capital.

46 Desbordes, "Representing 'Informal Empire' in the Nineteenth Century", *Media History* 14, no. 2 (August 2008): 121–139.

47 The Economist, "The False Argentine Telegrams," 19 December 1874, pp. 1510–1511.

48 The Economist, "The False Argentine Telegrams," 9 January, 1875, pp. 35–36.

ship. Both were elite and influential circuits of news, but it was the former that was being called to task for the harm to international investors and South American markets. The soul-searching in the articles seemed to suggest that there might be something in the temporal rhythms of news sent by the wires that was making markets jittery, while the slow pace of the printed word sent by ship comes across as reflecting higher standards of quality, deliberation and stability.

Reuters responded to the criticism in several ways. It initially tried to distance itself from them, first by claiming that it had been duped by others acting on its behalf in South America, namely the Agencia America Telegraphica.[49] One James W. Hall writing on behalf of that agency led an early defense of Reuters, acknowledging that Reuters' reports from the River Plate were *false*, but arguing that a single journalist, who had since been fired, was the source of the problem. Several letters to *The Times* and *The Economist* by Hall and the Secretary of Reuters, F. J. Griffiths followed. Over the course of this correspondence, Reuters appeared to move to the foreground in accepting responsibility, while the otherwise unknown Agencia America Telegraphica faded to a lesser role, likely because neither *The Times* nor *The Economist* seemed to be impressed by the initial attempts to "pass the buck." The letters mounted four more points in Reuters's defense: first, that the complexity and intrigue surrounding events in South America made them exceedingly difficult to grasp; second, that Reuters was the first to report on the events; third, that no costs had been spared in sending news to London despite "the very heavy expenses of sending telegrams from the River Plate."[50] Fourth, the letters ritualistically invoked impartiality and accuracy as the "gold standards" of international news reporting in a way that tried to re establish Reuters' credibility and adherence to modern "news values."[51]

The real impact of Reuters' and speedy "cable news" on the stability of South American markets is difficult to assess. However, the extent and content of the coverage of these events reveals that there was a wide-spread belief in influential circles that severe damage was being done. Reuters' response indicates as much, with senior people at the agency rising to defend it and,

49 The Economist, "The Argentine Telegrams," 16 January, 1875, p. 63; The Economist, "The False Argentine Telegrams," 9 January, 1875, pp. 1510–1511; The Economist, "The False Argentine Telegrams," 9 January, 1875, pp. 35–36.

50 The Economist, "The Argentine Telegrams," 16 January, 1875, p. 63.

51 The Times, "Money Market and City Intelligence," 19 April 1875, 9; also The Times, "Money Market and City Intelligence," 20 April 1875, 10.

crucially, to issue a *mea culpa* for the admittedly poor quality of its news service. The fact that the dispute was also drawn out for four months in the pages of two of Britain's most prestigious publications is one index of its importance. That it was given substantial emphasis in the *Annual Report* (1875) of *the* premier body representing international finance – the Council of Foreign Bondholders – points in the same direction. The dispute did not just revolve around the impact of the global media on financial markets, but entailed a clash between capital, on the one side, and Reuters and its clients in the press, on the other. In sum, the credibility of the global news *system*, at least in the eyes of elite audiences, appeared to be at stake. The case also revealed that elite audiences were not tied exclusively to the hyper-speedy flows of news sent by the wire, but also to the slower circuits of news culled from local sources and forwarded to Europe by ship. These two circuits of global news, thus, overlapped with the old and new temporal worlds of capital.

Economic historians have since studied these events, but with little, if any, attention given to the role of communications media in them.[52] Media historians, in contrast, have ignored them, focusing instead on political and military crises, not surprisingly given the tenets of the "struggle for control" model. This discussion is not exhaustive by any means, and is meant more to suggest possible new lines of inquiry and to highlight the ambivalent outcomes of an increasingly interdependent world.

Capital, Cosmopolitan Corporate Identity and Cartels in the Global Media Business

While financial crises are one overlooked aspect of the global media, their cosmopolitan character is another. Herbert Feis vividly captured this idea in his seminal book, *Europe: the World's Bankers*, observing that: "London was the center of a financial empire, more international, more extensive in its variety, than even the political empire of which it was the capital."[53] Of course, London was the base for many important cable firms, notably the British-owned Eastern Telegraph Company. This remained so well into the

52 For example, Feis, *Europe*, Kindleberger, *Manias, Panics and Crashes*; Marichal, *A Century of Debt Crisis in Latin America*.

53 Feis, *Europe*, p. 5; also see Jones, *Multinationals and Global Capitalism*, p. 282.

twentieth century. However, even this needs to be qualified given its alliances with other firms and that the influential local directors of its affiliates around the world. But London was also the cosmopolitan hub of world finance and business, and this meant that European and US investors flocked to the city, pooled their resources, and invested heavily in the cable-building boom that occurred between 1870 and 1910.

Many of the firms registered in London contained a mixture of British, European and North American capital, and their boards of directors reflected this fact. The French Atlantic Cable Company in the 1870s illustrates the point, combining French and British capital with two boards of directors, one in London, another in Paris. The Great Northern Telegraph Company and its affiliate, the Western and Brazilian Telegraph Company, were also excellent cases in point, mixing capital from Denmark, Britain and, in lesser amounts, other European investors. So too was the Commercial Cable Company. That company combined capital and directors from the U.S., Britain and Canada. There are others, but most studies obscure this by referring generically to firms as British, American, German, French, Japanese, and so on, with the inference that they were tools of the state. In fact, the multinational complexion of the business would have made the realist assumptions about the national basis of power difficult to achieve; state and corporate actors often pursued their interests independently of one another – although this was not always the case especially as World War I yoked the two closer together than ever before.

The Commercial Cable Company was a master at taking an opportunistic approach toward national identity, claiming to be British when it successfully sought subsidies for two "British" affiliates that it owned – the Halifax Bermudas Telegraph Company (1890) and Direct West India Cable Company (1898) – and as "American as apple pie" when it stood before the US Congress to claim why it should be chosen to lay the first US-owned cable across the Pacific (1904).[54] From its inception in 1884 until the early 1920s, the Commercial Company displayed its ambivalent attitude toward national identity by leading alliances with European cable companies in the trans-Atlantic market, while joining with an Eastern Company affiliate and the Anglo-Danish Great Northern Telegraph Company for its operations in the Far East. The Commercial Cable Company was far from exceptional. Indeed, with respect to the formation of multinational cartels, its actions

54 United States, House of Representatives, *Pacific Cable Bills*, Washington 1902.

were the norm. In fact, high capital costs, monopolistic concessions and a desire to avoid "ruinous competition" meant that a cartel controlled every major market. The northern trans-Atlantic market fell first to the cartels in 1868, the Far East in 1870, South America by 1873 and the Indo-European market in 1878. All were dominated by Eastern Company affiliates or closely allied firms – the Anglo American Company in the North Atlantic, the Western Telegraph Company in South America, the Eastern Company in the Indo-European market, and the Eastern Extension Company in Asia.[55] Rampant cartelization did not squelch rivalry entirely, however, and in the northern Atlantic two alliances, one led by the US-based Commercial Company and the other by the British-based Anglo American Company, competed after 1884. The South American cartel also faced competition from James Scrymser's US-based Central and South American Telegraph Company after their early joint-purse collapsed in 1891.[56] Yet, most importantly, the cartels were thoroughly multinational in nature, consisting of British, U.S., Chinese, Danish, French, German and Japanese firms – depending on the market.

Cartels were a defining feature *across* the global media. Building on earlier agreements going back to 1859, the "ring circle agreement" of 1870 expanded the terms of the global news cartel consisting of Reuters, Havas and Wolff (revised to include Associated Press in 1893).[57] Swiftly after World War I, the big four commercial wireless firms – Marconi, Compagnie Générale de Télegraphie sans Fil, RCA and Telefunken – adopted the same methods. Several conclusions arise from these observations. First, in contrast to Jill Hills' claim, cartels emerged far before their adoption in the 1920s in response to a crisis of over-capacity in the world communication system.[58] Second, historian Michael Hogan refers to cartels as "private structures of control." He focuses on their essential role in trans-Atlantic commerce and diplomacy after 1918, but these private structures, as this chapter shows, functioned globally, not just in the trans-Atlantic economies, and emerged earlier and with an even greater range of national interests than Hogan suggests.[59] Third, these cartels lasted well into the twentieth century, with the news cartel collapsing in 1934, the wireless cartel expiring in the mid-1940s, and cable cartels

55 Winseck and Pike, *Communication and Empire*, chap. 2 and pp. 169–176.
56 Ahvenainen, *The European Cable Companies in South America*, pp. 282–318.
57 Rantanen, The Struggle for Control of Domestic News Markets, p. 21.
58 Hills, *Struggle for Control.*
59 Hogan, *Informal Entente.*

continuing, albeit in different form, until the present. That is, cartels were a persistent and defining feature of the global media well into the twentieth century. All of this suggests that it is the "*globalization* of control," rather than the "*struggle* for control," that defined the early global media.

New Constellations of Media Power and the Empire of Capital

New technologies of communication also changed the spatial configuration of the early global system and gave rise to new constellations of media power. As James Scrymser observed, by 1885 a message could be sent from Buenos Aires to New York or London in about an hour and a half, but only 400 large businesses accounted for the vast majority of his firm's – the Central and South American Telegraph Company – revenues. The overwhelming majority of their messages, in turn, were sent to just 90 recipients in just 18 North American and European cities: 25 in New York, eleven in Paris, seven in Antwerp and Baltimore, six in Hamburg and Marseilles, four in Bordeaux, two in Bremen, and one each in Montreal, Christiana, Naples, Rome, and Venice (WTC, 1885).[60] Simon Potter similarly notes that new communication technologies had the potential to annihilate the spatial limits of geography, but in reality tended to promote concentration among news agencies and the press *within* the British Empire because of their high cost, while the direct, but slower, patterns of news circulation that had existed among the press in the early nineteenth century was marginalized. As Potter concludes, "telegraphs and cables did not simply reinforce existing imperial networks; rather, they shaped and, to some extent, restricted them." [61]

Potter's analysis of the press system *within* the British Empire applies *across* the global empire of capitalist modernity. As Walter Lippmann noted in 1920 in *Liberty and the News*, "the real censorship on the wires is the cost of transmission. This in itself is enough to limit any expansive competition or any significant independence."[62] The press occasionally took this as a good reason to criticize the cable firms, but news agencies and large press combinations also had good reason to keep silent on the issue. This was because, as

60 Western Telegraph Company, "Superintendant's Letters from Rio 1883–1893," July 12, 1885. Brazil files, Brazil, Box 1.

61 Potter, Webs, Networks and Systems, p. 630.

62 Lippmann, *Liberty and the News*, p. 43.

John Denison-Pender told a British Dominions Royal Commission in 1914, "the large merchant would [...] prefer to pay the higher rates and not bring in the small merchants."[63] With Reuters as one of his top clients, we can be sure that Denison-Pender had it in mind when making these comments.

Press associations allowed the cost of "cable news" to be shared among their members but equally importantly, such organizations limited competition in local and national news markets. This furthered media concentration in many markets, but also spawned press combinations powerful enough to curtail the influence of the global news agencies.[64] Peter Putnis illustrates the point well in his discussion of how Australian and New Zealand newspapers formed their own press associations quickly after the Eastern Telegraph Company's cables reached their shores in the early to mid-1870s.[65] The arrangements helped their members by allowing them to share the high cost of cables and the Reuters news service, while maintaining some competition between the two rival press combinations that had formed in Australia. However, mounting frustration led to the removal of even this limited rivalry in order to create the United Cable Association in 1895, which gathered and edited its own service in London on behalf of the big Australian metropolitan papers and the New Zealand Press Association. In the end, Reuters was left to a much reduced role serving small country papers in Australia but with its profitable private telegram and remittances services still intact.

Menahem Blondheim sees similar outcomes in the US, where the Associated Press(es) and Western Union's "double-headed monopoly" raised the ire of newspapers outside the AP umbrella, reformers and critics, as well as some members of Congress between 1866 and 1900, a period when the commercial media and media regulation took shape in the US.[66] These practices spilt over into Canada, but this time with a rival Canadian press group, the Western Associated Press, turning to regulators to break-up the strangle-hold on news distribution held by the Canadian Press and the Canadian Pacific Telegraph Company, both of whom worked together with AP and West-

63 Britain, Dominions Royal Commission, *Minutes of Evidence Taken in London in November 1913*, HMSO 1914.

64 Blondheim, Rehearsal for Media Regulation; Potter, *News and the British World*; Putnis, How the International News Agency Model Failed; Winseck and Pike, *Communication and Empire*; Rantanen, Foreign Dependence and Domestic Monopoly.

65 Putnis, How the International News Agency Business Model Failed, pp. 10–13.

66 Blondheim, Rehearsal for Media Regulation.

ern Union.[67] In sum, the "global media" bolstered the consolidation of "big media" *within* territorial-based empires and *across* the Empire of Capital. This did not occur in a simple one-way flow of influence, but via a complex struggle for dominance in global, national and city markets.

The Struggle for Control or the Globalization of Control?: The Case of South America

This complex interplay between competition and consolidation is often obscured in the "struggle for control" model. Latin America is often taken to exemplify this view, with the US government cast as standing four-square behind the existing cable companies, US news agencies – the Associated Press, United Press and International News Service – and the new technology of wireless in a bid to overthrow the hegemony of British cables and the French news agency, Havas, on the continent.[68] Many examples can be plucked from archival and published sources to support this view, but counter-examples are not hard to find either. The US-based Central and South American Telegraph Company formed a cartel with John Pender's group of cable companies in South America, for example, that lasted from 1878 until 1891 before cooperation collapsed into what can reasonably be called a "struggle for control" for South American cable markets that lasted until the 1920s. Reuters and Havas also had an agreement between 1869 and 1876 to jointly operate nine bureaus in South America, with Reuters concentrated on exporting news to Europe, while Havas focused on wholesaling the agencies' service to the South American press and its lucrative "private dispatches service."[69] This was part of their arrangements under the "ring

67 Canada, Board of Railway Commissioners, "The Western Associated Press versus The Canadian Pacific Railway Company's Telegraph and the Great Northwestern Telegraphs Company of Canada," *Sessional Papers of the Parliament of Canada*, Ottawa 1910.

68 Key examples of this line of research extend back into the 1920s. See, for example, Shreiner, *Cable and Wireless and their Role in the Foreign Relations of the United States*; Cooper, *Barriers Down*; Schiller, *Mass Communication and American Empire*; Tulchin, *The Aftermath of War*; Kennedy, Imperial Cable Communications and Strategy; Headrick, *The Invisible Weapon*; Hills, *Struggle for Control*. For a recent example, see Rantanen, Foreign Dependence and Domestic Monopoly.

69 Desbordes, Western Empires and News Flows, pp. 3–5; Rantanen, The Struggle for Control of Domestic News Markets, pp. 37–41.

circle" agreement, but it also reflected the high costs of news production and distribution. This did not mean that friction was eliminated, but it was precisely the role of cartels to manage conflict when it did arise. As some agreements fell apart, new ones were created. Thus, while the joint bureaus fell apart, the South American Bureau jointly run by the two agencies out of Reuters' headquarters in London continued from 1874 until 1916, when Havas moved the base of its South American operations to the Associated Press' offices in New York.

As for the cable companies, extensive monopolistic concessions held by the Central and South American Telegraph Company and John Pender on both coasts of Latin America, respectively, impeded others' access to these markets. Yet, in the case of the latter, these were leveraged as a means to manage "ruinous competition," with Pender handing off his company's exclusive concessions in the early-1870s to the Danish-Anglo-based Western and Brazilian Telegraph Company in return for a free hand in developing the southern trans-Atlantic cable market. Again, conflict was not eliminated, but managed, and when this did not succeed, Pender acquired the firm outright, while again turning to the familiar device of the cartel for the same reasons when the French-based South American Cable Company and German-based South American Telegraph Company introduced their own services in 1902 and 1911, respectively.[70]

US interests were involved in South America, but in a modest role before the turn of the twentieth century, with newspapers periodically sending journalists to cover important stories, as the *New York Herald* did during the crisis surrounding the Balmaceda reforms in Chile in 1891.[71] The Spanish-US War (1898) and the Associated Press's consolidation of its position as the pre-eminent news agency in the US led to changes to the "ring circle" agreement in 1903 that gave it exclusive rights in Central America, the Caribbean and the Philippines – essentially overlapping the territorial base of the new American Empire.[72] The "struggle for control" model, however, usually identifies the period from 1914 to 1918 when Britain and Europe were otherwise consumed by World War I, as the decisive moment when US media interests, backed by the US Government, from the State Department all the way

70 Ahvenainen, *The European Cable Companies in South America*, pp. 282–318.

71 All America Cables, *The Central and South American Telegraph Company: Via Galveston*, New York 1895, pp. 1–5.

72 United States, 66th Congress, 3rd Session, *Cable-Landing License Hearings*, Washingtons 1921, p. 4; Read, *Power of News*, p. 86.

up to President Wilson, opportunistically made their bid to overthrow the hegemony of British cables and the French news agency, Havas. Rantanen notes that aspects of this narrative that are "routinely repeated in the scholarly literature" are highly circumspect, but limits her critique to Kent Cooper's claim that he only became aware of how the European cartel blocked AP's access to Latin American newspapers in 1914.[73] I think that the problem is much deeper.

That the major US news agencies expanded their role in South America, first through the 1903 revisions, and subsequently during the war years is not at issue. Indeed, the United Press was the first to enter South American markets in 1916. Within four years, it was serving 23 dailies in Argentina, five in Rio de Janeiro and another in Sao Paulo. For its part, AP served 25 of the foremost papers in South America by 1919, and opened its own bureaus in the most important commercial cities.[74] What is at issue, however, is how the expansion of US media interests in South America is interpreted. The first point to be made is that United Press was not a party to the news agency cartel, and this allowed it to compete with AP and the European agencies in Europe, South America and Asia. While AP was impeded by the cartel's exclusivity provisions, it is also possible to see the cartel not so much an impediment to its ambitions in South America but as the device through which it gained access to markets with the *approval* of Havas, Reuters and Wolff, begrudgingly or not. Indeed, Melville Stone, President of AP at the time, made just this point when claiming that his aim was to expand AP's role in the cartel, not to dismantle it.[75]

This view is supported by the fact that the State Department not only negotiated special arrangements with the All-America Company to offer cheap rates to the US agencies, but that it shepherded negotiations between AP, Havas and other members of the cartel which led up to the 1919 revisions that gave AP access to South American markets to begin with.[76] Moreover, AP agreed to compensate Havas for "lost business" and to offer a condensed version of its service to Havas subscribers. Although the source material on this point is murky, it appears that the All-America

73 Rantanen, Foreign Dependence and Domestic Monopoly, p. 20.

74 United States, *Cable- Landing License Hearings*, pp. 50, 176, 282–284; Cooper, *Barriers Down*, p. 70–77.

75 Associated Press, *M.E.S. in Memoriam: A Tribute to the Life and Accomplishments of Melville E. Stone*, New York 1921.

76 United States, *Cable-Landing License Hearings*, pp. 6–7.

Company's cheaper press rates were extended, with State Department encouragement, to Havas. The main impediment to an enlarged role for any of the US agencies appears not to be the cartel so much as the fact that all cables to South America were groaning under the weight of wartime correspondence. This was aggravated by the poor state of telegraph networks within South America upon which the news services depended to get their services over the "last mile" to clients. This was critical during the War, and continued to be so several years later. William Hawkins, the President of United Press, for instance, told members of Congress that the All-America Company originally had good intentions to boost the amount of news it carried at the cheaper rates brokered by the State Department, but quickly came to view the press

> [...] more or less as a burden. I am sure that has come to be the case with the All-America Company in South America, because I have it on the authority of the officials of that company, and we are more or less in the attitude of supplications; the American press is practically in that attitude toward the cable companies of *the world* (*emphasis added).*[77]

As far as the US news agencies were concerned, they did not seem to care much one way or another who controlled the medium as long as the messages were carried affordably and reliably to and from wherever they wanted to send their services. But what about the position of the cable networks in the supposed "struggle for control" of South American media and communications markets? If the most widely circulated view in the literature is accepted, then they were deeply embattled in the "cable wars" of South America. The following statement of the All-America Cable Company, or more precisely, its legal counsel Elihu Root Jr., to Congress in 1920, appears to present an iron-clad case: [78]

> There is an international struggle going on [...] between the United States and Great Britain for the control of cable communications [...]. The British and American lines [...] get what they can for themselves and, where it is possible, break up the monop-

77 United States, *Cable-Landing License Hearings*, p. 10.

78 Elihu Root Jr., was the son of Elihu Root, Sr., the former U.S. Secretary of War under President McKinley (1897–1901) and of State in Theodore Roosevelt's Administration (1901–1909), both of whom oversaw the U.S.'s transformation into an imperial power after the Spanish-American War in the first place. Western Union, in contrast, was represented by William McAdoo, Secretary of the Treasury in both of the Wilson Administration (1913–1919).

olies of the rivals, and so far the British Foreign Office has actively assisted the British lines and the American State Department has assisted the American lines.[79]

Those who accept the "struggle for control" model typically view the jingoistic statement of the All-America Company as if it is *the* U.S. position in general.[80] It is not. In fact, the company's position was out of sync with other key US interests, notably Western Union, the Associated Press and United Press. Indeed, Western Union had an agreement with the British-based Western Telegraph Company to share certain Latin American markets. The gulf grew even wider when the All-America Company rejected overtures from the Western Telegraph Company and Western Union to join them in an expanded version of the South American cable cartel.[81] Newcomb Carleton, President of Western Union, was explicit that his firm's cables were instruments of commerce, not weapons of politics: "We will not permit ourselves to be used as a coercive measure."[82] The State Department did not stand in unison with the All-America Company, either, but wanted an *international agreement* that would eliminate *all* monopolistic aspects of the global media system.[83] Western Union, however, *opposed* the State Department and the Wilson Administration's attempt to internationalize the issues through the planned World Communications Conference that was to follow the Versailles Treaty negotiations, but which was ultimately never held.[84] Instead of using this case to demonstrate the struggle for control, perhaps a better interpretation focuses on two other aspects. First, it examines the tensions between that model versus the *globalization* of control perspective. Second, it considers the deep cleavages in US political culture between the "corporate nationalism" of the All-America Company, on the one side, versus the "corporate internationalism" of Western Union, on another, and the liberal internationalism of Woodrow Wilson and elements of the State Department, on yet another.

79 United States, *Cable-Landing License Hearings*, p. 140.

80 See Headrick, *The Invisible Weapon*, pp. 178–179; Hills, *Struggle for Control*, p. 185; Tulchin, *The Aftermath of War*, pp. 207–211; Shreiner, *Cable and Wireless and their Role*, pp. 136–137.

81 United States, *Cable-Landing License Hearings*, pp. 59–71; Western Union, *Annual Report*, New York 1920, 18.

82 United States, *Cable-Landing License Hearings*, p. 107.

83 United States, *Cable-Landing License Hearings*, p. 39.

84 Ibid.

Conclusion

In this chapter I have argued that there are several approaches to the study of global media history: "the struggle for control" model, the "circuits of news" approach, the "globalization of news" view and, lastly, one advanced here that situates the global media in the context of the empire of capitalist modernity. Each approach has its strengths, with the "struggle for control" perspective illuminating the connection between communication media, national security and territorial imperialism. Such ties shed light on how the control of media infrastructures plays a vital role in militarism and imperialism. The approach, however, places too much stress on communications media within the territorial boundaries of competing empires, neglecting how the "empire of capital" constituted a much larger terrain upon which the forces of capital, modernization, imperialism, crises and cosmopolitanism played out. The circuits of news approach makes a valuable contribution with its focus on the overlapping flows and temporality of communication and culture and how these worked to simultaneously foster the culture(s) of empire while yielding nationalist and anti-imperialist movements that challenging Innis's idea that the solidity of the imperial edifice was an index for the efficiency of its communications system. The circuit of news approach, however, appears to be confined to the study of these processes *within* territorially-based empires, and to politico-military and cultural crises. As this chapter has shown, in contrast, networks and news flows did not conform to the territorial limits of imperial power but rather flowed in ways that coincided with the composition of capital, cooperation among empires, and the more cosmopolitan sensibility of the age.

Moreover, I have shown that crises were also financial, and this revealed that interdependence did not just harmonize space but made it prone to the destabilizing force of politico-military *and* financial crises. The case of the dispute between Reuters and foreign investors with regard to the former's coverage of South America in the midst of a financial market meltdown revealed just that. Just as importantly, that case is also one which, to my knowledge, has not been investigated by other scholars, probably reflecting the overly-politicized narrative of the "struggle for control" model. The preliminary discussion here, it is hoped, reveals the importance of shining a light on the role of the global media in the context of *financial crises* but also how the "circuits of news" approach can usefully inform future studies of this topic.

Lastly, the globalization of news has made valuable contributions over several decades and that continues to be the case, especially in light of the recent turn to focusing on the complex interactions between the global news agencies, on the one hand, and competition and consolidation in metropolitan and national media markets, on the other. That said, however, there is still ambivalence in this approach about the nature of the historical period being covered, with some authors taking events *within* the bounds of territorial imperialism, notably the British Empire, as their framework while others tie *some* aspects of their work, unnecessarily in my view, to the "struggle for control" narrative while placing events in a more *cosmopolitan* frame of reference at other times.[85]

In short, the discourse of media and nation continues to loom large in the field, but with a view to how the linkages between the two played out on the international stage. Perhaps this is not surprising given the dominant role that the media-nation coupling has played in communication and media studies. This paper shows, however, that we should avoid juxtaposing things in an either/or fashion, as in imperialism *or* globalization, nationalism *or* internationalism, the state *or* capital, and so forth. This tendency is likely an outcome of what Ulrich Beck calls methodological nationalism, which he claims has dominated sociology for the last century and is rooted in dubious research methods that take the nation-state as the primary container of social action and the pivot upon which events in the "real" world are assumed to turn.[86] Beyond the mental boundaries of methodological nationalism, however, sit a far more dynamic set of interactions that shaped space and place and the vital role played by the global communications media in such a context. In my view, rather than "either/or" we need to develop perspectives that illuminate the interplay, as Schmidt and Long as well as David Harvey tell us, between all of the just-mentioned forces that defined the late nineteenth and early twentieth centuries.[87] So, not either/or, but "and" offers greater potential to understand how globalization and imperialism set the context within which the global media emerged and consolidated as a *system*. It is

85 For example, Simon Potter and Peter Putnis have undertaken most of their research within the framework of the British Empire (see n. 4 and n. 4 respectively). Terhi Rantanen ties aspects of her work to "the struggle for control" in "Foreign Dependence and Domestic Monopoly" but offers a more cosmopolitan approach in "The Cosmopolitianization of News."

86 Beck, *Power in the Global Age.*

87 Long and Schmidt, *Imperialism and internationalism*; Harvey, *The New Imperialism.*

also an approach that I believe offers more insights into the present, as well as the sinews that link us to the past.

Bibliography

Ahvenainen, Jorma, *The European Cable Companies in South America*, Finland 2004.

All America Cables, *The Central and South American Telegraph Company: Via Galveston*, New York 1895.

Arrighi, Giovanni, *The Long Twentieth Century: Money, Power and the Origins of our Time*, London 1994.

Associated Press, *M.E.S. in Memoriam: A Tribute to the Life and Accomplishments of Melville E. Stone, General Manager of the Associated Press, 1893–1921*, New York 1921.

Bayly, Charles, "Informing Empire and Nation: Publicity, Propaganda and the Press 1880–1920," in: Hiram Morgan (ed.), *Information, Media and Power through the Ages*, Dublin 2001, pp. 179–201.

Beck, Ulrich, *Power in the Global Age*, London 2005.

Blondheim, Menahem, "Rehearsal for Media Regulation: Congress Versus the Telegraph-News Monopoly, 1866–1900," *Federal Communications Law Journal*, vol. 56 (2004), pp. 299–328.

Boyd-Barrett, Oliver and Terhi Rantanen (eds.), *The Globalization of News*, London 1998.

Britain, Dominions Royal Commission, *Minutes of Evidence Taken in London in November 1913*, HMSO 1914.

Britain, Pacific Cable Committee, *Report, Minutes and Proceedings, etc.*, London 1899.

Britain, Second Report of the Interdepartmental Committee on Cable Communications, London 1902.

Britain, *The Telegraph Act, 1869*, in: *Public General Statutes*, 32–3 Vict., Chap. 73, London 1869.

Canada, Board of Railway Commissioners, "The Western Associated Press versus The Canadian Pacific Railway Company's Telegraph and the Great Northwestern Telegraphs Company of Canada," *Sessional Papers of the Parliament of Canada*, Ottawa 1910.

Carey, James, *Communication as Culture*, Boston, MA 1998.

Conant, Charles, "The Economic Basis of Imperialism," *The North American Review*, vol. CLXVII (1898), pp. 326–346.

Cooley, Charles, *Social Organization: A Study of the Larger Mind*, New York 1909/29.

Cooper, Kenneth, *Barriers Down*, New York 1940.

Cooper, Robert, "Why We still Need Empires," *The Observer*, London, 7 April 2002.

Desbordes, Rhoda, "Western Empires and News Flows in the 19th Century," Paper presented at the International Association of Media and Communication Research Conference, Porto Alegre 2004.
The Economist, "The False Argentine Telegrams," 19 December, 1874.
The Economist, "The False Argentine Telegrams," 9 January, 1875.
The Economist, "The Argentine Telegrams," 16 January, 1875.
Feis, Harold, *Europe: The World's Banker, 1870–1914*, New Haven, CT 1934.
Ferguson, Niall, *Empire*, New York 2003.
Gallagher, John and Ronald Robinson, "The Imperialism of Free Trade," *The Economic History Review*, vol. VI (1953), pp. 1–27.
Globe Telegraph and Trust (1872–1880) *Board Meeting Minute Book*, vol. 1, 25 April, 1873.
Globe Telegraph and Trust Company, *Memorandum of Articles of Association*, 11 July, 1873.
Globe Telegraph and Trust Company, *Minutes*, 1 January, 1874.
Hardt, Michael and Antonio Negri, *Empire*, Boston, MA 2000.
Harvey, David, *The New Imperialism*, Oxford 2003.
Headrick, Daniel, "Double-Edged Sword," *Historical Social Research*, vol. 35, no. 1 (2010), pp. 51–65.
—, *The Invisible Weapon*, Oxford, CA 1991.
—, *The Tentacles of Progress: Technology Transfer in the Age of Imperialism*, New York 1988.
Held, David, Anthony McGrew, David Goldblatt and Jonathan Perraton, *Global Transformations*, Stanford 1999.
Hills, Jill, *The Struggle for Control of Global Communications*, Chicago, ILL 2002.
Hoag, Christopher, "The Atlantic Telegraph Cable and Capital Market Information Flows," *The Journal of Economic History*, vol. 66 (2006), pp. 342–353.
Hogan, Michael, *Informal Entente*, London 1977.
Ignatieff, Michael, *Empire Lite*, Toronto 2003.
Innis, Harold, *Empire and Communication*, Oxford 1950.
Jones, Geoffrey, *Multinationals and Global Capitalism*, New York 2005.
Kaul, Chandrika (ed.), *Media and the British Empire*, London 2006.
Kennedy, Paul, "Imperial Cable Communications and Strategy, 1870–1914," *English Historical Review* (Oct. 1971), pp. 728–748.
Kindleberger, Charles, *Manias, Panics and Crashes*, London 2006.
Koskiennemi, Martti, *The Gentle Civilizer of Nations: The Rise and Fall of International Law, 1870–1960*, New York/Cambridge, U.K. 2004.
Lee, Kelly, *Global Telecommunications Regulation*, London 1996.
Lippmann, Walter, *Liberty and the News*, New Brunswick, NJ 1920.
Long, David and Brian Schmidt (eds.), *Imperialism and Internationalism in the Discipline of International Relations*, Albany, NY 2005.
Lukes, Steven, *Power: A Radical View*, London 1974.
Marichal, Carlos, *A Century of Debt Crisis in Latin America*, Princeton, NJ 1989.

Mattelart, Armand, *Mapping World Communication*, Minneapolis 1994.
O'Rourke, Kevin and Jeffrey Williamson (eds.), *Globalization and History*, Cambridge, MA 1999.
Potter, Simon, *News and the British World*, Oxford 2003.
—, "Webs, Networks and Systems: Globalization and the Mass Media in the Nineteenth and Twentieth Century British Empire," in: *Journal of British Studies*, vol. 46 (2007), pp. 621–646.
Putnis, Peter, "How the International News Agency Business Model Failed – Reuters in Australia, 1877–1895," *Media History*, vol. 12 (2006), pp. 1–17.
Rantanen, Terhi, "Foreign Dependence and Domestic Monopoly: The European News Cartel and U.S. Associated Presses, 1861–1932," *Media History*, vol. 12 (2006), pp. 18–35.
—, "The Cosmopolitianization of News," *Journalism Studies*, vol. 8 (2007), pp. 843–861.
—, "The Struggle for Control of Domestic News Markets," in: Boyd-Barrett, Oliver and Terhi Rantanen (eds.), *The Globalization of News*, London 1998, pp. 35–48.
Read, Donald, *The Power of News: The History of Reuters*, New York 1999.
Reinsch, Paul, *Public International Unions*, Boston, MA 1911.
Rosenberg, Emily, *Financial Missionaries to the World*, Cambridge, MA 1999.
Schiller, Herbert, *Mass Communications and American Empire*, Boston, MA 1969.
Shreiner, George, *Cable and Wireless and their Role in the Foreign Relations of the United States*, Boston, MA 1924.
Sklar, Martin, *The Corporate Structuring of American Capitalism, 1890–1916*, Cambridge, MA 1988.
The Times, "The Atlantic Telegraph and the Panic," 6 Oct 1873.
The Times, "Money Market and City Intelligence," 19 April 1875.
The Times, "Money Market and City Intelligence," 20 April 1875.
Tulchin, Joseph, *The Aftermath of War*, New York 1971.
United States, 66th Congress, 3rd Session, *Cable-Landing License Hearings*, Washington, DC 1921.
United States, House of Representatives Committee of Interstate and Foreign Commerce, *Pacific Cable*, Washington, DC 1902.
United States, House of Representatives, *Pacific Cable Bills*, Washington 1902.
Western Telegraph Company, "Superintendant's Letters from Rio 1883–1893", July 12, 1885.
Western Union, *Annual Report*, New York 1920.
Winseck, Dwayne and Robert M. Pike, *Communication and Empire*, Durham, NC 2007.

Globalizing the Telegraph: The ITU and the Governance of the First Globalization of Telecommunications

Léonard Laborie

The transnational circulation of knowledge and technologies was elemental to the early development of world-wide telegraph networks.[1] Yet, on top of this, the globalization of telegraphy also required the transformation of a patchwork of national networks into an interacting international system. This transformation process occurred under the auspices of the International Telegraph Union (ITU), a political, social and institutional framework designed for shaping technological and commercial interconnections, or, to put it differently, for channeling "increases in exchange, in interrelationships and in circulation" between telegraph networks.[2]

The dominant "struggle for control" interpretative model of international communications developments usually defines international communication networks as weapons of power subject to a struggle for control among nations.[3] At the same time, however, Daniel Headrick illustrates that 'rival' countries also cooperated and that "international cooperation preceded disagreements."[4] Similarly, Pascal Griset acknowledges in his analysis of competing companies over transatlantic communications that the ITU provided an "indispensable framework" for the expansion of international telegraphy.[5] Indeed, cooperation not only preceded, but also survived disagreements and coexisted with them. As a consequence, cooperation was by no means the opposite of rivalries. Whether we call it cooperation or the global governance of telegraphy, both make clear that the 'struggle for control' model falls short to fully interpret this larger historical phenomenon of global communication electric. Following other historians, I argue that this new kind of global

1 See for instance circulations between Russia, German States and Great Britain in the 1830s in Gouzevitch, L'essor.

2 Conrad, *Globalisation*, p. 4.

3 See Dwayne Winseck's chapter in this book.

4 Headrick, *The Invisible Weapon*, p. 12.

5 Griset, *Entreprise*, p. 144.

political arrangements marked a decisive turn in the longer world history of globalization.[6]

Dwayne Winseck and Robert Pike opened new vistas on the globalization of telegraph networks.[7] Focusing on all forms of arrangements between actors, from cartels to intergovernmental cooperation, they highlighted global trends and facts that were elided by analyses limited to national power strategies and methodological nationalism. In line with this transnational approach, this chapter considers the ITU from its origins to the 1900s as a major site of globalization. Such a focus helps us to rethink the role of network technologies in globalization. Telegraphy functioned as a globalizing agent not only because it "directly affect[ed] the very constituents of globalization: global exchanges, movements, transfers, flows," but also because telegraphy itself was globalized.[8] Actors involved in the ITU administered the world-wide expansion of telegraph networks and thus information flows. According to Jorma Ahvenainen, technical innovation, the pressure of demand and competition between companies lowered the cost of international telegraphy; together with "the common framework for the operation of the telegraph provided by the ITU," this enabled the globalization of the telegraph business.[9] Digging into this 'common framework' that multiple actors consciously built together allows us to understand how the resulting global telegraph network embedded socio-cultural and political relations, reflecting homogenization as much as delimitation and polarization. 'Common' is in no way equivalent to homogeneous nor so flat as the institution's rhetoric or analysis focusing only on the ITU's creation of connectivity might suggest.[10]

The chapter is organized in four sections. Section one explains why telegraph networks became interconnected and how multilateral cooperation progressively channeled the process. Section two scrutinizes how diplomats and European public telegraph administrations together made choices and negotiated interconnections in the long run, reconfiguring, not erasing, national boundaries and sovereignty. Section three introduces private interests into the picture. The governance of the global telegraph network soon had

6 Stearns talks about a "political globalization" going on with the emergence of first intergovernmental organisations, in *Globalization*, p. 6 and 159.

7 Winseck /Pike, *Communication and Empire.*

8 Wenzlhuemer, Editorial, p. 15.

9 Ahvenainen, The International Telegraph Union, p. 76.

10 Cowhey insisted similarly on the politics of international regimes in "The international telecommunications regime."

to perform the double and often contradictory task of protecting national public monopolies and facilitating the expansion of private companies that connected Europe to other continents. The cartel-like agreements reached after complex bargains reflect deeply entangled interests. Finally, section four discusses the multilevel connection between international cooperation and power stakes. It provides an overall interpretation of the globalization of the telegraph, which was a highly ordered and ordering process.

Interconnecting Telegraphs Networks: The Multilateral Way

Unconnected telegraph networks would have been very poor agents of globalization. But why and how did telegraph networks become interconnected?

The Power of Demand

There may be a "unifying tropism" of networks, namely that a network is ever more attractive as it connects to others and provides the opportunity to reach a greater number of destinations/people/information.[11] Known as Metcalfe's law, this tropism states that the value of a network increases with the square of the number of participants. Viktor Mayer-Schönberger and Deborah Hurley concluded their chapter on the "Globalization of Communication" at the end of the twentieth century with the typical idea that "consequently, digital networks feature an internal dynamic driving toward global interconnection."[12] This 'law' explained the rise of the internet. But similar mechanisms were also at play one and half centuries ago during the first era of globalization.

The law refers to 'participants' and so points to one crucial criterion, namely the demand and the openness of the network for everybody to use it. Here I stress the interconnecting power of demand. Telegraphs networks have not always been open to the general public. In France for instance, it was not until 1851 that the public was allowed to access a network primarily built for the government and the military. It then appeared to the government that, compared

11 Brun, La mise en réseau des réseaux, p. 329.

12 Mayer-Schönberger/Hurley, Globalization of Communication, p. 140.

with the optical telegraph, the electric telegraph could be opened to the public without compromising the security of its own communications. This was due to the invisible nature of the signals, the new traffic capacities and the ability to control and censor private dispatches in telegraph offices. The network had already been opened to the public by 1846 in Belgium, in 1849 in Prussia, in 1850 in Austria. As soon as the telegraph was commercialized, operators had a strong interest in coping with public needs to communicate beyond national borders – as it already sent ordinary postal mail beyond borders. Quicker than the telephone would do a few decades later, the telegraph soon allowed the transmission of information over long terrestrial distances, though at the expense of massive investment in copper wires of large section and later in relays. France had no international telegraph connection before the opening of the network to the public. It had several of them a few months after. Technology enabled this, but it was not the only force driving interconnection.

Interestingly, the first international connections occurred just after the transnational revolutions of 1848–1849. Austria and Prussia became interconnected in 1849. Revolutionary means of fast communication might have appeared to conservative regimes of the time as vital links for their common future. In any case, governments could interconnect their networks without any fear of their inner domestic communication becoming transparent to foreign intrusive ears or eyes. Lord Henry Palmerston best expressed early governmental expectations regarding international telegraph communication. For the British Foreign Minister, accessing foreign telegraph networks via a submarine cable connecting Great Britain to the continent was really a matter of tapping into a foreign distant resource. In his July 1851 note, "the free use of continental Electric Telegraphs" referred to the almost incredible possibility of using networks that did not belong to the British.[13] He envisioned three major benefits to that. It would accelerate governmental communication with its diplomatic apparatus abroad, enable "the rapid conveyance of intelligence of a political and military nature from China and from our Indian and other possessions in the East, as well as from our fleet and ports in the Mediterranean," and lastly, provide British newspapers with fresh information valuable to the government and the whole country.[14] The instantaneous transmission of intelligence to and from governmental authorities together with access to breaking news were attractive prospects

13 National Archives (Kew), FO 97 / 197: Foreign Office to the Board of Trade, 04/07/1851.
14 Idem.

when states representatives started considering international connections. In that sense, a key moment in the early interconnection and expansion of international telegraph lines was the Crimean War, with the erection in 1855 of an infrastructure enabling communication between London, Paris and the battlefield through the Austrian empire and the Black Sea.[15] Official communications had thus always benefited priority of transmission over all other types of communications according to the treaties regulating international telegraphy. The international convention signed in Paris in 1865 granted access to international lines to everybody (article 4) and guaranteed secrecy to all (article 5), but assigned a special status to state dispatches originating from head of state, ministers, commanders in chief of terrestrial or naval forces and diplomats (article 7, 11).[16]

Beside governments, businesses were also early advocates for and became principal consumers of international facilities. Railways, stock exchanges, banks, news agencies and the press were the most prominent groups.[17] At the end of the 1850s, French telegraph stations corresponded with about 2000 telegraph stations abroad and exchanged around 145000 dispatches with them per year.[18] This represented a little bit less than one third of the domestic correspondence, but supplied almost as much revenue. As section two will explain, international telegraphy was and would for a long time be a highly profitable activity for public telegraph operators.[19] Commercial and industrial agents provided 35 percent of the international traffic, stock exchanges 23 percent, newspapers 10 percent, the commerce of cereals and diplomacy 5 percent each. The rest, around 22 percent, was linked with family correspondences.[20] In the context of a liberalizing international economy and growing international flows of goods from the 1860s onwards, the need for international communication became stronger and the use of the telegraph intensified international exchanges.[21]

15 Headrick, *The Invisible Weapon*, p. 17.

16 *Documents diplomatiques de la conférence télégraphique internationale de Paris*, Paris, Imprimerie nationale, 1865, pp. 9–12.

17 See the chapter in this book by Volker Barth.

18 Laugel, Augustin, “La télégraphie électrique en France”, *Revue des deux mondes*, no. 31 (1861), pp. 892–914, p. 900.

19 In 1859, 2.07 and 1.95 millions francs for domestic and international correspondence respectively. Idem.

20 Ibid.

21 The telegraph lowered the cost of trade by increasing the capacity utilization of ships, which led to greater market integration, analysed Lew, B./Bruce Cater, The telegraph. Co-

Multilateral Cooperation

At the intersection of these two forces (technology and demand), international technical cooperation gathering diplomats and experts of the field shaped the globalization of telegraphy. By 1865, various multilateral treaties merged into a unique set of regulations, called the International Telegraph Union.

At the beginning, connections were made through the establishment of interface offices where dispatches were translated and manually forwarded from one network to the other.[22] Between Prussia and Austria, the exchange office with telegraphers from both states was in Öderberg. "Time losses and causes of errors that could be due to such transshipment" prompted the adoption of another solution, with direct through lines necessitating common end-line apparatus and code.[23] In this matter, the widespread adoption of the Morse system, with a "continental" European adaptation of the code by Friedrich Gerke, was crucial to the swift expansion of international telegraphy.[24] While France had taken quite a different technological path, the then new director of the telegraph administration, Henri de Vougy, took the decision to adapt the French network to this system in 1855.

Negotiations on operational, technological and commercial matters that went along with the establishment of such lines were almost immediately multilateralized. During the 1850s, three international telegraphic spaces emerged in Europe around Prussia and France.[25] The first gathered Prussia, Austria, Bavaria and Saxony, soon joined by other German states, in an Austrian-German Telegraph Union. The second was the result of a Belgian push for an agreement with the Austrian-German Telegraph Union and France. The resulting 'mixed' convention was signed in October 1852 after a year and a half of discussions. By this time, there were less than twenty telegraph stations in France. This country passed a new special multilateral convention with its neighbors (Belgium, Switzerland, Spain and the Kingdom of Sardinia) in December 1855, creating a third telegraph space. It is still unclear

---

operation reflected this practice, telegraph conventions taking provisions as soon as 1865 for international communication with ships at sea through coastal semaphore services. Article 43. *Documents diplomatiques*, p. 26.

22 Carré, Archéologie.

23 Bergon, L., "Du perfectionnement apporté par MM. Digney à l'appareil Morse", *Annales télégrahiques,* Juli–August 1858, pp. 7–15, p. 7.

24 Huurdeman, *Worldwide*, pp. 143–144.

25 Three spaces, and not only two as it is frequently presented. For a comparative analysis of the telegraph agreements signed during this period, see Laborie, *L'Europe*, pp. 72–81.

why France decided to pass this convention, though diplomatic tensions with Prussia and Austria caused by the Crimean War might have motivated it. A Belgian newspaper suggested that the French were frustrated by the latest negotiations with the Austrian-German Union in Berlin in June 1855 and wanted to offer more advantageous rates, based on the decimal system for price and distance measurement.[26] The convention's explanatory memorandum indeed exposed that it was drafted not only to offer uniform but "fair" tariffs.[27] The Netherlands, Belgium and Switzerland served as gateways between these three telegraphic spaces. Oriented toward the fostering of international flows of capital, information and goods, they constantly pushed for the coordination, if not the merger, of the telegraphic conventions on the basis of the "mixed" one.[28] The French government itself became receptive to the argument. On the eve of the conference convened in Brussels in 1858 to revise the "mixed" convention, it encouraged its diplomatic representative, "by mutual concessions, to take some more steps toward a definitive entente."[29] Governments finally reached a compromise a few years later in 1865. After 78 days of negotiations in Paris, they created the ITU.

This process and the resulting ITU cannot be presented as byproducts of competition between nation-states or inter-imperial rivalries. After all, Great Britain, the main imperial power of the time, was not a founding member. It joined the ITU only in 1871 once its telegraph networks had been nationalized. Rather, multilateral cooperation and the way it gained momentum in 1865 resulted from three main forces. First, the search for the reduction of transaction costs between national telegraph operators in the context of a growing international traffic. The increase of international demand made it more and more difficult for telegraph agents to handle the service correctly and for accountants to dispatch the taxes to the various participating countries. As a contemporary telegraph engineer observed, "a dispatch could be submitted to different rules along the different parts of its journey; there was even such portion of territory where it was just impossible to know which principle should be followed."[30] Turkey, Greece and Serbia for instance had

26 *L'indépendance belge*, 04/01/1856.

27 "Décret impérial portant promulgation de la Convention conclue entre la France, la Belgique, l'Espagne, la Sardaigne et la Suisse, pour régler la transmission des dépêches télégraphiques" (12/04/1856), *Lois et règlements*, tome III, 1856, 12 p., p. 1.

28 On the role of Switzerland: Grossi, Technologie, and Balbi G. et al.. Bringing.

29 Quoted in Bruley, *Le Quai d'Orsay*, p. 568.

30 Saveney, Edgar, "La télégraphie internationale. Les anciens traités, et la conférence de Paris", *Revue des deux mondes*, 15/09/1872, pp. 359–384, p. 366.

joined the mixed convention under its Berlin provisions (1855), but failed to follow its Brussels revision (1858).[31] Second, a major reform of national telegraph rates accelerated the process of standardization. Inaugurated in the Netherlands in 1858, the reform enforced a uniform rate for a dispatch whatever its journey within the country. Countries with comparatively larger territories followed suit at the beginning of the 1860s as a result of both technological improvements and the "postalisation" of the telegraph. The Hughes apparatus almost doubled the capacity compared to simple Morse apparatus. "Postalisation" was a tendency to align the management of the telegraph with the management of the Post Office as a public service. This made the telegraph more accessible in terms of rates and network spanning. In France, some observers saw this policy as a valuable follow-up to free trade options taken by Napoleon III by 1860. In their eyes, France had to improve its own transport and communication infrastructures to be able to cope with international competition.[32] Thus domestic dynamics largely fueled by international trends quickly impacted multilateral settings. New national rates promoted a new wave of agreements on international rates, at first with neighboring countries. This complicated the picture in Europe and created both the need and the opportunity to take new uniform international provisions. Third, the renewal of a "European concert" diplomacy prompted the creation of the ITU.[33] In the aftermath of the Italian war, Napoleon III and his minister Drouyn de Lhuys tried to restore dialogue in Europe on issues of common interest. They unsuccessfully invited a congress in Paris to deal with the Polish question in 1863. They also tried to enlarge the concert's scope both spatially and thematically. Navigation, communication and finance were to be on the agenda of several meetings linking diplomacy with technical issues. International telegraphy fitted into this pattern. Indeed a French telegraph engineer in 1865 suggested that "the telegraphic side may be not the most important side of the international conference."[34]

Cooperation at play in interconnection was thus not a technologically but socially driven process. Hereafter I stress the actors and effects of this social construction.

31 Ibid.

32 De Siorac, "La télégraphie française sous l'administration de M. le vicomte de Vougy," *Annales télégraphiques*, tome 8, July–August 1865, pp. 461–466, p. 462.

33 Bruley, Le concert, p. 149.

34 De Siorac, p. 463.

Interconnected Telegraph Networks: On Some Durable Lines of Tensions and Asymmetries

Some reporters saw the 1865 international telegraph convention as the legal framework that turned the various national networks into "a kind of common good that the people of the old continent can use."[35] Actually, coordination and standardization produced connections and convergences, which in turn globalized telegraph networks. But this new commonality involved tensions and gave birth to or reinforced asymmetries like every process of globalization. Debates and provisions turned around two lasting issues: enlarging or deepening the networks integration and articulating national sovereignty with international regulation. They were at play in each of the three types of norms designed through international cooperation: operating rules, technical standards, commercial rates.

Operating Rules

The topics debated at the conventions included, as we have already seen, the priority of messages, censorship, the languages permitted, codes and ciphers as well as the maximum number of words permitted, the opening hours of offices and the very services provided to customers. Negotiations shaped in detail the substance of the global network.

Among the novelties introduced in 1865, delegates decided that secret codes and ciphers would no longer be reserved to state dispatches. Each contracting state nonetheless had the liberty to ban this facility for private correspondence from or to its territory. Only Austria and Spain would use this provision. Delegates also agreed on the enforcement of a permanent service between capital cities of contracting states – main offices would be open day and night. Capital cities thus reinforced their role as nodes of the international telegraph network. This became the basis of a problem ten years later, when Austria again opposed the inscription into the convention of the possibility to sign bilateral or multilateral agreements offering special subscription fares to newspapers and news agencies for the use at night of

35 So spoke the French journalist and republican politician Alfred Darimon, former secretary of Joseph Proudhon: "La télégraphie universelle," *La Presse*, 28/10/1865, p. 2.

international telegraph lines.[36] The first agreement had been signed between France and Great Britain. Those opposed to its generalization brought up concerns about the cost and the health of the workforce necessary to provide the service. The Austrian delegates officially expressed their fear that their press would follow the example of other nations to ask for such a service. But this time Belgium, France, Germany, Great Britain, the Netherlands and Switzerland gained support from the majority. The director of the Italian telegraph administration, Ernesto d'Amico, explained to the Austrians in an ironic voice that foreign examples should not prove a problem to countries where good governmental reasons could be opposed to public claims.[37]

Registered dispatches were no longer inscribed in the list of basic international services after 1871. Discussions between governmental agents over this case are another example of the tensions surrounding the globalization of telegraph services. The service was too expansive for some operators. The majority preferred to side line it in order to gain participation to the general agreement from the largest number of countries possible.[38] This case exemplifies that in the early process of telegraphic globalization, participants prioritized enlarging over deepening cooperation. This would afterwards motivate the promulgation of restricted conventions producing spaces more deeply connected around political or economic affinities.

Technical Standards

In their work on telecommunication standardization of the last decades of the twentieth century, Suzanne Schmidt and Raymund Werle explained that "by no means do the participants simply execute what an invisible 'techno-logic' prescribes. Standardization is a process of choosing among several alternatives or options, none of which is necessarily the 'best'."[39] In the late nineteenth century, the ITU was no more politically or economically neutral

36 *Documents de la conférence télégraphique internationale de Saint-Pétersbourg publiés par la bureau international des administrations télégraphiques*, Berne, Rieder et Simmen, 1876, pp. 78, 408–410, 476–481.

37 Idem, p. 481.

38 Saveney, Edgar, "La télégraphie internationale. Les conférences de Vienne et de Rome", *Revue des deux mondes*, 01/10/1872, p. 551–583, p. 578.

39 Schmidt /Werle, *Coordinating*, p. 305.

in its choices about terminal equipment, unit measurements or technical features of international telegraphic lines.

The fact that all countries agreed on adopting the Morse system was decisive for the early stage of international cooperation. But further technological developments caused debates. It was in particular discussed whether all member countries should have to equip their network with the latest and most expensive devices, such as the Hughes apparatus. As with the registered dispatch, the answer was negative. One major limit to the homogenization of the networks was the necessity to preserve financial equilibriums of the more fragile, or cautious, participants to international cooperation.

The discussions around the Italian proposal to create a common school for the training of telegraph networks managers at the beginning of the 1870s sheds light on another aspect of the tension between member nations. According to Ernesto d'Amico, there was a severe lack of human resources in Italy and in most of the countries to manage the growing telegraph networks.[40] He saw pooling training resources as the solution. The director of the ITU's international bureau passed on the proposal to the member countries but he was not convinced.[41] To him, difficulties arose on three seemingly practical but in fact political issues. The first two touched upon the double choice of a teaching language and of a city for hosting the school. If a big European capital city were more appropriate than Bern to host all the necessary professors and equipment, then "scientific neutrality" might not be guaranteed.[42] The core problem was the probable reluctance of governmental administrations to take any risk by allowing the development and operation of their domestic network to occur along the same technical lines as the one learnt at school by a kind of new cosmopolitan technical elite. Although the Spanish, Ottoman and Russian administrations responded positively, the majority proved to be reluctant indeed.[43] Belgian and German partners proposed intermediate solutions in the form of exchange of officers, quarterly meetings or short training sessions.[44] Finally, the project was simply

40 ITU archives (Berne), Correspondence of the international bureau, year 1872, 132 / 1: E. d'Amico to C. Lendi, 06/07/1872.

41 Idem, 132 / 3: Charles Lendi, minute to the administrations, 24/07/1872.

42 Ibid.

43 Ibid., 132 / 10: direction générale des télégraphes au directeur du bureau international, Moscou, 14/08/1872; 132 / 13: direction générale des télégraphes au directeur du bureau international, Madrid, 11/09/1872; 132 / 12: ministère des Postes et des Télégraphes au directeur du bureau international, Constantinople, 05/09/1872.

44 Ibid., 132 / 3: international bureau, minute 24/11/1872.

abandoned. Opponents and proponents alike agreed on the fact that this emerging field was and would be crucial in the future for national sovereignties. But opponents had the feeling that the project endangered more than strengthened them.

Rates

Rates were the subject of the most heated debates. At the Paris 1865 conference, "the discussion, up to that point calm and regular, [became] tumultuous and fuzzy," reported an observer.[45] On transit taxes, members like Spain who proposed a uniform charge per country regardless of the size of the territory opposed those like France, which thought that such an equation would provide too great an advantage to small countries. Those members proposing a uniform terminal due throughout the member countries opposed members like Russia that wanted each operator to be able to fix it through case by case agreements and to make it vary depending on the distance to the border of the emitting or receiving station.

At some point during the discussions, France proposed a uniform international rate of six francs for any international dispatch, regardless of the distance and the number of transit countries.[46] This system would have been advantageous for long distance relations, which would then be subsidized by short-distance communications. Motivated by simplicity and smoothness in international communication, uniformism favored a flat world, or at least a flat continent, where distances and political discontinuities might have become abstractions that decreasingly influenced information flows. The abstraction of distance in international pricing was first imagined for postal correspondence and became the rule within the Universal Postal Union at the end of the 1870s. In the telecommunication world instead, realism – in fact a different kind of vision of what the telegraph was and its functions – took over and imposed another path. Terminal dues would depend on the area of each country and on mutual concessions. This complicated international accountability considerably, as pricing differences from or to a same

45 Saveney, La télégraphie internationale. Les anciens traités, p. 371.

46 The conference adopted the franc as the account currency and French as the common language for discussions during conferences, writing of the convention, and exchange of service dispatches. With the strong support of Switzerland and France, French and franc both became symbolic wheels of the globalization of the telegraph.

country depended on the origin or destination of the dispatch. Bilateral bargains monitored the system and meant that geography (distances) and even geopolitics (political distances) remained lasting factors in pricing.[47]

In the long term though, the most important asymmetry resulting from the ITU was the non-parallel evolution of national and international rates. European traffic became a cash cow to finance the constant lowering of national rates and expansion of domestic networks.[48] The explanatory memorandum of the international convention signed in Paris and revised periodically explained that the delegates wanted primarily "to guarantee to the telegraphic correspondences [...] the advantages of a simple and reduced rate."[49] International rates were indeed reduced and simplified, though not as much as some had hoped. But whereas national rates continually fell during the following decades, international rates almost stagnated.[50] From 1865 to 1903, a twenty-word domestic dispatch would halve in price in France, while a twenty-word dispatch between Paris and Lisbon only reduced in price by 20 percent and that was only thanks to a special agreement between these two countries. Between Switzerland and Russia, the price even increased from 9 francs to 9.3 francs. In the meantime, a transatlantic cable from Bern to New York dropped from 540 francs in 1866 to 30 francs in 1897. Emile Frey, director of the International Telegraph Bureau in Bern, pointed out the paradox that domestic correspondence was charged at a minimum with a maximum cost price, while international correspondence was being charged at a maximum with a relatively lower cost price.[51] To put it simply, profitable European international services subsidized loss-making domestic services.

Nations tended to defend or even build their sovereignty and infrastructures in and through cooperation. Transnational dynamics did emerge with collective decisions affecting national settings. All in all, the process of global interconnection of the telegraph networks did not erase national boundaries but reconfigured them around a new kind of common good that both made nations more interdependent and people more conscious of this. It soon

47 *Documents diplomatiques*, p. 37–40.

48 The situation kept unchanged until the deregulation of the 1980s–1990s. Zacher /Sutton, *Governing*, p. 166.

49 *Documents diplomatiques*, p. 1.

50 Figures below come from Frey, Emile, *La télégraphie et la téléphonie internationales*, Berne, [1908], 22 p., p. 15.

51 Idem, p. 6.

became clear that this common good had no continental frontier and that private actors should also be considered.

Aligning Public and Private Interests: From a Modus Vivendi to a Global Cartel

Private international submarine cables companies were key players in the globalization of the telegraph. Some contemporaries campaigned for the internationalization of submarine cables, arguing that they were too important for the future of mankind to depend on private initiatives and stay in private hands exclusively. Aristide Dumont, a French public works engineer, created a Committee for Universal Telegraphy during the 1860s with the purpose of convincing governments to foster and guarantee the laying of multiple submarine cables worldwide.[52] His idea was not to ban private investments, but to integrate them within an intergovernmental program. His public-private partnership scheme responded to the recent failure of several private initiatives. In 1866 though, a private company succeeded in laying and operating a transatlantic cable. Dumont's committee most probably disappeared. But what was the collective position of governmental agencies toward market forces? I argue that ITU provisions toward private companies were another form of public guarantee for their expansion and hence, again, a major site of globalization of the telegraph.

At first, the ITU members actually seemed to take a rather defensive position. In 1865, cable companies were far away from being as powerful as they would become a few years later. But the point was already raised. It seemed unfair to many heads of telegraph administrations that private companies offering very expensive services could take advantage of increases in telegraph traffic due to low and simple rates on public continental networks. Telegraph administrators did not believe that the convention should enable public agencies to, though indirectly, subsidize private undertakings. They decided that they would "impose as much as possible this convention's rules to land or undersea cables concession companies and to negotiate with the existing companies a reciprocal lowering of rates if needed."[53] This principle

52 Darimon, p. 2.

53 *Documents diplomatiques*, p. 34.

became even more explicit under the Vienna (1868) regulations which stated that "companies [would] be admitted to the advantages of the Convention, subject to their accession to all its obligatory clauses and after notification by the conceding State."[54] Members were furthermore authorized to double, triple or even quadruple their rates to connect their networks to the companies and public agencies that would not be part of the convention and that would have excessive commercial claims.[55] This was a way to enforce all operators throughout the globe to comply with European public provisions. But at least three interrelated issues complicated the matter, with long lasting debates over the decade from 1865 to 1875 dividing ITU members.

Protecting Public Monopolies: Rates Competition under Control

Given that a dispatch could follow several routes, some public, others private, or a combination of the two, the question of a *modus vivendi* soon arose. Competition between public telegraph administrations themselves was the first important problem to solve. Member countries could be tempted to calculate their rates in order to gain international market shares. Traffic between Europe and India opened the debate. In 1868, there were eight different routes between London and Karachi through the European continent toward the Persian Gulf. Delegates gathered in Vienna this year managed to fix an equal rate (61.5 francs for a simple dispatch), whatever the route taken.[56] This cartel agreement banned fare wars and its members proclaimed that competition should take place in terms of "quality of the service provided and not of exaggerated discounts."[57] A related issue was whether two member countries were allowed to create a direct submarine cable between them that would offer competing rates compared to the parallel (public) land route. A typical example was the traffic between Great Britain and Norway. A submarine cable was planned that could overnight eliminate the traffic previously sent through Denmark and German states. The answer was that public networks were free to concede such liaisons, but were also free to adapt their rates to this unavoidable competition.

54 *Documents de la conférence télégraphique internationale de Vienne*, Vienne, Imprimerie royale et impériale de la cour et de l'Etat, 1868, 459 p., p. 33.

55 Idem and Saveney, La télégraphie internationale. Les conférences, p. 564.

56 *Documents de la conférence télégraphique internationale de Vienne*, pp. 46–48.

57 Idem, p. 296.

The inauguration of a direct London to Bombay submarine cable in 1870, to be extended to Australia, China and Japan, revived the matter, as did preparation a few months later to open the Great Northern line connecting Great Britain to the Far East (China, Japan) via Scandinavia and Russia. The new competing route toward India questioned rates set in Vienna. Future traffic with China and Japan, via India or via Russia, raised the question of the adoption of another common rate. At the beginning, the Eastern and Associated Telegraph Companies operating the submarine line to and from India carefully followed the Vienna regulations. But it changed its strategy with the arrival of competition from the Great Northern. Its strategy was then to increase rates on the Indo-European part of the line in order to compensate for reducing rates to and from the Far Eastern to sustain competition. Theoretically, it was not free to do so. But it still increased the rate. From January 1871 onwards, a simple dispatch from London to Karachi cost 100 francs instead of 61.5 francs previously. Obviously, the Eastern asked the numerous operators involved in other Indo-European routes to align with their rates in order to safeguard its traffic and revenues. The International Bureau in Bern did not manage to find a solution. It then mobilized the "amphictyonic system" to be set up in case of serious problems, with a special commission meeting in Bern in September 1871. State delegates decided there to admit private companies' representatives as consultative participants. Managers of public and private operators knew each other. Some even crossed the lines. Louis Curchod for instance headed the Swiss administration when he was appointed as the first director of the International Bureau; after a few months, he left office and went to the Société du Câble Transatlantique Français, which he represented in Bern in 1871, before returning to the ITU. Inviting private companies would be the rule for the following conferences. Above all, delegates agreed to align with the companies operating the submarine line. The Turkish transit rate doubled, though Turkey's delegates tried to oppose this provision in order to increase its market share. It was indeed "a piquant feature" to see public administrations increase their rates to comply with private interests.[58] This is how competition over the traffic between Great Britain and China modified European public transit rates toward India. Networks regulations were truly entangled. The delegates also abandoned the idea of setting a common rate for traffic to China and

58 Saveney, La télégraphie internationale. Les conférences, p. 574.

Japan. They agreed though to ask for 61.5 francs for the transit along the Indian land routes toward the Far East.[59]

The Eastern had won an important decision on rates, but lost on another point. Delegates created provisions that enabled governmental operators to dispatch the traffic between different routes at any bifurcation point. The company consequently decided to quit the telegraph conference convened in Rome just two months after the Bern commission had broken up, complaining that a "desire of uniformity" had prevailed over the companies' liberty and necessity of flexibility.[60] To settle the problem, in 1875, Heinrich von Stefan, the influential head of the German post and telegraphs administration, proposed a cartel-like organization between public and private operators. His scheme was based on an agreement on rates, a clause of non-competition and of revenue sharing.[61] His proposition was not adopted but laid the groundwork for the new principle of a price floor, i.e. an imposed limit on how low a rate could be for a given service. Delegates decided in the meantime that "accession (to the convention) must be imposed on the operators that are linking two or several contracting States."[62] This integrated public and private undertakings more deeply.

Protecting Public Monopolies: Land Lines Concessions under Control

In 1868, the North German Confederation and Russian telegraph administrations allowed the Indo-European Telegraph Company (Siemens) to construct a landline between Emden, Warsaw, Odessa and Tiflis. The line provided a service that did not exactly follow international regulations. The case was immediately discussed in the ITU. A French delegate, one of the few diplomats of the assembly, phrased his discontent rather directly. To him, this was a "serious slur" on the spirit of international cooperation.[63] The case was not about the liberty for a government to concede whatever landline it wished. It was about the normative power of international regulations. It took many years to reach an agreement. In 1875 in St Petersburg, a provision finally stated that "private telegraph operators working within one or several

59 Ibid.

60 Ibid., p. 581.

61 Ibid.

62 *Documents de la conférence télégraphique internationale de Saint-Pétersbourg*, p. 86.

63 *Documents de la conférence télégraphique internationale de Vienne*, p. 302.

contracting States' borders and participating to international service are considered, as far as this service is concerned, as part of the telegraph network of these States."[64] This definitively secured national monopolies and their international organization.

Facilitating the Expansion of Private Companies

If landline concession companies could not operate outside of international regulations, then international regulations had to be changed. This, at least, was the idea of the North German Confederation and Russian administrations when they advocated a revision of international rates that would be advantageous to the Indo-European Telegraph Company. They asked for a special rate to be introduced for dispatches no more than ten words long – whereas until then the entry rate was for a twenty-word dispatch. Two dispatches of ten words each would make more money than one of twenty, but individually, they would be more affordable for the public. Private business was risky and needed commercial flexibility to find new customers, they argued. Furthermore, there was a growing business of packing very short messages up to the maximum number of words of a simple dispatch. The "packers" made money buying a simple dispatch to the company and re-selling it into small pieces.[65] After lengthy discussions, delegates aligned with the Belgian proposition to authorize private companies to offer this kind of fare, but not to generalize it throughout public European services.[66] In a way, they had found a win-win arrangement. Both public and private operators won on the back of the consumer. A ten-word dispatch from Vienna to New York would pay the price of a twenty-word dispatch for its public landline transit, and then access the private submarine cable at an affordable but remunerative price. This provision was completed in 1871 with a rate per word over ten words on private lines. "This was a great inconvenience, a great complication for the European offices," but this was the price to pay for both facilitating companies' expansion and protecting administrations' routines and revenues.[67]

64 *Documents de la conférence télégraphique internationale de Saint-Pétersbourg*, p. 85.
65 Saveney, La télégraphie internationale. Les conférences, p. 561.
66 Idem, p. 562.
67 Ibid., p. 580.

In 1875, an "extra-European regime" gathered all exemptions granted to private and public operators which were not formally part of the Union; notably, cable companies were granted a new, more flexible, rate by the word that would also be applied on the European lines' portion of the journey.[68] The globalization of the telegraph hence resulted in different, though clearly articulated, telegraphic spaces. A distinct European space emerged within a broader scope. But this European space could stretch extensively, including Tunisia and Senegal for instance.[69] Sending a message to or from Tunisia was like telegraphing within Europe and followed the same pricing rules. It was indeed up to the European mother countries to choose whether "the lines that they have outside of Europe and which they have made part of the convention belong to the European or extra-European regime."[70] Over time, this major pricing difference with the European regime eroded. The rate by the word, favored by cable companies, became universal following the conference in Berlin in 1885. The extra-European regime had taken over the formative European regime as part of the process of globalization.

All these issues and the way the telegraph administrations settled them show that the emergence of powerful private operators complicated international cooperation. Tensions arose and the telegraph convention and regulations were adapted to, but not immediately nor systematically aligned with, private companies' interests. Generally speaking though, the ITU framed provisions both protecting national public monopolies and enabling, as the convention said, "the progressive development of communications" throughout the world.[71]

European Order and Global Power

The globalization of the telegraph was neither anarchic nor the expression of the hegemony of one world power. Power relations were at stake but not in the classical sense of a competition between nations for control. Instead,

68 *Documents de la conférence télégraphique internationale de Saint-Pétersbourg*, p. 11 and 26.

69 "Tableau A des taxes du régime européen," in *Documents de la conférence télégraphique de Londres*, London, 1903.

70 *Documents de la conférence télégraphique internationale de Saint-Pétersbourg*, p. 85.

71 Idem, p. 11.

the regulatory process structuring the international communications order in the long term empowered European nations and empires simultaneously.

Networks, Nation-State Building and International Cooperation

The historical roots of monopolization through public agencies were two-fold. On the one hand, this process was linked with concerns on sovereignty (state fiscal monopoly and state security), on the other hand, this progressively mixed with concerns about providing a public service (equal access, low rates, economic and social development of the country).[72] Post and telegraphs services thus became "public services of sovereignty" during the nineteenth century, contributing to the rise of modern nation-states.[73] Co-operation initiated by the public agencies in charge of these services did not intend to weaken this dynamic. On the contrary, as we have seen with how rates were set, interconnection provided material (money) and symbolical (legitimacy) resources to the participants, reinforcing their own domestic position. Contemporaries saw the ITU as a "syndicate of European administrations."[74] More than one century later, Eli Noam talked about a "bastion of the traditional national public monopolies."[75] During this long period, "The international telecommunications regime provided a multilateral framework that reinforced domestic monopolies and bilateral agreements in the global market and thereby created one of the most lucrative and technologically significant international cartels in history," summarized Peter Cowhey.[76]

The ITU was essentially an intergovernmental treaty, periodically revised by an assembly of experts gathered into international conferences. During the 1865 conference, the French delegation proposed the creation of a permanent body composed of one delegate per member country, with its seat in the country where the conference last met – hence for the following years, in Paris. Its tasks would include drawing maps of international lines, updating rates and conducting studies of general interest. The proposition was rejected because it was politically unacceptable to many to transform Paris into the capital city of European telegraphs, the symbolic brain of Europe, even tem-

72 Millward, European governments and the infrastructure industries, c. 1840–1914, p. 19.
73 Laborie, *L'Europe*, p. 49.
74 Saveney, La télégraphie internationale. Les conférences, p. 564.
75 Noam, *Interconnecting*, p. 151.
76 Cowhey, The international telecommunications regime, p. 169.

porarily.[77] In 1868 instead, delegates decided to set up a very small but still permanent international bureau in Bern, under neutral Swiss supervision. War between Prussia and Austria had shown two years before that international politics could put technical cooperation at risk. The international bureau of telegraph administrations was not a supranational player. It was voluntarily limited to coordinative tasks: facilitating the circulation of information between members and organizing the conferences. Again, national sovereignty was not endangered by cooperation.

Institutions, Norms, Values: A European Order

To many contemporary observers, growing international economic and technological interdependences contrasted with increasing political differentiation between nations. This mix, they thought, could be highly explosive and necessitated a reengineering of the international political system.[78] Governing interdependence posed a new challenge. This was just what telegraph experts were doing actually. Communication networks both embodied the building of the nation-states and the rising needs accompanying transnational economic integration, materialized by their interconnection. Apparently modest administrative cooperation was in fact a decisive link between political fragmentation and economic integration. To Emile Saigey, a telegraph engineer of the time, technical cooperation thus produced something more than just technological interdependence. "Let us multiply such agreements," he stated, "and the States of Europe would be federated by the simple force of things."[79] It is not surprising then that such eminent reformers of international law like Louis Renault, who would be very much involved in the conferences on the protection of cables, the Hague Peace Conferences and would receive the Nobel Peace prize in 1907, studied these questions closely.[80] According to Louis Renault, the International Telegraph Union was, with the Universal Postal Union (1874), the first institution of its kind, as it was the product of social forces rather than international law and

77 Saveney, La télégraphie internationale. Les anciens traités, p. 381.

78 These reflections preceded the birth of the League of Nations. Arcidiacono, Les projets de réorganisation, p. 11.

79 Saveney, La télégraphie internationale. Les anciens traits, p. 360.

80 Griset, *Entreprise*, pp. 145–148; Winseck /Pike, *Communication and Empire*, pp. 46–48; Laborie, *L'Europe*, p. 137–138.

held a chronologically and geographically broad mandate, as its role lay not only in solving problems of the past, but also designing the future.[81] "Post and telegraph administrators who created the Postal Union and Telegraph Union, the name of whom are ignored by the public," he said, "did more for civilization and entente among peoples than many famous diplomats."[82]

Emile Saigey's or Louis Renault's interpretation of international cooperation was typical of pre-functionalist thinking and nascent technocratic internationalism – a mix of pragmatism, apolitical attitude, and faith in the ability of experts to negotiate national and universal issues.[83] In my view, they were active participants of a transition towards a new kind of international order in Europe. No longer resulting simply from the balance of power, the stabilization of international relations resulted in the nineteenth century from an institutionalized continual dialogue among the Great Powers and their pursuit of common goals. From the Congress of Vienna (1814–1815) onward, the European Concert generated common institutions and common legal norms, cemented by shared values.[84] Technical international institutions such as the ITU were at the forefront of this ongoing transition, all working together and generaling each other in a spill-over dynamic.[85] For Georges-Henri Soutou, "still limited in scope, they nonetheless contributed to structure the European order […] early announcing the European construction of the twentieth century."[86]

The delegates who composed the ITU stood at the crossroad of national interests and the service of a transnational public. In order to manage the tensions emerging there, they believed in "mutual concession," a notion they often referred to. A strong feeling of human solidarity, based on the interdependence of technical networks and shared professional knowledge, made the delegates open to negotiation, despite their national allegiances. They shared four common beliefs and values, pillars of their epistemic community. First, they respected national sovereignty. This was clear in such varied provisions as the liberty for each member country to stop international ser-

81 Renault, Louis, "Les Unions internationales," *Revue générale de droit international public*, no. 28 (1896), pp. 14–26, p. 15.

82 Renault, Louis, *Études sur les rapports internationaux. La poste et le télégraphe*, Paris, 1877.

83 Schot, J./Vincent Lagendijk, Technocratic Internationalism, p. 198.

84 On the notion of international order in the nineteenth century European and world context: Schroeder, *The Transformation*; Soutou. Le Concert européen; Conrad /Sahsenmaier, *Competing*.

85 Norman Weiß, Institutionalised co-operation in International Communication, p. 74.

86 Soutou, Le Concert européen, p. 120.

vice within its territory at any moment – revealing perceived threats of an information invasion or evasion —, or to authorize or prohibit the use of ciphers. Second, they shared the natural monopoly thesis, which held that network industries naturally tended to be organized through the institution of a monopoly. Third, they championed the idea that communications and trade were vectors of peace and civilization. The more complex individual or social organisms were, be they animal or human, the more developed their nervous systems. Following this idea, popularized by the philosopher and sociologist Herbert Spencer, communications seemed to be the essence of civilization. Postal and telegraph lines were the nervous system of more advanced societies. Cooperation itself was thought to be Europe's signature, embodying its superiority and collective mission. The delegates celebrated it as a key connection between technological progress and social, humanitarian progress. "It increases the number of links and makes any rupture harder," assumed the Austrian technical delegate, Brunner de Wattenwyl while the Austrian Foreign Minister, de Beust, declared that "the European character of our association should not hide the fact that the civilizing mission that Europe had always gloriously embraced calls for the expansion and not restriction of its action's realm."[87] As we have seen, the delegates agreed on the fact that cooperation should embrace the planet, not only one region, as technology had no predefined limit. Fourth, they thought that "rationally" acting technical delegates were best equipped to handle international cooperation and build consensus. This meant that they should be freed from diplomats' supervision. In this perspective, the third international telegraph conference in St Petersburg in 1875 made an important move. The delegates split provisions into two parts. On the one hand, the political convention could only be revised by plenipotentiary delegates (i.e. diplomats). On the other hand, the administrative regulations for international service could be revised by non-plenipotentiary delegates (i.e. experts). Future conferences would then be plenipotentiary or administrative depending on the text they would have to deal with. In the meantime and more importantly, they transferred most of the provisions from the convention to the regulations.[88] This had a direct but usually invisible consequence that international regulation of the telegraph was now almost officially out of professional diplomats' scope. This, in a way, facilitated the globalization of the Union.

87 *Documents de la conférence télégraphique internationale de Vienne*, p. 343 and p. 93.
88 *Documents de la conférence télégraphique internationale de Saint-Pétersbourg*, pp. 3–87.

Global Power

The ITU was an open institution. A simple diplomatic notification was sufficient to become a member. Even if it was not as comprehensive as its twin organization, the Universal Postal Union, it counted 52 members in 1908 compared to 20 in 1865.[89] Overseas offices often joined the ITU after private lines had linked their country to Europe. For instance, Japan was connected in 1871 and sent an observer in Roma the same year, before becoming a member in 1879. Brazil was connected in 1874 and sent a representative to the London conference in 1879; Siam became a member in 1883 and sent a delegation in 1885. South American countries joined afterwards (Argentina, 1889; Uruguay, 1902; Bolivia, 1907; Chile, 1908; Guatemala, 1914).[90] Persia (1869) and Egypt (1875) had joined earlier. Many of these countries were more or less part of what historians have called European "informal empires."[91]

Formal empires were not automatically part of the Union when it was created. First regulations explicitly excluded colonies and territories that were located "outside Europe," notably the Asian parts of Russia and Turkey, upon the request of these powers.[92] This was because these countries were concerned about the prospective cost of reducing rates to and from these remote parts of the country. But this changed swiftly, as Asian Russia and Turkey were included into the convention and regulations in 1867 and 1868 respectively. The provinces of British India became part of the Union as early as 1868, even before Great Britain itself. Other British colonies became part of the Union once Great Britain had joined and after the significant revision of the convention in St Petersburg in 1875 had distinguished the two types of conferences for the future. "British colonies" formed a delegation during the London administrative conference (1879). In subsequent conferences, this ensemble was split into separate delegations: South Australia, New South Wales, Tasmania and later Cape Province, Natal, New Zealand, etc. France delegated representatives from Cochinchina, Senegal and Tunisia to the Berlin conference in 1885. Spain and the Netherlands did the same at the next conference, in Paris (1890).

89 Codding, *The International*, p. 30 and p. 35.

90 Codding /Rutkowski, *The International*, pp. 333–336.

91 Crouzet, *De la supériorité*, p. 374.

92 *Documents diplomatiques*, p. 78 and 160.

Daniel Headrick quoted a note from the Director-General of French Post and Telegraph on the reasons for enlarging the Union to non-sovereign territories: "The adherence of our colonies is of great interest, because every administration that joins has the right to vote in international conferences, and every new adherence of one of our colonies has the effect of giving one more voice to France in the voting which follows the debates."[93] The more colonies, the more voting power. Integrating colonies within international organizations became a kind of a race. But this quote forgot one important element. When Great Britain joined in 1871, it was immediately felt, and agreed, that its delegates would have, with India, two voices. The Netherlands took advantage of the situation and asked for the same treatment regarding its Dutch East Indies colony. But the members refused.[94] Non-colonial nations who composed the majority were especially reluctant to embark on this trend. It was the convention signed in St Petersburg that finally granted individual voting rights for the future to administrations dependent on the same government, provided that permission to participate had been obtained through diplomatic channels from the host country and that these administrations were each represented by their own delegation.[95] The new distinction between plenipotentiary and administrative conferences enabled this change. It then became possible to give voting right and full membership power to administrations of colonial territories in an international (administrative) conference.[96] Voting procedures within the new International Radiotelegraph Union (1906) even went further following Universal Postal Union's procedures, with a "plural vote" system giving to colonial powers up to six voting rights each irrespectively of the conference nature (plenipotentiary or administrative). By the 1912 radiotelegraph conference, Germany, France, Great Britain and the United States each had five voting rights. The epistemic community directly involved in ITU meetings was relatively narrow. Stable between 1865 and 1875, at around 35 attendees, it almost doubled in 1879 with 65 attendees, and doubled again between 1879 and 1890; at its height, it gathered 139 people in 1903

93 Letter of March 24, 1894 to the minister of Colonies. Quoted in Headrick, *The Invisible*, p. 14.

94 Saveney, La télégraphie internationale. Les conférences, p. 576.

95 Article 16. *Documents de la conférence télégraphique internationale de Saint-Pétersbourg*, p. 10.

96 De Lapradelle, Le droit de vote, p. 408.

in London. In London again in 1912, 154 people attended the radiotelegraph conference.[97]

As we can see, cooperation was not exclusive to classic inter-national power stakes. From a global perspective though, the ability of Europeans together to fix the rules of the game developed as a kind of collective power. Imperialist countries went hand in hand through this process. Cooperation was hence of a co-imperialist nature. It both gave extra power to imperial countries within this international organization and enabled the tighter integration of colonies within the global telegraph network, under the close control of their colonizer. The system also operated as a collective mediator toward non-members. The latter had to comply with some of the provisions or at least negotiate their interconnection. They could stay independent but still had some interests in aligning with international regulations. As the third section showed, this was the case with submarine cable companies, many of which became part of the agreement. Thanks to some adaptations, the European international order of communication thus included not only public but private actors, which in turn made it stronger and reinforced its capacity to extend worldwide.

Conclusion

This chapter surveyed the origins, channels and principal consequences of the dynamics that made local telegraph networks part of a globalized ensemble. Seen from the perspective of the ITU, the globalization of the telegraph first resulted from a tacit agreement that provided national loss-making public monopolies with important income from international (and mainly trans-European) traffic. Here we see how the modern nation-state was built on transnational flows. Second, private cable companies operating intercontinental routes benefited from European endeavors to integrate their continental networks and succeeded conversely in influencing European regulations. National monopolies and private companies' interests became more and more entangled. Third, a kind of European co-imperialism was at play, underpinned by shared values of communication and trade as vectors of civilization and peace, which provided Europeans with national sovereignty and

97 Laborie, En chair et en normes.

global hegemony through the setting and control of global rules within the context of increasing interdependencies.

Beginning with the twentieth century, the period that followed increasingly focused on maintaining this order, preserving it from technological (wireless) and business (Marconi) pressures, and incorporating a newcomer, the United States. Maintaining the order required new rounds of mutual concessions and careful negotiations that propelled a new phase in the history of the governance of global communication, not its disruption.

Bibliography

Ahvenainen, Jorma, "The International Telegraph Union: the cable companies and the governments," in: Finn Bernard/Daqing Yang (ed.), *Communications under the Seas. The Evolving Cable Network and its Implications*, Cambridge, MA 2009, pp. 61–77.

Arcidiacono, Bruno, "Les projets de réorganisation du système international au XIXe siècle (1871–1914)," *Relations internationales*, no. 123 (2005), pp. 11–24.

Balbi, Gabriele/S. Calvo, S. Fari, G. Richeri, "Bringing together the two large electric currents that divide Europe: Switzerland's Role in Promoting the Creation of a Common European Telegraph Space, 1849–1865," ICON, no. 15 (2009), pp. 61–80.

Bruley, Yves, "Le concert européen à l'époque du Second Empire," *Relations internationales*, no. 90 (1997), pp. 145–163.

—, *Le Quai d'Orsay impérial.* Histoire du Ministère des Affaires étragères sous Napoléon III, Paris 2012.

Brun, Gérard, "La mise en réseau des réseaux: matériaux pour une étude morphologique, in: Gemelli, Ginliana (ed.), *Big Culture. Intellectual Cooperation in Large Scale Cultural and Technical Systems: an Historical Approach*, Bologna 1994, pp. 327–355.

Carré, Patrice, "Archéologie d'une Europe des télécommunications 1865–1939," *Revue française des télécommunications*, no. 70 (1989), pp. 72–83.

Codding, George A., *The International Telecommunication Union. An Experiment in International Cooperation*, Leiden 1952.

Codding, George A./Anthony M./Rutkowski, *The International Telecommunication Union in a Changing World*, Deedham 1982.

Conrad, Sebastian/Dominic Sachsenmaier (eds.), *Competing Visions of World Order. Global Moments and Movements, 1880s-1930s*, London 2007.

Conrad, Sebastian, *Globalisation and the Nation in Imperial Germany*, Cambridge, MA 2010.

Cowhey, Peter, "The international telecommunications regime: the political roots of regimes for high technology," *International Organization*, vol. 44, no. 2 (1990), pp. 169–199.

Crouzet, François, *De la supériorité de l'Angleterre sur la France. L'économique et l'imaginaire, XVIIe-XXe s.*, Paris 1985.

De Lapradelle, Paul, "Le droit de vote aux conférences des telecommunications," *Revue juridique internationale de radioélectricité*, no. 56 (1933), pp. 398–424.

Gouzevitch, Irina, "L'essor de l'enseignement électrotechnique en Russie : genèse, filières et aboutissements (1832–1917)," in: Efmertova, Marcela/André Grelon (eds.), *Un monde progressivement connecté. Les électrotechniciens au sein de la société européenne au cours des 19e et 20e siècles*, Brussels, 2013.

Griset, Pascal, *Entreprise, technologie et souveraineté : les télécommunications transatlantiques de la France (XIXe-XXe s.)*, Paris 1996.

Grossi, Verdiana, "Technologie et diplomatie suisse au XIXe siècle," *Relations internationales*, no. 39 (1984), pp. 287–307.

Headrick, Daniel, *The Invisible Weapon. Telecommunications and International Politics, 1851–1945*, New York 1991.

Huurdeman, Anton A., *The Worldwide History of Telecommunications*, New York 2003.

Laborie, Léonard. *L'Europe mise en réseaux. La France et la coopération internationale dans les postes et les télécommunications (années 1850-années 1950)*, Bruxelles 2010.

—, "En chair et en normes. Les participants aux conférences de l'Union internationale des télécommunications, de sa fondation à sa refondation (1865–1947)," *Flux*, no. 74 (2008), pp. 92–98.

Lew, Byron/Bruce Cater, "The telegraph, co-ordination of tramp shipping, and growth in world trade, 1870–1910," *European Review of Economic History*, vol. 10/2 (2006), pp. 147–173.

Mayer-Schönberger, Viktor/Deborah Hurley, "Globalization of Communication," in Nye, Joseph S. Jr./John D. Donahue (eds.), *Governance in a Globalizing World*, Cambridge, MA/Washington DC 2000, p. 135–155.

Millward, Robert, "European governments and the infrastructure industries, c. 1840–1914," *European Review of Economic History*, vol. 8 (2004), pp. 3–28.

Noam, Eli, *Interconnecting the Network of Networks*, Cambridge, MA 2001.

Schmidt, Susanne K./Raymund Werle, *Coordinating Technology: Studies in the International Standardization of Telecommunications*, Cambridge, MA 1998.

Schot, Johan/Vincent Lagendijk, "Technocratic Internationalism in the Interwar Years: Building Europe on motorways and electricity networks," *Journal of Modern European History*, vol. 6, no. 2 (2008), pp. 196–217.

Schroeder, Paul, *The Transformation of European Politics 1763–1848*, Oxford 1994.

Soutou, Georges-Henri, "Le Concert européen, de Vienne à Locarno," in Bérenger J./G.-H. Soutou (eds.), *L'ordre européen du XVIe au XXe siècle*, Paris 1998, pp. 117–136.

Stearns, Peter N., *Globalization in World History*, New York 2010.

Weiß, Norman, "Institutionalised Co-operation in International Communication: the International Administrative Unions as a Means of Governing Globalisation Processes", in: Löhr, Isabella/Roland Wenzlhüemer, Heidelberg 2013, pp. 65–82.

Wenzlhuemer, Roland, "Editorial - Telecommunication and Globalization in the Nineteenth Century," *Historical Social Research*, vol. 35, 2010/1, pp. 7–18.

Winseck, Dwayne/Robert Pike, *Communication and Empire. Medias, Markets and Globalisation (1860–1930)*, Durham, NC 2007.

Zacher, Mark W./Brent A. Sutton, *Governing Global Networks. International Regimes for Transportation and Communications*, Cambridge 1996.

The Wiring of the Working Class: On the Interdependence of Telegraphy and Social-Revolutionary Discourses in the Nineteenth-Century

Martin Doll

Research in the field of global media studies so far typically confines the impact of telecommunication in the nineteenth century to the advancement of imperialist politics and the global capitalist economy. In his study *Geopolitik der Literatur* (*The Geopolitics of Literature*), for example, Nils Werber provided an overview of perspectives on the connection between media technology and the development of a global communality.[1] His analyses of relevant German prose works and of utopian conceptions led him to conclude that communication technologies in the nineteenth century only served to fuel geopolitical expansionism. These conclusions are widely shared by most studies concerning the impact of communication technology in the nineteenth century.[2] Other authors concentrate on the effect of communication technologies to advance the world economy.[3]

In contrast, or rather in addition to these approaches, my chapter addresses other – less commonly considered – political interpretations of communication technologies, namely, from socialist and communist points of view.[4] I will show that the 'wiring of the world' not only created a global capitalist economy but – from a Marxist perspective – was also understood

1 Werber, *Die Geopolitik der Literatur.*

2 For the imperial side see Headrick, *The Tools of Empire*; Kittler, "Geschichte der Kommunikationsmedien"; "Lakanal und Soemmering"; Flichy, *Dynamics of Modern Communication.*

3 The capitalist economy perspective is explored by: Fischer, "'Touch Someone': The Telephone Industry Discovers Sociability"; Marvin, *When Old Technologies Were New*; Winseck/Pike, *Communication and Empire*; Wobring, *Die Globalisierung der Telekommunikation im 19. Jahrhundert.*

4 One exception is the book *Media Marx* edited by Jens Schröter, particularly his own article where he reflects upon the relation between Marxism and transmission media. Cf. Schröter, "Übertragung und Explosion – Telegraphie/Telephonie/Transport."

as a powerful element of its dialectical reversal by offering the promise of the universal connectibility of the international working class.

In order to investigate these connections, the first part of this chapter will consider Benedict Anderson's ground-breaking theory of 'imagined communities.' On this basis, I will ask whether media technology in the eighteenth century not only prepared the ground for an 'official nationalism' in the nineteenth century, as he argued, but whether media technology – especially in the form of telegraphy – might also be understood as one source for nineteenth-century social-revolutionary ideals and concepts of global social-revolutionary communality.

Accordingly, the second part of my analysis will investigate how (imagined) structures of communication affected nineteenth-century communist conceptions of political subjects and collectivities, or more precisely, how global communication networks were imagined as immanent political possibilities for 'changing the world' – as 'agents' for the emancipatory organization of global society. I will support this with reference to Marx's and Engels' theories of societal change such as in *The Communist Manifesto* from 1848 or the *Grundrisse. Foundations of the Critique of Political Economy* from 1857/58. If one takes a closer look at their texts, one can show that – in contrast to established interpretations – Marx and Engels based their analyses not only on industrial technology particularly on the machine, but also on the field of communication. Against this background it will become apparent that the telegraph was perceived by them to play a relatively important role in the global process of social transformation.

The Technologically Driven Formation of (Bourgeois) 'Imagined Communities' (Benedict Anderson)

In his work *Imagined Communities*, Benedict Anderson showed that the growth of the printing industry and the reception of newspapers on a massive scale in the eighteenth century led to readers' historically new experience that they constituted a homogeneous linguistic community. This created new forms of imagined communities. The readers "gradually became aware of the hundreds of thousands, even millions, of people in their particular language-field [...]. These fellow-readers, to whom they were connected through print, formed, in their secular, particular, visible invisibility, the embryo of the nationally imagined community [...]."[5] Anderson called these new practices of newspaper reading a "mass ceremony," because "each communicant is well aware that the ceremony he performs is being replicated simultaneously by thousands (or millions) of others of whose existence he is confident, yet of whose identity he has not the slightest notion. [...] What more vivid figure for the secular, historically clocked, imagined community can be envisioned?"[6] In order to understand this argument in its historical context, one might supplement this account by explicitly mentioning two additional aspects of Anderson's argument: first, he insists that the impact of the printing press cannot be seen in isolation from the historically specific socio-political situation, namely the logic of the market in the development of capitalism. Hence for Anderson, imagined communities were only made possible by the fact that the "accumulation of technological innovations in the fields of shipbuilding, navigation, horology and cartography" between the sixteenth and the end of the eighteenth century was "mediated through print-capitalism."[7] In other words, Anderson in no way subscribed to media or technological determinism. Second, the experience of community is necessarily the experience of a "community in anonymity."[8]

At this point, I would like to raise the question of whether this kind of 'imagined community' not only led to an official nationalism in the nineteenth century, but also formed the embryo of a certain sense of global communality. Interestingly enough, even if Anderson touches upon technological developments like sailing ships and steamships, railways, motor transport and aviation – as well as, in footnotes, radio broadcasting and television – there is not a single passage in which he even mentions telegraphic communication.[9]

Nevertheless, Anderson drew some interesting observations from the impact of written language, which he termed, presumably to stress the power of the printing press, "print-language" – an impact which I wish to apply to

5 Anderson, *Imagined Communities*, p. 44.

6 Ibid., p. 35.

7 Ibid., pp. 43–44;188. My emphasis.

8 Ibid., p. 36. Anderson explains: "It became conceivable to dwell on the Peruvian altiplano, on the pampas of Argentina, or by the harbours of 'New' England, and yet feel connected to certain regions or communities, thousands of miles away, in England or the Iberian peninsula. One could be fully aware of sharing a language and a religious faith (to varying degrees), customs and traditions, without any great expectation of ever meeting one's 2 partners." Ibid., p. 188.

9 Ibid., pp. 54, 61, 115 and 151.

the case of telegraphy. According to Anderson, the factory-owner in Lille was connected to his counterpart in Lyon "only by reverberation. They [...] did come to visualize in a general way the existence of thousands and thousands like themselves through print-language."[10] From this Anderson concluded that "in world-historical terms bourgeoisies were the first classes to achieve solidarities on an essentially imagined basis."[11] Starting from this idea of imagining, we can pursue two theoretical paths: first, if print-language has as strong an impact as Anderson argued, telegraphy must at least be considered as a sort of amplifier, a reinforcer of the particular imagined communities produced by the new forms of print-language. I will return to this idea later on. Second, and more strikingly, if we consider the emerging codes of telegraphy in the nineteenth century as constituting a new abstract and *universal language*, we can extend Anderson's conception of regional linguistic communities to the idea of a truly universal principle of global community.

On the one hand, one might argue on a concrete level that telegraphic codes were restricting and controlled by those who had access to them and could read them. On the other hand, in spite of this factual institutional limitations, the mere availability of the universal communication via the telegraph (even if it was actually often not used) fuelled the imagination of the people and propagated a process that disseminated an abstract idea of humankind, fully in line with the project of the Enlightenment and French Revolution: And, indeed, even the earlier establishment of optical telegraphy in France after the French Revolution can be linked to contemporary attempts to homogenize and standardize the calendar as well as to projects to reform the system of weights and measures according to principles of rationality, simplicity and universal validity.[12] The abstract (visual) code of optical telegraphy was explicitly tied to the idea of a universal language.[13] Ignace Chappe, a member of the Legislative Assembly and the brother of the inventor Charles Chappe, would later make a similar argument. He claimed that the recent technological development forged not a "characteristica universalis" as sketched out by Gottfried Wilhelm Leibniz but at least a language "that becomes almost universal in that it indicates combinations of

10 Ibid., p. 77.

11 Ibid.

12 Flichy, *Dynamics of Modern Communication*, pp. 12–17; Kula, *Measures and Men*; Baczko, *Utopian Lights*, pp. 159 and 170–173.

13 Flichy, *Dynamics of Modern Communication*, pp. 11–15. See also: Ozouf, *L'école de la France*, p. 27.

numbers instead of words, and that the manner of expressing these numbers is generally known and can be applied to the words which compose all dictionaries."[14] Nevertheless, for Chappe, this was more a question of efficiency than of universal understanding, because in connection with telegraphic language, the "aim is not to find a language which is easy to learn without a dictionary (Leibniz's expression in his letter to M. Renard) but to find the means to express many things with few signs [...]."[15]

Through this process of reducing and abstracting language to a series of coded numbers or signs, the medium of telegraphy created a sense of similarity, quick connectibility and supra-regional communality. This might remind us of McLuhan's famous elaborations in *Understanding Media* that the organic character of electricity led to a worldwide co-presence: "The simultaneity of electric communication, also characteristic of our nervous system, makes each of us present and accessible to every other person in the world [...]."[16] But I would like to stake out a position unlike McLuhan's fusion of organicist positions with religious or, rather, Christological arguments. Rather than understanding telegraphy as an "extension of man," in accordance with McLuhan's hypothesis – an extension which would enrich and supplement the living body – telegraphy should be interpreted as a medium of abstraction. Indeed, it must be seen as first creating the abstract language that rendered the very idea of a general or universal public conceivable. Electric telegraphy – with its specific experimental codes – abstracted from all particularities of the human communicative apparatus (e.g. acoustic signifiers like regional accents and tonality or visual signs as skin colour or gender) and consisted instead of a purely mathematical code. This mathematical abstraction permitted the idea of a human being *per se* to come to the fore, in accordance with the Enlightenment vision of universal human reason and rationality. This process of abstraction engendered in Walter Benjamin's words "humanity's whole constitution" (*Verfassung der Menschheit*).[17]

Through this technological abstraction, a new type of supra-individuality was brought into play, an idea reflected in Marx's use of Feuerbach's concept

14 Leibniz, "On the General Characteristic," pp. 221–228 and Chappe, *Histoire de la télégraphie*, pp. 136, English translation quoted from: Flichy, *Dynamics of Modern Communication*, p. 15.

15 Ibid.

16 McLuhan, *Understanding Media*, p. 333.

17 Benjamin, "The Work of Art in the Age of Its Technological Reproducibility [Second Version]," p. 27.

of "species being" (*Gattungswesen*). In his early *Economic and Philosophical Manuscripts* from 1844, Marx defined man as a species-being "in the sense that he makes the community (his own as well as those of other things) his object both practically and theoretically, but also (and this is simply another expression for the same thing) in the sense that he treats himself as the present, living species, as a universal and consequently free being."[18] In this context, Marx, with a nearly humanistic idealization (that only occurs in his early works), articulated the idea that the test of man's level of development is "how far man's needs have become human needs, and consequently how far the other person, as a person, has become one of his needs, and to what extent he is in his individual existence at the same time a social being [*Gemeinwesen*]."[19]

At the beginning of the twentieth century, the German philosopher Helmuth Plessner diagnosed the internationalist cosmopolitan communism developed in Marx's later writings as appealing to a "supra-national world-brotherhood", a "unity out of and above all differences [...]."[20] According to Plessner's interpretation, Marx based this "unity of all persons" simply on their humanity and this produces a sufficient condition "to found, at least, a community of ideas, justice and conviction."[21] Plessner concluded from this that this "form of the communist ethos points to [...] *a community of the ideal* [*Gemeinschaft der Sache*] through partnership in one and the same value."[22] This form of communism – according to Plessner deeply rooted in the eighteenth century and in the Enlightenment – "is resolutely rational and intellectual. It operates with an abstract universality [...]."[23] Plessner connected this abstract universality to a conception of non-coercive persuasion operating through reason and understanding. This form of persuasion provides the principle of universal communality.[24]

My point here is to demonstrate that nineteenth-century media such as telegraphy – in combination with infrastructural developments including the establishment of railway tracks and steamboat lines, the precondition of global traffic (in German: *Weltverkehr*, sometimes translated as "world

18 Marx, "Economic and Philosophical Manuscripts [Excerpts]," p. 100.
19 Ibid., pp. 126–127.
20 Plessner, *The Limits of Community*, pp. 91–93.
21 Ibid., p. 92. Plessner's emphasis.
22 Ibid. Plessner's emphasis.
23 Ibid., p. 93.
24 Ibid.

intercourse") – provided a necessary supplement to the concept of abstract humanity by offering the *promise* of a universal connectibility. The word promise has to be stressed here in order to distinguish analytically between imagined and actual connectivity (the latter prominently assumed by McLuhan). At least McLuhan's already quoted stark statement about the accessibility of "each of us [...] to every other person in the world" has to be put into perspective, because in the beginning, technologies like telegraphy were far from mass media intended for everyone's use.[25] However, in an imaginary realm, each individual human being *seemed to be* reachable and addressable. An abstract general public, a public that would embody a principle of solidarity, first became conceivable on the basis of this idea of universal addressability.[26] In other words, technologies of communication such as telegraphy represented – regardless of their actual use – a medium addressing not the individual human being in his particularity but mankind as such. Because humanity was conceived of as universally accessible, each individual human was understood to be "reachable" through universal principles of persuasion. In Marx's terms, a "socialized humanity" (*gesellschaftliche Menschheit*) came into being.[27]

Anderson underlined that "the most important thing about language is its capacity for generating imagined communities, building in effect particular [bourgeois, M.D.] solidarities."[28] For Marx and other communists, however, the economically and technologically driven imagined solidarity of the bourgeoisie became at a certain historical point in the nineteenth century the principle for the development of its counterpart: the universal demands of the newly constituted proletariat. Engels wrote 1870 that in "proportion as the bourgeoisie develops its industry, commerce and means of communication, in the same proportion it increases the numbers of the proletariat

25 McLuhan, *Understanding Media*, p. 333.

26 This abstract general public has to be distinguished from Rudolf Stichweh's "global public sphere" or "world public." Whereas Stichweh's concept is linked to distributive mass media spreading opinions to – for the most part – the communicative fiction of a passive homogeneous 'audience' "not doing the acting but being observers," the abstract general public focused on here is rather linked to how communication media fueled the ideal of political mobilization, of permitting all actors to communicate mutually and of actively involving them in societal change (cf. Stichweh, "The Genesis of a Global Public Sphere," pp. 26–29).

27 Marx, "Theses on Feuerbach," pp. 145.

28 Anderson, *Imagined Communities*, p. 133.

[…]."[29] To support this abstract thesis, let me refer to other specific passages in which Marx explicitly mentioned recent developments in technologies of communication. These citations will also show that Marshall McLuhan was fundamentally wrong when he wrote in *Understanding Media*: "Marx and his followers […] reckoned without understanding the dynamics of the new media of communication. Marx based his analysis most untimely on the machine, just as the telegraph and other implosive forms began to reverse the mechanical dynamic […]."[30] This more or less unsubstantiated criticism can also be found a few pages later, when McLuhan concluded that "[w]edded as they are to nineteenth-century industrial technology as the basis of class liberation, nothing could be more subversive of the Marxian dialectic than the idea that linguistic media shape social development, as much as do the means of production."[31] This critique on the exclusion of "language, signs and communication" from the realm of productive force is later adopted from Jean Baudrillard in his famous essay "Requiem for the Media." But Baudrillard even pushed the argument forward, when he accused Marx not only of neglecting to conceptualize "railroads as 'media,'" but also of ignoring "technological evolution in general, except from the point of view of production."[32] When Baudrillard wrote that only Hans Magnus Enzensberger in his 1970 article "Constituents of a Theory of the Media" managed to include the field of signification and communication into "the analysis in terms of productive forces" and thus addressed the blind spot and "the immense retardation of classical Marxist theory," he was as wrong as McLuhan.[33] On the contrary, Marx and Engels themselves already theoretically undermined this naïve grasp of the Marxian dialectic – relegating the domain of communication to the superstructure (*Überbau*) – as they were well aware of the massive impact of the new technologies of communication, notably telegraphy.

29 Engels, "Preface [to the Second Edition of 'The Peasant War in Germany']," p. 97.
30 McLuhan, *Understanding Media*, p. 58.
31 Ibid., p. 74.
32 Baudrillard, "Requiem for the Media," p. 164.
33 Enzensberger, "Constituents of a Theory of the Media," pp. 13–36.

Discourses on the Economically and Technologically Driven Formation of a Worldwide Union of the Proletariat (Marx and Engels)

In their writings Marx and Engels constantly referred to the global expansion of the Bourgeoisie, a development that is not only driven by economic constraints to constantly enlarge the market but also massively propelled by newly invented communication technologies like telegraphy. In the *Communist Manifesto* (1847/48), Marx and Engels wrote:

> The need of a constantly expanding market for its products chases the bourgeoisie over the entire surface of the globe. It must nestle everywhere, settle everywhere, establish connections everywhere. [...] The bourgeoisie, by the rapid improvement of all instruments of production, by the *immensely facilitated means of communication*, draws all, even the most barbarian, nations into civilisation. [...] The bourgeoisie, during its rule of scarce 100 years, has created more massive and more colossal *productive forces* than have all preceding generations together. Subjection of nature's forces to man, machinery, application of chemistry to industry and agriculture, steam-navigation, railways, *electric telegraphs*, clearing of whole continents for cultivation, canalisation of rivers, whole populations conjured out of the ground – what earlier century had even a presentiment that such *productive forces* slumbered in the lap of social labour?[34]

It is worth noting that in this passage Marx and Engels subsumed electric telegraphs under productive forces – productive forces that help to improve or rather to accelerate the circulation of commodities. Later, in the 1857/58 *Grundrisse*, Marx wrote explicitly: "The improvement of the means of transport and communication likewise falls into the category of the development of the productive forces generally."[35] And already in 1855 when Marx wrote about "electric telegraphs transforming the whole of Europe in one single Stock Exchange" it is crystal-clear that he did not in the least underestimate the social and economic effects of the new communication technology.[36]

Instead, Marx commented on its immense socio-economic impact inasmuch as it strongly accelerated the circulation of people, capital and signs.

34 Marx/Engels, *The Communist Manifesto*, pp. 38–40. My emphases.

35 Marx, *Grundrisse*, p. 523.

36 Marx, "The Commercial Crisis in Britain," p. 37. In this context one should recall that already in the beginning the most frequent practical public use of the electric telegraph was to communicate price trends on the stock exchange. Cf. Flichy, *Dynamics of Modern Communication*, p. 50.

For example, in the third volume of *Capital*, far less consulted than the first volume, Marx further elaborated on the interrelation of telegraphy and the socioeconomic sphere. He explained:

> The main means of cutting circulation time has been improved communications. And the last fifty years have brought a revolution in this respect that is comparable only with the industrial revolution of the second half of the last century. On land the Macadamised road has been replaced by the railway, while at sea the slow and irregular sailing ship has been driven into the background by the rapid and regular steamer line, the whole earth has being girded by *telegraph cables*.[37]

A similar argument concerning the massive impact of recent "technologies of communication" with a focus on the rapid acceleration of economic progress can be found in a text by Engels from 1850. In the context of his discussion of the "growth of German industry and commerce, of railways, telegraphs and ocean steam shipping since 1848" Engels wrote: "[T]his progress […] is unprecedented for Germany and has accomplished more in twenty years than was previously done in a whole century."[38]

Just as a side note to these quotations. While Marx and Engels serve as an object of study rather than as a theoretical point of departure, I fully agree with the argument they implicitly invoke here that the social impact of telegraphy cannot be examined in isolation from other technological developments, such as the construction of railway tracks and steamboat lines. Against this background one also has to bear in mind that the word "communication" in the nineteenth century – in English, French (*communication*) and German (*Kommunikation*) – was used in a far broader sense than it is today, comprising every means of connecting at least two things. At the time, it was commonly employed to describe any kind of traffic and connection and thus even to denote architectural connections within complexes of buildings.[39] Accordingly, telegraphy as a means of communication is for Marx and Engels subsumed under means of transport as only one element within the bigger picture of modern technology. Against this background, historians who write about the socio-economic impact of telegraphy might be well advised not to lose sight of the wider context of their object of investigation.

37 Marx, *Capital*, vol. 3, p. 164. My emphasis.
38 Engels, "Preface [to the Second Edition of 'The Peasant War in Germany']," p. 96.
39 Cf. "communication, n.," in: Oxford English Dictionary.

In the context of an analysis of how Marx and Engels conceived the effects of communication technology, in particular telegraphy, on society one much debated general question has to be addressed here concerning whether both are closer to a social and economic determinism or to a technological determinism – Marx is constantly accused of the former. A closer look at the passages just quoted seems to imply that the difficulty cannot be resolved conclusively, since one can interpret the relevant passages in either way.[40] On the one hand, one is confronted with sentences like "electric telegraphs transforming…" which presuppose a more media-materialist approach, in particular implying that the telegraph (along with other communication technologies) propelled the development of a European stock exchange and of a world economy. On the other hand, one can find in the paragraphs already quoted phrases stating that "The bourgeoisie, during its rule of scarce 100 years, has created […] electric telegraphy."[41] These imply a certain predominance of social forces, notably that the "bourgeois capitalists" invented and further developed these new technologies for their economic purposes. As a way out of this obvious paradox, one is inclined to suppose that Marx and Engels had a much more balanced view, namely that both of the determining forces, social as well as technological, are intrinsically interrelated. This would lead to the conclusion that as a whole, for Marx and Engels, technological development, as well as the use and the socio-economic effects of the productive forces of communication technologies, have to be understood as non-intentional effects *both* of the economic need of an ever increasing surplus value *and* of the mere existence of technological "discoveries that have not been called for […]."[42] For this paper, this means that we have to question the common denunciation of Marx as solely adhering to an over-simplified economic determinism as regards his elaborations on telegraphy.

40 However, Hanno Pahl tries to figure out this problem by arguing historically and reasoning that Marx's technological determinism that can yet be found in the *German Ideology* later, in the *Grundrisse*, gave way to a kind of social determinism, assuming that the development of technology can be solely put down to the social. Cf. Pahl, "Marx, Luhmann, Kritische Gesellschaftstheorie [Interview, Teil 2]," no page.

41 Marx/Engels, *The Communist Manifesto*, p. 40. A similar argument can be found in *Capital* where Marx elaborates on the use of productive forces "in order to consume the forces of nature productively." In this context he mentions for example that, once discovered, the knowledge of "magnetization of iron by electricity" is free of cost, but its appliance for "the purposes of telegraphy etc. necessitates a costly and extensive apparatus" (Marx, *Capital*, vol. 1, pp. 508–509).

42 Brecht, "The Radio as an Apparatus of Communication," p. 51.

A passage from the *Grundrisse* about the influence of the new possibilities of economic information exchange on people's individual actions, as well as the influence of the latter on the former, might serve as evidence for this hypothesis:

> [I]nstitutions emerge whereby each individual can acquire information about the activity of all others and attempt to adjust his own accordingly, e.g. lists of current prices, rates of exchange, interconnections between those active in commerce through the mails, *telegraphs* etc. (the *means of communication* of course grow at the same time). (This means that, although the total supply and demand are independent of the actions of each individual, everyone attempts to inform himself about them, and this knowledge then reacts back in practice on the total supply and demand. Although on the given standpoint, alienation is not overcome by these means, *nevertheless relations and connections are introduced* thereby which include the possibility of suspending the old standpoint.)[43]

Since the last sentence already invokes a certain dialectical development of a socio-economic backlash or countermovement, I shall turn to my crucial meta-Marxist argument for the necessity of a dialectical reversal of bourgeois technologies of communication. Bearing in mind the argument from the *Communist Manifesto* (that the bourgeoisie is urged to spread its trade relations over the entire surface of the globe), one might not be surprised that Marx dialectically predicted its necessary politico-historical consequence. In the same paragraph of the *Grundrisse* from which the preceding quotation is derived, he argued that, together with the formation of the global capitalist market system, the circumstances to transcend it were gradually being produced: "In the case of the *world market*, the connection of the individual with all, but at the same time also the *independence of this connection from the individual*, have developed to such a high level that the formation of the world market already at the same time contains the conditions for going beyond it."[44] In other words, for Marx, the fast economic developments of capitalism fuelled inter alia by communication technologies necessarily lead to the vast aggravation of social conditions and thus prepare the conditions of its own downfall. In a footnote in the third volume of *Capital*, Engels argued in the same vein dialectically that "[t]he colossal expansion of means of communication – ocean-going steamships, railways, *electric telegraphs*, the Suez canal – has genuinely established the world market for the first time. [...] All these things mean that most of the former breeding-grounds

43 Marx, *Grundrisse*, p. 161. My emphases.

44 Ibid. Marx's emphases.

of crises and occasions for crisis formation have been abolished or severely weakened."[45] But according to Engels these measures to combat recent crises only pave the way for even worse economic disasters to come, because "each of the elements that counteracts a repetition of the old crises, conceals within it the nucleus of a far more violent future crisis."[46]

Against this background, the "means of communication" played an important role not only in shortening the circulation process, but also – as a result of this – in sharpening class antagonisms:[47] According to Engels, along with machines like steam-engines or the mechanical weaving looms, the railways and electric telegraphs "are gradually and remorselessly destroying all the relics of feudal and guild conditions and are reducing all the petty social contradictions surviving from former times to the one contradiction of world-historical significance: that between capital and labour."[48] In *The Communist Manifesto*, a similar argument claimed that the "epoch of bourgeoisie" had simplified class antagonisms by increasingly dividing society "into two great classes directly facing each other – bourgeoisie and proletariat."[49] But this diagnosis does not have to be interpreted as mere resignation because this intensified situation according to Marx is the one necessary and therefore imminent precondition for the revolutionary reversal of society. So, it would be just a question of time until the class contradictions, full of suspense, would unfold: "[W]ithin bourgeois society [...] there arise *relations of circulation* as well as of production which are so many mines to explode it."[50] Again, the very relations of transportation and exchange that were indirectly produced by capitalism serve as precondition for its destruction: "[I]f we did not find concealed in society as it is the material conditions of production and the corresponding relations of exchange prerequisite for a classless society, then all attempts to explode it would be quixotic."[51] This development of explosive forces is caused by communication technology among other

45 Marx, *Capital*, vol. 3, pp. 620. My emphasis.

46 Ibid, p. 621.

47 In "Value, Price and Profit" Marx speaks about the "shortening of time and space by means of communication and transport." Marx, "Draft for the Final Passage of 'Value, Price and Profit'", p. 165. And in another text from 1862/63 he precisely lists the specific means "to shorten the circulation process, like railways, roads, navigation, telegraphs." Marx, *Theories of Surplus-Valve*, p. 491.

48 Engels, "The Prussian Military Question and the German Workers' Party," p. 70.

49 Marx/Engels, *The Communist Manifesto*, p. 34.

50 Marx, *Grundrisse*, p. 159. My emphasis.

51 Ibid.

factors; it is, as Engels emphasized – rigorously following the concept of historical materialism – a historical necessity,

> [b]ecause large-scale industry, the development of machinery, *communications* and world trade are assuming such gigantic proportions that their exploitation by individual capitalists is becoming daily more impossible; because the mounting crises of the world market are the most striking proof of this; because the productive forces and the *means of exchange* [*Verkehrsmittel*] which characterise the present mode of production and *exchange* [*Verkehr*] are daily becoming increasingly more than individual exchange [*Austausch*] and private property can manage; because, in a word, the moment is approaching when communal management of industry, of agriculture and of exchange [*Austausch*] will become a material necessity for industry, agriculture and exchange [*Austausch*] themselves — for this reason private property will be abolished.[52]

In other words, if communication technologies as a productive force play a pivotal role in the expansion of the capitalist economy and hence of the bourgeoisie, from a Marxist perspective, this technology must play the same role for the proletariat as long as they succeed in gaining access to it through a revolutionary act, thus acting according to historical necessity.[53] Engels wrote in 1876–78: "In making itself the master of all the means of production, in order to use them in accordance with a social plan, society puts an end to the former subjection of men to their own means of production. [...] The old mode of production must therefore be revolutionised from top to bottom."[54] One communist method to achieve this goal was to socialize the "means of production" including first of all the technologies of communication. As Engels underlined: "This necessity for conversion into State property is felt first in the great institutions for intercourse and communication – the post office, the telegraphs, the railways."[55] This quotation perfectly shows how wrong McLuhan and Baudrillard were, when they prematurely

52 Engels, "The Communists and Karl Heinzen," p. 304. My emphasis.

53 This logic is also the core of Enzensberger's later argument (cf. Enzensberger, "Constituents of a Theory of the Media").

54 Engels, "Anti-Dühring ," p. 279. Noticeable in 1867, Marx also takes into account the amount of workers in the so called new industries: "According to the census of 1861 for England and Wales, we find in the gas industry (gas-works, production of mechanical apparatus, servants of the gas companies, &c), 15,211 persons; in telegraphy, 2,399; in photography, 2,366; in steam navigation, 3,570; and in railways, 70,599." Marx, *Capital*, p. 573.

55 Engels, "Socialism: Utopian and Scientific," p. 711.

condemned Marx and Engels for allegedly underestimating the relevance of contemporary communications institutions for shaping social development.

If successful, a kind of industrial paradise awaited the protagonists. The feral productive forces formerly terrorizing the industrial society would thus be tamed and in the future be put to the service of mankind, as Engels put it poetically in the same text: "But when once their nature is understood, they can, in the hands of the producers working together, be transformed from master demons into willing servants."[56] Like Marx in *Capital,*[57] Engels elaborates on the means of utilizing the forces of nature productively with reference to the telegraph given that it does not stay in the hands of the bourgeoisie but that it will be socialized in order to follow the demands of the people:

> The difference is as that between the destructive force of electricity in the lightning of the storm, and electricity under command in the telegraph and the voltaic arc; the difference between a conflagration, and fire working in the service of man. With this recognition, at last, of the real nature of the productive forces of today, the social anarchy of production gives place to a social regulation of production upon a definite plan, according to the needs of the community and of each individual.[58]

Yet in the context of the revolutionary process, the very technological conditions of the bourgeois capitalist market served for Marx and Engels as the concrete means to help "WORKING MEN OF ALL COUNTRIES [...] UNITE!" Again, in the *Communist Manifesto*, Marx and Engels spoke of "immensely facilitated means of communication" and specified that "the ever expanding union of the workers [...] is helped on by the improved means of communication that are created by modern industry, and that place the workers of different localities in contact with one another. It was just this contact that was needed to centralise the numerous local struggles, all of the same character, into one [national] struggle between classes."[59] Like their counterpart, the bourgeoisie which, as already quoted, "must nestle everywhere, settle everywhere, establish connections everywhere" communists make use of the very same technology but instead with reversed aims. In place of the Bourgeoisie's pursuit of individual profit based on private property, communists promote international understanding based on common property: "Finally, they labour everywhere for the union [*Verbindung*] and

56 Ibid., p. 712.

57 See footnote 41.

58 Marx, *Grundrisse*, p. 712.

59 Marx/Engels, *The Communist Manifesto*, pp. 46–47.

agreement [*Verständigung*] of the democratic parties of all countries."[60] In a later text from 1894 Engels specified retrospectively the importance of communication technologies like the telegraph for that purpose: "It took a whole quarter of a century from the foundation of the International before [...] unity could be established at least in respect of most general economic viewpoints. And that with our *means of communication* – railways, *telegraph*, giant industrial cities, the press, organized people's assemblies."[61]

The means of communication – railways, steamboats and telegraphy – functioned in this context as no more than global facilitators of the worldwide simultaneous unification of the working class. Communication media were conceived as liberatory, as Baudrillard rightly summarized the Marxist scheme, but in order to unfold their full potential "it is necessary to liberate them."[62] Thus to exploit global communication for emancipatory political goals, it had to be brought immediately and entirely under communist control,[63] as Marx states in the *The German Ideology*: "Empirically, communism is only possible as the act of the dominant people 'all at once' and simultaneously, which presupposes the universal development of productive forces and the world intercourse [*Weltverkehr*] bound up with communism."[64]

From a metaperspective, this technologically facilitated global social-revolutionary communality in Marx and Engels' texts might also be interpreted by reference to Anderson's thoughts about 'imagined communities.' Due to telegraphy and its promise of universal connectivity, the working class gradually became aware of the hundreds of thousands, even millions, of fellow-workers all over the world. Hence, the attributed political agency of communication media like telegraphy was less driven by actual emancipatory media practices but more or less by their mere possibility. Nevertheless, it fuelled the idea of bringing working men of all countries further together to form the 'imagined community' of a communist international, of a globally organized proletariat. So, even if the connectivity existed only potentially, one might argue, this promise in and of itself had a huge historical influence on the development and strengthening of the communist movement.

60 Ibid., p. 84.

61 Engels, "On the History of Early Christianity," pp. 457–458.

62 Baudrillard, "Requiem for the Media," p. 168.

63 For a detailed account of the significance of the concept of simultaneity in nineteenth century media history see: Uricchio, *Media, Simultaneity, Convergence.*

64 Marx, *The German Ideology*, p. 56.

Conclusion

Others have since built on Marx's and Engels' approach to media technology and have taken on both their belief that media were in the wrong hands and their aspirations to take over the media in order to exploit their vast political possibilities. In his famous essay "The Radio as an Apparatus of Communication," Brecht argued that radio had to be changed from a medium of entertainment and top-down distribution of information into a medium of communication, a medium that enabled exchange. In particular, radio had to become a medium that should bring the participant "into a relationship instead of isolating him […]."[65] Brecht described radio – a discovery that had "not been called for" – in the tradition of Marx as a technology that was more advanced than society urging the people to strive for radical social change by overthrowing the existing institutions:

> When I say that the radio or the theatre 'could' do so-and-so I am aware that these vast institutions cannot do all they 'could,' and not even all they want. But it is not at all our job to renovate ideological institutions on the basis of the existing social order by means of innovations. Instead our innovations must force them to surrender that basis. So: For innovations, against renovation! […] Impracticable in the current social order, practicable in another one, these suggestions – which, after all, only form the natural consequence of technical development – serve for the propagation and formation of this *other* order.[66]

Later, in 1970, Enzensberger would react to this approach by sketching out a "socialist media theory" which – based on the assumption that the "new media are egalitarian in structure" – focused on taking the "promises of the media" seriously. In order to fulfil the "collective wishes […] to take part in the social process on a local, national and international scale" Enzensberger demanded a more "emancipatory use of media."[67] In line with Marx's and Engels' ideas about the seizure of communication technology and institutions due to their socio-political impact, Enzensberger argued about "the mobilizing power of the media" in a very Brechtian way: "One immediate consequence of the structural nature of the new media is that none of the

65 Brecht, "The Radio as an Apparatus of Communication," p. 51.

66 Ibid., pp. 52–53. The last part is omitted in the English version of the text and translated from German from: Brecht, "Der Rundfunk als Kommunikationsapparat," p. 557.

67 Enzensberger, "Constituents of a Theory of the Media," pp. 14, 20 and 25–26.

régimes at present in power can release their potential. Only a free socialist society will be able to make them fully productive."[68]

These arguments have been influential for all types of techno-optimists up to the present day. But what often remains unacknowledged are their Marxist roots. Most strikingly, late twentieth century discourses about the emancipatory aspects of the internet display an odd mingling of rather disparate thoughts like social utopist, Marxist and liberalist ideas. Either these considerations are connected to imagined virtual communities, which serve as prime examples of a future liberated society as in Howard Rheingold's writings or the new information technologies are conceived of as emancipatory and liberatory per se through references to Brecht and Enzensberger.[69] Alexander Galloway concisely summed up the techno-optimism of our day by calling the qualities Enzensberger attributed to an emancipatory use of media as "a perfect foreshadowing of today's most optimistic descriptions of the Internet."[70] Accordingly, the decentralized forms of internet communication are for the most part comprehended as models of striving for liberty and independence on a larger scale, as in Jeff Jarvis' eulogy of publicness *Public Parts*: "The net [...] is our tool of disruption, a catalyst that breaks old bonds and sets us loose to explore our natures anew. [...] This transformation also takes momentous form: revolutions, dead industries, economic upheaval. We atomize. We re-form."[71] Wael Ghonim, one of the major activists in the Egyptian uprising puts it more briefly: "[I]f you want to liberate a society just give them the internet."[72] These discursively constructed political impacts of new media occasionally come under scrutiny.[73] But

68 Ibid., p. 21.

69 See for example: Rheingold, *The Virtual Community*.

70 Galloway, *Protocol*, p. 58.

71 Jarvis, *Public Parts*, p. 10. One might even go as far as to say that the phrase about the exploration of human nature is a kind of echo of Marx's idea of species-being as already described above.

72 Wael Ghonim in: Cohen, "Google's Wael Ghonim Thanks Facebook for Revolution;" cf. for a general account of the revolutionary impact of new media and the internet: Shirky, "The Political Power of Social Media;" Shirky, *Cognitive Surplus*. For general discourses about the liberating power of the internet see: Galloway/Thacker, *The Exploit*; Atton, *An Alternative Internet*; Hardt/Negri, *Multitude*; Rheingold, *Smart Mobs*. The perspective on social media is explored in: Jarvis, *What Would Google Do?*; Fuchs, *Internet and Society*. Fuchs/Tapscott/Williams, "Wikinomics [Review];" Benkler, *The Wealth of Networks*; Lessig, *Code*. Cyber communism is reflected in: Barbrook, "The::Cyber.com/munist::Manifesto" (1999).

73 See for example: Reichert, *Amateure im Netz*; Lovink, *Zero Comments*; id., *My First Recession*; id., *Dark Fiber*; Barbrook/Cameron, "The Californian Ideology;" Barbrook, *Media Freedom*; Kroker/Weinstein, *Data Trash*.

the corresponding observations are rarely substantiated by reference to historical analyses. Whereas some recent publications trace back the internet ideals or utopias of freedom and global communality of the 1990s to the 1960s, this article might contribute to their genealogy and show that they are already deeply rooted in nineteenth-century discourses about 'new' communication technology like telegraphy.[74]

Bibliography

Anderson, Benedict, *Imagined Communities. Reflections on the Origins and Spread of Nationalism*, London 1983.

Atton, Chris, *An Alternative Internet*, Edinburgh 2004.

Baczko, Bronislaw, *Utopian Lights. The Evolution of the Idea of Social Progress*, New York 1989.

Barbrook, Richard, "The::Cyber.com/munist::Manifesto" (1999), accessed June 20, 2010. http:www.nettime.org/Lists-Archives/nettime-l-9912/msg00146.html.

—, *Imaginary Futures. From Thinking Machines to the Global Village*, London 2007.

—, *Media Freedom. The Contradictions of Communications in the Age of Modernity*, London 1995.

Barbrook, Richard/Cameron, Andy, "The Californian Ideology," *Science as Culture*, vol. 6, no. 26 (1996), pp. 44–72.

Baudrillard, Jean, "Requiem for the Media," in: *For a Critique of the Political Economy of the Sign*, St. Louis, Mo 1981, pp. 164–184.

Benjamin, Walter, "The Work of Art in the Age of Its Technological Reproducibility [Second Version]," in: Michael William Jennings/Brigid Doherty/Thomas Y. Levin (eds.), *The Work of Art in the Age of Its Technological Reproducibility, and other Writings on Media*, Cambridge, MA 2008, pp. 32–68.

Benkler, Yochai, *The Wealth of Networks. How Social Production Transforms Markets and Freedom*, New Haven, CN 2006.

Brecht, Bertolt, "Der Rundfunk als Kommunikationsapparat," in: Werner Hecht et al. (eds.), *Werke. Große kommentierte Frankfurter und Berliner Ausgabe*, vol. 21, Frankfurt/M. 1992, pp. 552–557.

Brecht, Bertolt, "The Radio as an Apparatus of Communication," in: John Willett (ed.), *Brecht on Theatre. The Development of an Aesthetic*, New York 1964, pp. 51–53.

Cohen, Jackie, "Google's Wael Ghonim Thanks Facebook for Revolution," in: *All Facebook* (11.3.2011), accessed November 11, 2012. http://allfacebook.com/googles-wael-ghonim-thanks-facebook-for-revolution_b32315.

74 Barbrook, *Imaginary Futures*; Turner, *From Counterculture to Cyberculture*.

"communication, n.," in: *Oxford English Dictionary*. 3. edition, September 2009; online version June 2011, accessed August 4, 2011. http://oed.com/view/Entry/37309.

Engels, Friedrich, "Anti-Dühring, Herr Eugen Dühring's Revolution in Science," in: *Collected Works*, vol. 25, Moscow/New York/London 1987, pp. 5–312.

—, "The Communists and Karl Heinzen," in: *Collected Works*, vol. 6, Moscow/New York/London 1976, pp. 291–306.

—, "On the History of Early Christianity," in: *Collected Works*, vol. 27, Moscow/New York/London 1990, pp. 445–469.

—, "Preface [to the Second Edition of 'The Peasant War in Germany']," in: *Collected Works*, vol. 21, Moscow/New York/London 1985, pp. 93–100.

—, "The Prussian Military Question and the German Workers' Party," in: *Collected Works*, vol. 20, Moscow/New York/London 1985, pp. 37–79.

—, "Socialism: Utopian and Scientific," in: Robert C. Tucker (ed.), *The Marx-Engels Reader*, New York 1978, pp. 683–717.

Enzensberger, Hans Magnus, "Constituents of a Theory of the Media," *New Left Review*, no. 64 (1970), pp. 13–36.

Fischer, Claude S., "'Touch Someone:' The Telephone Industry Discovers Sociability," *Technology and Culture*, vol. 29, no. 1 (1988), pp. 32–61.

Flichy, Patrice, *Dynamics of Modern Communication. The Shaping and Impact of New Communication Technologies*, London 1995.

Fuchs, Christian, *Internet and Society. Social Theory in the Information Age*, New York 2008.

Fuchs, Christian/Tapscott, Don/Williams, Anthony D., "Wikinomics: How Mass Collaboration Changes Everything [Review]," *International Journal of Communication*, no. 2 (2008), pp. 1–11.

Galloway, Alexander R., *Protocol. How Control Exists after Decentralization*, Cambridge, MA 2004.

Galloway, Alexander R./Thacker, Eugene, *The Exploit. A Theory of Networks*, Minneapolis, MN 2007.

Hardt, Michael/Negri, Antonio, *Multitude. War and Democracy in the Age of Empire*, New York 2004.

Headrick, Daniel R., *The Tools of Empire. Technology and European Imperialism in the Nineteenth Century*, New York 1981.

Jarvis, Jeff, *Public Parts. How Sharing in the Digital Age Improves the Way We Work and Live*, New York 2011.

—, *What Would Google Do?*, New York 2009.

Kittler, Friedrich, "Geschichte der Kommunikationsmedien," in: Jörg Huber/Alois Müller (eds.), *Raum und Verfahren*, Basel/Frankfurt/M. 1993, pp. 169–188.

—, "Lakanal und Soemmering. Von der optischen zur elektrischen Telegraphie," in: Brigitte Felderer (ed.), *Wunschmaschine Welterfindung, Kat. Kunsthalle Wien*, Wien et al. 1996, pp. 286–295.

Kroker, Arthur/Weinstein, Michael A., *Data Trash. The Theory of the Virtual Class*, Montreal 1994.
Kula, Witold, *Measures and Men*, Princeton 1986.
Leibniz, Gottfried Wilhelm, "On the General Characteristic," in: Leroy E. Loemker (ed.), *Philosophical Papers and Letters*, Dordrecht 1989, pp. 221–228.
Lessig, Lawrence, *Code. Version 2.0*, New York 2006.
Lovink, Geert, *Dark Fiber. Tracking Critical Internet Culture*, Cambridge, MA 2002.
—, *My First Recession*, Rotterdam 2003.
—, *Zero Comments. Blogging and Critical Internet Culture*, New York 2008.
Marvin, Carolyn, *When Old Technologies Were New. Thinking about Electric Communication in the Late Nineteenth Century*, New York 1990.
Marx, Karl, *Capital. A Critique of Political Economy*, vol. 1, London 1992.
—, *Capital. A Critique of Political Economy*, vol. 3, London 1993.
—, "The Commercial Crisis in Britain," in: *Gesamtausgabe (MEGA)*, vol. 14, Berlin 2001, pp. 37–41.
—, "Draft for the Final Passage of 'Value, Price and Profit'," in: *Gesamtausgabe (MEGA)*, vol. 20, Berlin 2003, pp. 140–186.
—, "Economic and Philosophical Manuscripts [Excerpts]," in: Erich Fromm, *Marx's Concept of Man*, New York 1961.
—, *The German Ideology. Part One and Selections from Parts 2 and 3*, New York 1970.
—, *Grundrisse. Foundations of the Critique of Political Economy*, London/Harmondsworth 1973.
—, *Theories of Surplus-Value (Volume IV of Capital. Part II)*, Moscow 1963.
—, "Theses on Feuerbach," in: Robert C. Tucker (ed.), *The Marx-Engels Reader*, New York 1978, pp. 143–145.
Marx, Karl/Engels, Friedrich, *The Communist Manifesto*, London 2008.
McLuhan, Marshall, *Understanding Media. The Extensions of Man*, Corte Madera, CA 2003.
Ozouf, Mona, *L'école de la France. Essais sur la Révolution, l'utopie et l'enseignement*, Paris 1984.
Pahl, Hanno, "Marx, Luhmann, Kritische Gesellschaftstheorie [Interview, Part 2]," *trend onlinezeitung*, no. 3 (2008), accessed November 10, 2012. www.trend.infopartisan.net/trd0208/t010208.html.
Plessner, Helmuth, *The Limits of Community. A Critique of Social Radicalism*, New York 1999.
Reichert, Ramón, *Amateure im Netz. Selbstmanagement und Wissenstechnik im Web 2.0*, Bielefeld 2008.
Rheingold, Howard, *Smart Mobs. The Next Social Revolution*, Cambridge 2002.
—, *The Virtual Community. Homesteading on the Electronic Frontier*, Reading, MA 1993.
Schröter, Jens, "Übertragung und Explosion – Telegraphie/Telephonie/Transport," in: id. (ed.), *Media Marx. Ein Handbuch*, Bielefeld 2006, pp. 201–212.

Shirky, Clay, *Cognitive Surplus. Creativity and Generosity in a Connected Age*, London 2010.

—, "The Political Power of Social Media," *Foreign Affairs* (January/February 2011), accessed February 13, 2011. www.foreignaffairs.com/articles/67038/clay-shirky/the-political-power-of-social-media.

Stichweh, Rudolf, "The Genesis of a Global Public Sphere," *Development*, vol. 46, no. 1 (2003), pp. 26–29.

Turner, Fred, *From Counterculture to Cyberculture. Stewart Brand, the Whole Earth Network, and the Rise of Digital Utopianism*, Chicago 2006.

Uricchio, William, *Media, Simultaneity, Convergence: Culture and Technology in an Age of Intermediality*, Utrecht 1997.

Werber, Niels, *Die Geopolitik der Literatur. Eine Vermessung der medialen Weltraumordnung*, München 2007.

Winseck, Dwayne Roy/Pike, Robert M., *Communication and Empire. Media, Markets, and Globalization, 1860 - 1930*, Durham, NC 2007.

Wobring, Michael, *Die Globalisierung der Telekommunikation im 19. Jahrhundert. Pläne, Projekte und Kapazitätsausbauten zwischen Wirtschaft und Politik*, Frankfurt/M. 2005.

Agents|Actors

From Partnership to Confrontation: Japan and the Great Northern Telegraph Company, 1871–1943

Daqing Yang

On May 4, 1940, H. S. Poulsen, the Far Eastern manager of the Danish telegraph company Great Northern (GN) arrived at the Japanese Ministry of Communications (MOC) in Tokyo. In the presence of the Minister, he signed a document that would put an end to the company's landing rights in Japan. Since both of the company's two major submarine telegraphy cables in East Asia connected in the western Japanese city of Nagasaki, the event was a major blow to the company's long-standing presence in the entire region. Understandably, it was seen in an entirely different light in Japan. In the words of Tamura Kenjirô, Japan's Director-General of Telecommunications, this event bore "witness to a powerful Japan making great strides" and laid "a founding stone for the construction of the new communications order in East Asia."[1] The mass-circulation national daily *Mainichi* described the event as "a dream coming true:" with the present position and power of Nippon, "we have at last secured complete communication autonomy."[2] Such views are still echoed in Japan today. One recent history of Japan's international communications describes GN's presence as "colonial" and considered 1940 a key turning point.[3]

Although Japan's era of unequal treaties is generally considered to have ended by the beginning of the twentieth century, the retreat of a foreign communications company like GN from Japanese soil might be considered the real endnote of that lamentable era.[4] Indeed, at a roundtable discussion

1 Tamura Kenjirô, Taihoku Denshin Kaisha menkyojo no kaitei ni tsuite, pp. 2–9. Some of the materials in this chapter have appeared in my *Technology of Empire: Telecommunications and Japanese Expansion in Asia*.

2 *Mainichi shimbun* (May 3, 1940).

3 Ishihara Fujio, *Kokusai tsûshin no Nihon: Shokuminchika kaishô he kutô no 99-nen*.

4 The "unequal treaties" usually refer to treaties signed between Japan and foreign countries during the late Tokugawa era (1853–1868). The only remaining issue after Japan's successful treaty revisions was the permanent rent of land in areas once designated as foreign settlement in Japan. In 1937, the Japanese government reached agreement with the eight remaining countries that such rights would expire in April 1942.

convened after the signing of the new agreement, a member of the Japanese Upper House, the House of Peers, who had previously served in the Ministry of Communications, described GN's presence in Nagasaki as "the last special foreign privilege in Japan since the Meiji era (1868–1912)."[5]

Since the beginning of Meiji, GN was the harbinger of modernity that first brought Japan into the age of worldwide telecommunication. Why did the relationship between GN and the Japanese government that had weathered so many changes since the 1870s come to such an abrupt end? This paper explores the factors that transformed the relationship between GN and Japan from that of cooperation to conflict. I argue that while technological changes and Japan's quest for autonomy gradually undermined the once-cooperative relationship, Japan's new ambition in Asia and its bureaucratic politics as well as GN's own business calculations also played important roles. In doing so, the paper challenges the simplistic portrayal of the history of international communications in Japan as a continuous struggle for autonomy and sheds light on the complex structure of international telegraphy in the first half of the twentieth century.

The Age of Partnership

The early history of the Great Northern in Japan has been told competently elsewhere and only merits brief recounting here. In August 1870, the new Japanese government reached an agreement with the recently incorporated Danish firm, whereby GN would lay submarine telegraph cables between Shanghai and Nagasaki, Nagasaki and Vladivostok, and Nagasaki and Yokohama. The company would provide financial papers of its operations should the Japanese government desire to purchase the above cables. In return, the Japanese government offered most-favored status to the Great Northern in case Japan were to grant concessions to other companies. Japan also gave in to the Danish demand that this agreement remain in effect for thirty years. In 1871, the Great Northern completed the submarine cable between Shanghai and Nagasaki and opened it to public service. This was followed by the Nagasaki-Vladivostok line a few months later. As a result, by the

5 Zandankai: "Daihoku Denshin Kaisha" no sesshû o kataru (1), *Teishin kyôkai zasshi* 382 (June 1940), p. 17.

beginning of 1872, Japan was linked to Europe and beyond through the worldwide telegraphic network for the first time in history.[6] This was a time when Japan had not yet introduced international mail service. Its domestic telegraph had barely started and was limited to the Tokyo-Yokohama and Osaka-Kobe areas.

International telegraphic communication required both advanced technology and enormous financial resources. Having just emerged out of two centuries of isolation and embarked on modernization, Japan had neither. It had no choice but to rely on GN. Still, the Japanese government strove to prevent foreign control of Japan's domestic telecommunications. It placed high priority on building its own domestic telegraph network and declared a state monopoly over all telegraphic operations within Japan. Its own swift construction of the Nagasaki-Yokohama landline in 1873 deprived the Great Northern of its opportunity to lay a parallel submarine cable between the two port cities as stipulated in the 1870 treaty.[7]

After that point, GN and Japan were able to fashion a cooperative partnership largely thanks to their asymmetrical interests and capabilities. Based in a small northern European country that largely stayed out of world politics, GN desired above all to increase its revenue and protect investment. For Japan, the GN international telegraph link was a major boost to its overseas trade and diplomatic relations. All government telegrams were charged half price. Exchange of information between Japan and the US, its largest overseas market for silk export, was reduced from several months to a matter of hours.[8]

Moreover, Japan's international communication began to assume strategic significance. As Japan's rivalry with China in Korea intensified at the beginning of 1880s, the need for swift communication with the peninsula grew urgent. Lacking necessary funds as well as technology, Japan had no choice but to turn to the Great Northern to lay the first submarine telegraph cable across the Korean Strait in 1882. After further negotiations with the Korean

6 For the early history of GN presence in Asia, see Ahvenainen, *The Far Eastern Telegraphs;* Lange, *Partnere og rivaler.* The Japanese text of the agreement is reprinted in Nihon Kagakushi Gakkai comp., *Nihon kagaku gijutsu shi daikei,* vol. 19, p. 49. For a Japanese study that emphasizes international factors, see Nagashima, Daihoku denshin kaisha no Nihon shishutsu to sono haikei.. As GN and the Great Eastern Telegraph Company pooled their cables between Shanghai and Hong Kong, telegrams could also travel via the southern route.

7 Jacobsen, In the Service of the Nation.

8 Nakayama: "Zandankai: 'Daihoku Denshin Kaisha' no sesshû o kataru (2)," p. 23.

government, Japan was allowed to build a short landline from Great Northern's landing site to the Japanese settlement in Pusan, where Japan would establish its first overseas telegraph office. For the first time, Japanese-language telegrams could be received and sent outside the Japanese islands after GN completed the cables to Pusan. This was of considerable importance to the growing Japanese presence in Korea.[9] The Great Northern also agreed to add new cables between Nagasaki and Shanghai and Nagasaki and Vladivostok to handle increased traffic. In addition, the company promised not to raise cable rates without the consent of the Japanese government and to ensure that all submarine cables remain in working order. In exchange for Great Northern's timely service of building the new cable, the Japanese government granted the Danish company a twenty-year monopoly over Japan's international cable communications, to be extended to 30 years if Chinese and Russian governments also extended such privileges for a comparable period.[10] Acting on the principle of no foreign involvement in its domestic communications, Japan acquired the right to operate part of this cable from Japan proper to the Tsushima Island in the Korean Strait.[11]

The GN's telegraph cable network in East Asia also proved to be a critical asset for Japan's military operations in Asia, albeit sometimes in ways unknown to the company. After the outbreak of the Russo-Japanese War in 1904, the Japanese government needed to build new submarine cables to connect with its army advancing toward the Russian stronghold in Manchuria. Unable to purchase new cables in a timely manner, it secretly removed sections of the two GN Nagasaki-Vladivostok cables which temporarily went out of service and put them to its own military use. It quietly reconnected the cables after the hostility ceased, apparently without being detected by GN.[12]

9 For details of the Korea cable, see Yang, *Technology of Empire*, chap. 1.

10 Ahvenainen, *The Far Eastern Telegraphs*, 67. A complete text of the agreement is reprinted in Ishikawa comp., *Kokka sôdoin shi: Shiryô hen* III, pp. 696–701.

11 Nihon Denshin Denwa Kôsha Kaiteisen Shisetsu Jimusho, *Kaiteisen hyakunen no ayumi*, pp. 179–181. Later, as Japan annexed Korea after driving out both China and Russia, the Japanese government purchased the remaining portion of the Korea cable – between Tsushima and Pusan – from the Great Northern in 1910 on the same grounds.

12 This episode was first revealed during the Pacific War by a former Japanese army officer who had participated in the clandestine operation. Its details were provided in the authoritative history of Japan's submarine cables published in 1970. On July 4, 1904, Japan's cable ship *Okinawamaru* captured 5.8 n.m. and 20 n.m. respectively of no.1 and no. 2 GN's Vladivostok cables. Nihon Denshin Denwa Kôsha Kaiteisen Shisetsu Jimusho, *Kaiteisen hyakunen no ayumi*, 152, 158.

One indication of the Japanese government's appreciation of GN was the numerous official awards bestowed upon its personnel. Between 1898 and 1918, at least nine Danish employees of GN received awards from the Japanese government. Several of them were cited for their services such as the timely repair and maintenance of the cables from Japan as well as assistance in censorship matters during times of overseas conflict such as the Sino-Japanese War, the Boxer Rebellion in China and the Russo-Japanese War. In 1910, the Japanese government awarded the Imperial Order of the Rising Sun 3rd Class to two senior GN executives – head of the Far Eastern department Peter Michelson and Far Eastern manager J.J. Bahnson – for successfully concluding the agreements to sell the remaining portion of the Korea cable to Japan. Several of them repeatedly received awards with one technician receiving it as many as four times.[13] At the very least, such awards indicated the desire of the Japanese government to ensure continued cooperation of this important partner.

However, the Japanese government was also aware of the cost of GN's cooperation. For instance, gaining telegraphic access to Korea came at a price to Japan's own autonomy, as the 1882 arrangement led to granting GN permanent landing rights in Japan. In 1902, Japan had no choice but to extend GN's concession by ten years while its own efforts to secure alternative routes and to reduce telegram rates only had limited success. In 1906, Japan managed to open a direct route to the U.S. via the Commercial Pacific Cable Co., an American company jointly set up by GN and the Great Eastern Extensions Co. Japan's attempt to establish an independent telegraph link with the Asian continent would encounter stubborn opposition from the Danish firm, even after Japan achieved a de facto link after building a military line to southern Manchuria during the Russo-Japanese War. Given the fact that GN's concession would expire by the end of 1912, Japan's Ministry of Communications (MOC) began exploring ways to end GN's monopoly of telegraphic traffic with China. One problem, as an MOC official found out while visiting London, was that cable companies managed to have different expiration dates of their landing rights with governments on two ends of the cable, thus making it difficult for either government to end the concession unilaterally.[14] After negotiation with both GN and the Chinese government, Japan was able

13 Records for these awards are in Japan's National Archives and accessible through Japan Center for Asian Historical Records (abbrievated as JACAR). The award for Michelson and Bahnson is under A10112703900. In other documents, Michelson and Bahnson are referred to as chief of GN's secretariat and principle agent in the East respectively.

14 Nakayama, Zandankai: "Daihoku Denshin Kaisha" no sesshû o kataru (2), pp. 22–23.

to realize its much-cherished goal in 1913 of laying its own cable to Shanghai. In return, the Japanese government agreed to share with GN all revenue on telegraphic traffic between Japan and China, whether it traveled on GN lines or not. The construction of Japan's own telegraph cable to Shanghai represented yet another comprise between these two asymmetrical actors despite sources of irritation and potential conflict.

The cooperation continued in the 1920s. For instance, when the new Nationalist government in China attempted to recover its "lost rights" from foreign countries, Japan and Great Northern cooperated to defend the status quo. Before GN began negotiating with China over the renewal of its landing rights, Bahnson held talks with Japanese officials and proposed that GN and Japan exchange information in secret while maintaining separate stances in dealing with China. Bahnson even alerted the Japanese government that China might refuse Japan's landing rights altogether.[15] In April 1933 the Chinese government reached an agreement with the three foreign cable companies – GN, Eastern Extension, and Commercial Pacific – whereby China terminated landing rights and monopoly concessions and instead granted landing license for a 14-year period from 1931 till 1944.[16] The end of GN's permanent landing rights in China would impact its relationship with the Japanese government, as there would be no more need for the united front in China.

Wireless and Frictions

A serious challenge to GN's global business as well as its presence in Japan was the new technology of wireless. First appearing in the beginning of the twentieth century and undergoing rapid technological advance, wireless threatened the monopoly of the telegraph cable in international communications. As Kurt Jacobsen and others have shown, GN acted like other international cable companies and refused to fully embrace the new communications technology. Instead, it bought a wireless patent with the sole intention of preventing it from being used on rival routes to Asia.[17]

15 "Denmukyokuchô-Bahnson kaidan yôryô," Ministry of Communications (MOC) Papers, Tokyo.

16 *Zhongguo jindai yudian shi*, pp. 179–181. Citing internal GN documents, the author accused GN of bribing Chinese officials, including the Minister of Communication and Director of Telecommunications, into extending landing rights.

17 Jacobsen, Wasted opportunities?, pp. 231–250.

In contrast, the Japanese saw wireless as a godsend. After World War I, when Japan sought to further expand its influence on the Asian continent, Japanese government leaders as well as opinion makers began calling for measures to increase Japan's international communications capability as well as its autonomy. The Army, Navy, and Communications Ministries agreed on the urgent need to "quickly establish Japan's superior position in international communications in East Asia" and conducted joint investigations on high power wireless as a new solution for overseas communications. After conducting careful studies, the three-ministry joint committee proposed that wireless telegraph would be given precedence over cables in the future expansion of Japan's international telecommunications facilities. It also called for direct wireless communications links with foreign countries without transmission via third countries.[18]

To overcome perennial problems of public funding in Japan, the Army-Navy-Communications joint committee recommended that a private enterprise in Japan be allowed to build long-distance wireless facilities and submarine cables in the future while keeping operation and management in government hands. The Foreign Ministry also found reasons for such a formula.[19] In 1925, the Japan Wireless Company was founded by a group of businessmen to construct wireless facilities for overseas communications.[20] The company, capitalized at 20 million yen, was authorized to open direct wireless communications with Europe, the South Seas islands, and East Asia. A few years later, the private International Telephone Company was founded to operate wireless telephone links to Japan's colonies as well as other locations in Asia. Eventually, the two companies would be merged, at the government's insistence, into the International Telecommunications Company in 1937.

Beside strategic implications, there were mounting economic reasons for Japan to reduce reliance on foreign telegraph cables. In the wake of the worldwide economic depression, Japan's export declined sharply. As a result, the Japanese government adopted a policy of reducing hard currency payments in order to ensure import of vital resources. In 1935 alone, however,

18 "Rikugun daijin, kaigun daijin, teishin daijin ni teishitsu taigai dai musen keikaku ni kansuru sanshô kankei kanri no chôsa hôkokusho," (received on August 12, 1924), MT 3.6.11.23, Japan Ministry of Foreign Affairs Archives Microfilm.

19 Memo of five-ministry meeting on private operation of wireless (May 29, 1924). Ibid.

20 For a Japanese study, see Sunaga Noritake, Senzenki Nihon no taigai tsûshin jigyô no tokushitsu to tsûshin shuken, pp. 35–54.

over ten million yen were paid to GN for international telegraphic traffic. This was partly due to the 1912 agreement with the Great Northern, which stipulated that Japan was to pay the company for one third of all traffic on all submarine cables between Japan and China, including Japan's own cables. As Japan rigorously promoted wireless for its overseas telegraph communication, the share of cable telegrams declined. After 1935, the majority of Japan's overseas telegrams were sent over wireless. An important reason was that, at the encouragement of MOC, the Japanese business community was mobilized to promote the use of wireless in their international communications. International wireless promotion associations were set up at major commercial centers such as Kobe and Osaka.[21]

Business Frictions in Manchukuo and at Home

Japan's aggressive promotion of international wireless telegrams was bound to cause frictions with GN in Japan as well as in its newly expanded empire. After Japan invaded Manchuria (northeastern China) in 1931, it established the Manchurian Telegraph and Telephone Company (MTT) in 1933 to operate telecommunications and broadcasting in the puppet state Manchukuo. As a result, the new MTT had to compete with foreign cable companies for the lucrative international traffic. The ensuing dispute between the cable companies – GN in particular – and MTT was as much a result of latter's aggressive business practice as the emerging competition between cable and wireless communication technology.

According to previous agreements with foreign cable companies in China, outgoing telegrams from the port city Dalian (Port Arthur, in the Japan-administered Kwantung Leased Territory) and other places in Manchuria, if marked "VIA Northern" or "VIA Eastern," were to be handed over to the respective foreign cable companies. Making use of its powerful wireless facilities in Shinkyo (Xinjing or Hsinking in Chinese, present-day Changchun), MTT launched an aggressive campaign for international traffic to increase its overall revenue. Soon the cable companies discovered that many of these routing requests had been disregarded in the telegraph offices in the area, leading to their declining revenue. The companies dispatched F.B. Pfordten, Divisional Manager of the Eastern Extension Company, to

21 *Nihon musen shi*, vol. 6, pp. 559–560.

Figure 1. Wireless vs. Cable Telegrams from Japan, 1929–1941

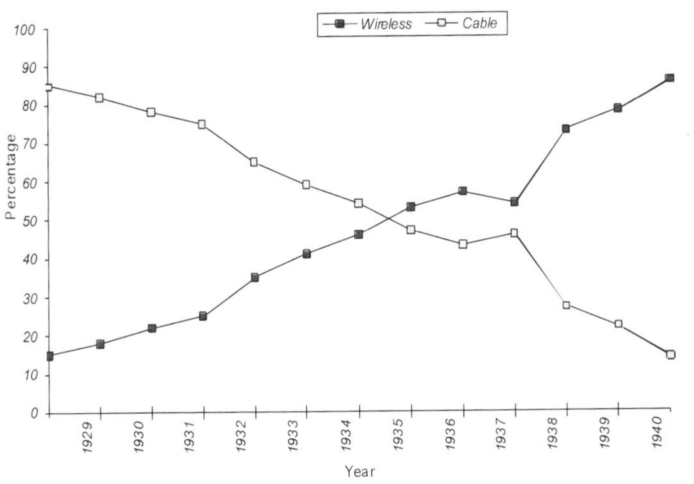

Figure 2. Japanese Payment for Overseas Telegrams (1926–40)

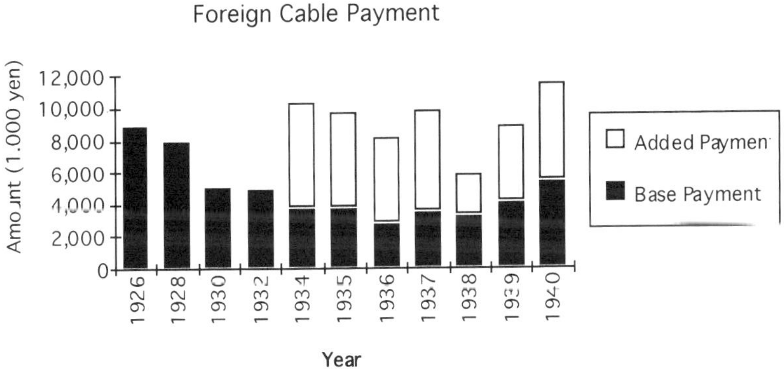

Source: Nihon musen shi, vol. 5, p. 560.
Note: Added payment due to devaluation of yen.

Dalian in September 1934 to bring the matter up with the Japanese Director of Communications there. The latter acknowledged the irregularities and promised Mr. Pfordten to straighten them out.

It was not before long, however, that the companies had to lodge new complaints with fresh evidence throughout 1935, though without much effect. On February 18, 1936, H.P. Krogh, Chief Traffic Agent of the Great

Northern Company, dispatched four telegrams from Dalian - two at the Central Telegraph Office, two at the Post Office on Yamagata-dori, all distinctively marked "VIA NORTHERN" or "VIA EASTERN." As he had suspected, they were not forwarded to the cable companies. Moreover, Krogh observed personally that clerks at the counter erased the routing requests for cable and substituted with "HSIN," marker for the MTT-operated "Hsinking Radio." In addition, he obtained similar information from other Western firms in Dalian. Despite these complaints, MTT apparently did not relent. Its First District Bureau Chief Meeting in 1936 specifically discussed measures of attracting international telegrams to Hsinking Radio.[22]

During the next few months, GN and other cable companies filed more complaints and requested meetings with Japanese officials in Dalian, but to little avail. The companies finally decided to ask for diplomatic intervention, but in such a way as not to cause an imminent showdown with the Japanese authorities. The British Consul General in Dalian was informed of these developments. The situation continued to deteriorate, however, with the cable companies receiving a total of two telegrams from Dalian to Europe and America during the first quarter of 1937. In their meeting at Japan's Communications Bureau in Dalian, with its director absent, Pfordten and Krogh brought along the British Consul and let it be known to the Japanese that the consular authorities would have to report to the British Ambassador and Danish Minister in Tokyo if situation did not improve as a result of the latest representations.[23] On top of all this, H.S. Poulsen, Great Northern's General Manager for the Far East, made a visit to Tokyo and held talks with officials in the Ministry of Communications. Poulsen presented them with a list of 57 telegrams dispatched in March 1937 that had requested to be routed via Eastern or Northern by the senders, none of which had been handed over to the companies. Officials in the MOC Telecommunications Department complained that it was inappropriate [*omoshirokarazu*] to take a business matter to diplomatic channels and that a proper adjustment should instead have been negotiated between concerned parties.[24]

22 *Manshû Denshin Denwa Kabushiki Kaisha 10-nen shi*, p. 612.

23 Memo (April 26, 1937) in Nomura (Denmukyoku) to Ichihashi (MTT Gaishin) (May 6, 1937s) G1452, MOC Papers.

24 Nomura (Denmukyoku) to Ichihashi (MTT Gaishin) (May 6, 1937s, G1452), MOC Papers.

Behind the scenes, the Communications Ministry had referred this matter to the Manchurian Telegraph and Telephone Co. (MTT) each time they received a protest. MTT's behavior must be understood in the context of its emphasis on increasing revenue and fierce competition for international telegraphic traffic as well as its quasi-autonomous status. As part of their business practice, foreign cable companies distributed pre-printed forms to various firms, about 110 in Dalian overall. As all Japanese firms and "the majority" of foreign firms were sending their telegrams via Hsinking Radio, MTT concentrated their efforts on the remaining few sticking with foreign cable companies. MTT claimed that many of these foreign firms were simply using those pre-printed forms marked with via-cable without intending to go through cable companies.[25] In May 1937, MTT's chief of foreign telecommunications told an MOC official that "telegrams from foreign firms absolutely refusing Hsinking Radio are sent as requested." However, he also revealed that "in local branches, the enthusiasm for increasing company business was somewhat excessive, leading to recent frequent protests from the cable companies." Acknowledging that "the continuation of the aggressive line only complicated matters," he suggested that "from now, about half of the telegrams sent by the two or three foreign firms with special relationship with the foreign companies will be handled as requested, whereas the remaining shall go through Hsinking Radio under the pretext that the connecting line to the requested route is crowded [...] it will placate things somewhat."[26] This halfway measure seemed to have worked, prompting Poulsen in a letter dated June 30 1937 to thank MOC for the "distinctive improvement in the situation since May 13." He noted that routing requests were still disregarded to some extent and presented a list of 71 requests made after May 13, of which only 31 had been handed over to the companies.[27]

Japan's aggressive push for the use of wireless telegraphy for international communications threatened the cable companies. GN sought to counter it by its own effort but soon ran into trouble with the Japanese authorities. On December 10, 1936, the Communications Ministry sent a letter to Poulsen,

25 [Enclosure] "Encouragement of Use of Hsinking Radio," by MTT Representative in Dalian (July 30 1937), in Ichihashi to Nomura (August 9, 1937 sent, received August 14) 624. MOC Papers, vol. 183.

26 Ichihashi to Nomura (May 3, 1937s #373), MOC Papers.

27 Poulsen to Nomura (June 30, 1937). Of these 71 telegrams, 47 were bound via Northern (27 sent).

GN's General Manager in Shanghai, accusing the company to be violating the concession:

> We have learned from certain telegraph users and others in this country that some officials of your Company [sic] engage in direct dealings with the public in this country, by handling messages regarding repetitions, inquiries, etc., which should be handled by this Administration as paid service messages. Judging from the fact that code messages are frequently exchanged between your officials and your Nagasaki Station, the above information seems likely to be true.[28]

The letter was vague on the details of the allegations as well as the implications:

> In view of the long-established relations with your Company, we hope that there might not occur any such irregularities nothing short of infringement of the Concession. But if by any chance it may prove true, the matter will be of grave nature and so it is requested that you would take proper steps at once not to allow them any longer to interfere with the Administration's proper business.[29]

In his reply, H.S. Poulsen simply acknowledged the receipt, with an equally vague promise that "strictest instruction have been issued."[30] The lack of any denial from the Great Northern must have confirmed the suspicions in the minds of Japanese officials.[31]

New Geopolitics and Collision Course

Japan's Plan for Communications Hegemony in East Asia

No less serious a challenge to GN and other foreign communications companies in Asia than the new wireless technology was Japan's new empire-building strategy in East Asia. Following the 1931 Manchurian Incident, when

28 Director General of Telecommunications (Hirasawa) to General Manager in the Far East (December 10, 1936), vol. 610, GN Archives, Copenhagen.

29 Ibid.

30 General Manager in the Far East to Director General of Telecommunications (Hirasawa) (December 17, 1936). Ibid.

31 Kobayashi Takeji, then MOC Chief of International Communications, later commented that GN appeared to admit the allegations. Japanese government began stationing censor in GN Nagasaki station, which served as a check on its activities. Zandankai: "Daihoku Denshin Kaisha" no sesshû o kataru (3), *Teishin kyôkai zasshi* 383 (August 1940), p. 50.

Japan's Kwantung Army occupied Northeast China and set up the puppet regime of Manchukuo, Japan embarked on building an autarkic "bloc economy" in Northeast Asia. In August 1936, the Cabinet adopted the Fundamentals of National Policy at the Five Ministers' Conference, proclaiming Japan to be the stabilizing power (read: leader) in East Asia. Japan would seek to forge close ties between the three countries of Japan, Manchukuo, and China against the influence of Britain and the United States.[32]

Developments in the area of telecommunications paralleled this overall policy shift. In 1935, Japan's Ministry of Communications and the recently established Manchuria Telegraph and Telephone Company launched an ambitious project of building a Japan-Manchukuo long-distance cable network. Simultaneously, MOC technocrats began to formulate an "East Asian telecommunications policy," whose centerpiece was an East Asian cable telecommunications network linking Japan, Manchukuo and China by Japan's "unique, advanced technology of non-loaded cables."[33]

In July 1937, full-scale war broke out with China. As the new Director of the Telecommunication Bureau, Tamura Kenjirô, put it, the new conflict "ushered in a new era of Sino-Japanese cooperation."[34] MOC officials were dispatched to central and north China to work with the advancing Japanese army to take over Chinese telecommunication operations. In December 1937, MOC set up a Committee on Administration of Communications in China to coordinate its activities and to stay in control of the new developments. In less than a year after the outbreak of the war, under the joint sponsorship of MOC and the military, Japan set up three telecommunications enterprises in Inner Mongolia, north and central China.[35]

The new opportunities for overseas telecommunications expansion on the continent were equally important to the military and other bureaucracies involved in China. For the first time, an inter-ministerial Committee on Telecommunications was established in September 1938 to serve as the advisory body on overall telecommunications policy. One of the objectives of the committee was "to develop a solid East Asian telecommunication bloc centered on our country:"

32 For a complete translation, see Lebra, *Japan's Greater East Asia Co-Prosperity Sphere in World War II*, pp. 62–64.

33 For a detailed discussion of the East Asian telecommunications policy as well as Japan's development of non-loaded telephone cable, see Yang, *Technology of Empire*, chap. 4.

34 Tamura, Denki tsûshin seisaku ni tsuite, p. 31.

35 For details, see Yang, *Technology of Empire*, chap. 7.

> [In] view of the new development in East Asia and taking note of telecommunication policies of other countries, to firmly establish a communication policy in our country that unifies East Asia areas and to make our country the nexus of telecommunication in East Asia, to bring about comprehensive development of facilities in all areas, should be the urgent task at this moment.[36]

Japan began to pursue a two-pronged approach in its external telecommunications, by continuing to rely on wireless expansion in international communication with distant countries, and adding cable for its own imperial network in East Asia.

Frictions in China

After full-scale conflict broke out between China and Japan in July 1937, all three foreign cable companies – GN, Eastern Extension, and Commercial Pacific – continued to operate in China from their bases in Shanghai. As the fighting spread to the Shanghai area in August, these companies felt the pinch of the ongoing conflict. The pressure increased in October after Japanese forces overcame fierce Chinese resistance and occupied all of Shanghai except the International Settlement. The cable companies found themselves in an unenviable position: on the one hand, they had to honor legal and financial obligations to the Chinese Nationalist government, which had withdrawn to Hankow. On the other hand, they had to deal with the Japanese as the de facto occupying authority in Shanghai.[37]

On November 20, 1937, representatives of three cable companies met with Japanese officials at the Japanese Consulate General. Among the issues discussed, two proved to be particularly difficult: Japanese insistence that telegrams sent by foreign cable companies in Shanghai be subject to censorship and that remittance of revenue from cable operations to the Chinese government be stopped.[38] The cable companies, in order to continue their

36 "Explanations concerning the establishment of the Committee on Telecommunications," and other related documents are to be found in *Kôbun Ruiju* 62-hen, 1938, Volume 4–3, Kokuritsu Kobunshokan [National Archives], Tokyo.

37 To curry favor with the new Japanese occupation authorities, one American executive even sent in a personal donation to the Japanese military via MOC. MOC Papers I.

38 Poulsen to Danish Consul General (November 20, 1937), vol. 415, Great Northern Archives, Shanghai.

business operation, tried to be cooperative with the Japanese authorities. In a telegram from Copenhagen, H. S. Poulsen was instructed that

> in view of the de facto situation at Shanghai and of the importance to all interests of the continued effective operation of the cable services of the port without unnecessary disturbance of the existing agreements, there is no alternative but to suggest that the companies concerned make the best arrangement possible with the Japanese authorities in the present difficult circumstances.[39]

In the end, all three cable companies agreed to the Japanese demands on censorship. They also agreed not to pay terminal fees to the Nationalist Chinese government as they had done before, but the Japanese demand to deposit these funds into a Japanese bank met with resistance. GN's dilemma in Shanghai seemed to be well understood by Copenhagen, as seen in a cable sent to Poulsen:

> difficulties are 1) do not wish offend either Chinese or Japanese 2) do not wish suffer loss revenue 3) cannot easily find reasonable explanations vis-à-vis Chinese for opening account and vis-à-vis Japs (sic) for not opening account. Reasons for and against: "1) opening account might assist cable negotiations Japan (pf seems very doubtful but thinks refusal open may react our disadvantage) 2) might protect us against repressive measures directed cable service Shanghai Japanese being well aware of traffic handled via cable 3) might assist us in present accounting and remitting difficulties 4) if do not open account Japanese will probably forward traffic for Great Britain via Berlin Paris or New York.[40]

GN relented only after considerable delay. On April 21, 1939, Poulsen finally informed the representative of the Chinese Nationalist government that "in accordance with a written demand from the Japanese Authorities, the Companies have paid today into the Suspense Account opened with the Yokohama Specie Bank the sum of $925,142.02 being the balance due to your Administration in respect of the General Balance Sheets for the months of March to December 1938."[41] GN tried to deduct from it the debt payments owed by the Chinese government, but without success.

In early 1938, an MOC official met with Poulsen in Shanghai and suggested the need to revise the previous arrangement to meet the new situation.[42] Understandably, GN was trying hard to stay ahead of these dras-

39 Scheel to Poulsen (December 7, 1937), Ibid.

40 Copenhagen to Nordiske Shanghai (March 6, 1939), vol. 418. GN Archives, Shanghai. Abbreviations were used in the original.

41 Poulsen to Sha (April 21, 1939), Ibid.

42 Zandankai: "Daihoku Denshin Kaisha" no sesshû o kataru (1), p. 53.

tic changes. During his visit to Japan in June that year, Poulsen met with senior officials in the Ministry of Communications, who responded only vaguely to his attempts to sound out Japanese intentions.[43] By August, the situation in central China crystallized somewhat: the Japanese founded the Central China Telecommunications Company (CCTC). Similar to MTT in Manchuria, CCTC was set up as a Chinese-Japanese joint venture aimed at taking over and developing the telecommunications enterprises in central China.[44] The Japanese made no secret of their intention of expanding the new telecommunications company in the area including Shanghai. In his letter of introduction to Poulsen, the head of CCTC noted that

> This Company being born during this critical moment of hostilities, our mission is to revive and reconstruct the telecommunications enterprises in Central China. In order to fulfill our important errand, we must concentrate our whole attention and devote our full energy to the business so that our budding hopes for future aggrandizement of our organization may be realized.[45]

For cable companies, Japanese competition in central China was already on the rise, even without the CCTC. For instance, the Japanese government set up a new wireless service between Shanghai and Hong Kong, which cut directly into the business of Great Northern and Great Eastern. In response, GN adopted a number of counter-measures. It hired Japanese business representatives and managed to maintain good business relationship with Japanese companies and banks in Shanghai, apparently with some success to the dismay of Japan's own telegraph office in the city.[46] Despite their conciliatory position in negotiations, therefore, the Great Northern Telegraph Company continued to be a headache for Japan's objectives by posing direct competition.

On the other hand, GN's problems were not limited to Shanghai. At a meeting with other foreign cable company executives, held at the British Consulate General in 1938, H.S. Poulsen sounded resigned: He noted that the cable companies were vulnerable not only in Shanghai, but also in their vital interests in Japan and in communications under the Japanese military control elsewhere. Although the British representative reminded the participants that Japan was dependent in some measure on the cable companies, Poulsen pointed out that it had radio communication at its disposal and

43 Poulsen to Copenhagen (June 4, 1938), vol. 416, GN Archive, Shanghai.
44 Yang, *Technology of Empire*, 114–119.
45 Fukuda to Poulsen (August 1, 1938), vol. 416, GN Archive, Shanghai.
46 Matsunaga Hangoro, Shanhai ni okeru denki tsushin no kakuchiku, p. 16.

had been trying to reduce the importance of the cable companies as much as possible.[47]

The GN Nagasaki Office: Irritant of a Foreign Enclave

Since 1871, GN had operated a telegraph office in Nagasaki, the Japanese city that had been the sole window to the outside during the 250-years of isolation before the mid-nineteenth century. The GN Nagasaki office played a pivotal role in connecting GN's two submarine cables in the Far East. It was also an object of envy and displeasure for the Japanese government.

At its peak, GN's Nagasaki office employed over 90 people, mostly Japanese. GN had strict office rules and punished operational errors with heavy fines, but otherwise gave its employees considerable freedom. Partly as a result of that, the Nagasaki office was able to provide high-quality service that excelled in speed and accuracy. At the same time, while Japanese employees at the office received considerably less pay than their Danish counter-parts and had to use separate restroom facilities, theirs were among the highest salaries anywhere in Japan, often exceeding that of the highest-ranked Japanese government communications officials in the Nagasaki area. For instance, a Japanese messenger carrying telegrams for the GN Nagasaki office was paid roughly the same salary (including bonus) as a Japanese official in charge of an entire telegraph office. In fact, these Japanese employees were paid so well that some of them became coveted customers in the expensive restaurants and geisha houses in town.[48]

Figure 3: Employees' Pay at GN Nagasaki Office (1939) (unit: yen)

Category	Number	Total salary	Total bonus	Average salary	Average bonus
Employee-foreign	9	124,915	5,845	13,879	649
Employee-Japanese	62	168,946	22,179	2,725	358
Messenger-Japanese	19	7,226	820	38	43
Pension-Japanese	8	12,086	1,511	1,510	189
Death compensation -Japanese	1	75		75	

Source: Japanese Ministry of Communication (MOC) Papers II.

47 "Minutes of a Conference of August 15, 1938," Ibid.

48 *Aoi umi o hashiru*, pp. 42–43.

As time went by, GN's presence in Nagasaki became in a sense anachronistic, a reminder of Japan's dependency status during the bygone Meiji era. One ranking MOC official later recalled that when visiting Nagasaki in 1914, he was indignant at the "third-class status" of Japan in telecommunications while the country as a whole had become first-class. Such a view was echoed by Hanaoka Kaoru, a young official in the ministry who was to play a key role in dealing with GN. He described the GN office as a foreign enclave. It was perhaps no coincidence that Tamura Kenjirô, who became a leading driving force behind the elimination of GN's rights in Japan, had previously served as the director of communications in the nearby Kumamoto prefecture. His tenure gave him ample opportunity to observe GN's presence at close range.[49]

After the outbreak of the Sino-Japanese conflict in 1937, there were new concerns among Japanese officials that, even with close surveillance and censorship by the Japanese government, the existence of a foreign communications facility near military installations and heavy industries in northern Kyûshû posed a major security threat. Japanese authorities in Nagasaki sent periodic reports on GN activities. The fact that GN agents had visited many "important firms throughout Japan" fed into their suspicion of an espionage network.[50]

Showdown

Preparations for Showdown

As Japan continued its expansion in Asia, complete control of communications at home became a strategic priority. At home, as the Japanese government sought greater autonomy in its industry and economy in the 1930s, many foreign-owned companies in Japan either withdrew or were bought by Japanese ones. After becoming Director of the Bureau of Telecommunications, Tamura Kenjirô began pushing for solving the Great Northern prob-

49 Zandankai: "Daihoku Denshin Kaisha' no sesshû o kataru (1)," p. 17–21.

50 Teishinshô, "Daihoku Denshin Kaisha mondai shitsugi ôtai shiryô" (March 1940), MOC Papers II–332. See also such reports from the governor in Nagasaki, in JACAR.

lem, by coercion if necessary. In documents prepared by MOC for lobbying other government ministries, MOC officials described GN as a threat to national security and an obstacle to realizing Japan's communications policy.

Japanese records suggest that MOC began addressing the GN issue in early 1937. The Ministry first examined the previous exchanges between the Foreign Ministry and the company.[51] Due to the outbreak of the war in China and the flurry of Japanese telecommunication's expansion there, it was not until February 1939 that the GN issue was placed on the table again and various concrete questions were discussed. On what basis could Japan eliminate this source of discomfort? Although hardliners in the Ministry such as Tamura even considered the suspicion of espionage as sufficient justification for an outright takeover, an internal memo admitted that the suspicion would be difficult to substantiate despite constant surveillance efforts. Vice Minister Owada Teiji suggested tackling the problem from Great Northern's business activities. The best way to "eliminate the trouble from its roots" was to focus on the concession granted by the Minister of Communications. In addition, MOC offered other less attractive alternatives such as purchasing the GN cables.[52]

To their delight, MOC officials discovered a technicality in their favor: a prohibition of direct contact with customers was written into the 1912 agreement, ironically enough, at the suggestion of GN. MOC then decided on the tactic of accusing the Great Northern of illegal business practice in Japan and compiled a detailed list of over sixty cases where GN had directly worked with customers. Almost all of these infringements had occurred in 1939. As MOC sought legal grounds to revoke GN's concessions, the 1936 exchange proved crucial: GN had been warned and now it had to face the consequence.[53]

Before actually confronting the Great Northern, however, MOC had to obtain full cooperation from other government agencies in Tokyo. This was by no means assured. Although the effort to revoke Great Northern's privileges in Japan enjoyed increasing support in the Japanese government, there were considerable differences over how to achieve such a goal and how fast

51 Horiuchi (Vice Minister, MOFA) to Tomiyasu (MOC Vice Minister) (February 6, 1937). MOC Papers.

52 For a detailed account by a Japanese participant, see Hanaoka, *Kaitei densen to Taiheiyô no hyakunen*, 191–212. See also *Kaiteisen hyakunen no ayumi*, pp. 346–352.

53 "Kaiteisen shiyô appaku ni kansuru ken" (March 1939), JACAR B10074928700; "Daihoku Denshin Kaisha ni kansuru ken" undated, MOC Papers II–332.

it should be done. The Army, Navy, and Home Ministries had no objection to the MOC's hardline approach. In contrast, however, the Foreign Ministry was much more cautious. Both Foreign Minister, Arita Hachirô, and Director of the Eurasian Bureau, Nishi Haruhiko, were against the aggressive policy. Officials in its Treaty Bureau initially even demanded that the negotiations over the GN issue be carried through the Foreign Ministry, not MOC. Others, including Vice Minister, Tani Masayuki, and Yamada Yoshitaro, a section chief in the Eurasian Bureau, only came over to the MOC side after the persuasive efforts of the MOC Vice Minister. It was not until March 12, 1940 that a consensus was reached within the Japanese government. As a compromise, MOC agreed not to revoke landing rights immediately but rather to adopt a gradualist approach.[54]

MOC justified the need for immediate and resolute action against GN citing indications that GN was considering pre-emptive measures, such as a proposal to shore up cable traffic and wireless service between Japan and Denmark. Since GN's rights in China would not expire until 1944, the company seemed to have every reason to delay any change to its status to the same time as those of British and American cable companies in China. Therefore, MOC argued, Japan must prevent the formation of a united front of all three foreign countries by tackling Danish interests ahead of Anglo-American rights. The MOC study specifically explained that the GN issue was different from the Commercial Pacific Cable Company. Even though MOC recognized that GN had been trying not to provoke the Japanese government recently, suspicions deepened after reports of frequent meetings between the company and British and American consular officials.[55]

Partly to convince hesitant government officials and partly to be cautious, MOC made careful preparations ahead of the final showdown. In a classified document issued in March, MOC officials considered numerous questions related to the revoking of GN landing rights. A sample list indicated some of Japanese concerns:

1. Are there any concerns over counter-measures overseas against our moves?

32. Will Denmark bring the case to arbitration?

33. Will Denmark bring a civil suit?

---

54 Zandankai: "Daihoku Denshin Kaisha" no sesshû o kataru (1), p. 18; Hanaoka, *Keitei densen*, 194–199.

55 "Daihoku Denshin Kaisha mondai shitsugi ôtai shiryô," MOC Papers II–332.

Moreover, although Japan possessed over 30 direct international wireless telegraphic circuits, MOC had to prepare for the temporary interruption of telegraphic service to several areas such as Hong Kong, Northern Borneo, and British West Africa, where Japan had to rely on submarine cables owned by GN. It came to the conclusion, however, that risks were low. For instance, telegrams to Hong Kong, the most important of these locations, could be relayed in Shanghai.[56]

Showdown 1940

Completely unaware of the MOC's moves, GN seemed to have gone on the offensive over its grievance with Japan's aggressive wireless policy. In a memo dated March 11, 1940, the Danish Legation in Tokyo made another protest to the Japanese government over the latter's "illegal anti-cable propaganda." It called attention to the case made by a Danish firm in Kobe that the "(Japanese) Telegraph Administration do [sic] not even refrain from using intimidation in their efforts to persuade business firms to send their telegrams by wireless instead of by wire."[57]

Exactly one week later, the Ministry finally struck. Just days after reaching the consensus within the Japanese government, on March 18, MOC officials handed an ultimatum to H.S. Poulsen, GN Far Eastern Manager who happened to be in Japan to discuss local taxation problems.[58] It accused the company of continuing illegal business practices in Japan such as direct contact with customers and demanded a quick reply within three weeks. Realizing the gravity of the matter, Poulsen asked for an extension to five weeks, citing the Easter Holiday in Denmark. In the end, a compromise of four weeks was achieved.[59]

Internal Great Northern documents show that Poulsen seemed prepared to give up the fight. In his cable to London on April 3, Poulsen evoked the

56 "Daihoku Denshin Kaisha ni kansuru ken" undated, MOC Papers II–332.

57 The Danish report alleged a Japanese official from the Kobe Central Telegraph Office pressured Messrs. H.A. Hansen to send telegrams via Anten (wireless) or otherwise "be prepared to receive a bad service of their telegrams."

58 The lightening speed on the part of the Japanese came at a price, for one Japanese official at least, who lost his only son due to illness just before Poulsen's arrival in Tokyo. Zandankai "Daihoku Denshin Kaisha" no sesshû o kataru (1), p. 20.

59 Teishinshô Denmukyoku, *Tai Daihoku Denshin Kaisha tokkyôjô fuku tokkyôjô jôkan kaitei keika* June 1940), MOC Papers II.

"old maxim of eastern that [it is] rarely good policy to fight a powerful concessionary govt [sic] on its own ground."[60] After examining the legal grounds of the matter, Great Northern's board in London also deemed it futile. As a result, the Great Northern Company beat retreat on all issues. The timing could not have been better for MOC: its pressure on the company coincided with the deterioration of the Danish position in the European War, which the Japanese used to its advantage to the full extent. When Poulsen returned to Tokyo a month later for further negotiation, Tamura tactfully informed him of the German invasion of Denmark which had taken place during his journey before negotiation began.[61]

The negotiation lasted over a dozen rounds, surprisingly without major hurdles. Citing the advanced technology and skills required to operate long-distance communication by submarine cables with Riga and Helsinki, Poulsen wondered whether Japanese operators could rise to the task. Director Tamura assured him that he needed not worry because "Japan's technology has progressed."[62] The new agreement was finalized at the end of April 1940. Instead of a "concession," a term that had the aura of a foreign settlement, the Great Northern would now receive a "license" from the Japanese government for its business activities in Japan. Moreover, the operation of its Nagasaki facilities would be taken over by the MOC beginning June 1, the date when the new agreement would take effect. Telegrams between Japan, Manchukuo, and China would no longer be handled by the Great Northern. The Great Northern's landing rights in Nagasaki, the once-permanent foundation of its whole operations in Japan, would be finally relinquished on April 30, 1943.[63] This three-year grace period constituted a concession on MOC's part to appease other Japanese bureaucracies that had opposed the complete takeover.

In a letter to the Japanese officials after the agreement was signed, Poulsen expressed "'my co-operators' and my own heartiest appreciation of, and thanks for, the unfailing courtesy shown to us, and the cordial and harmonious atmosphere which has been prevailing all through our negotiations."[64] This may be true to some extent. Although some Japanese compared

60 Poulsen to London (April 3, 1940), vol. 610: "Forhandling om ny overenskomst," GN Archive, Copenhagen.

61 Zandankai: "Daihoku Denshin Kaisha" no sesshû o kataru (1), p. 19.

62 Hanaoka, *Kaitei densen to Taiheiyô no hyakunen*, 204.

63 For a brief summary see Tamura, Taihoku Denshin Kaisha menkyojo no kaitei ni tsuite, pp. 8–9.

64 Poulsen to Tamura (May 4, 1940). MOC Papers II.

GN's presence to a "long time cancer in international communications," not everyone on the Japanese side was so hostile to GN per se; some even displayed a degree of sympathy for "Denmark being trampled by the iron hoofs of Germany."[65] Owada Teiji, a MOC Vice Minister, praised GN negotiators as "mature, honest" and even suggested GN's retreat from Japan to be an honorable departure after accomplishing its mission of providing cable communication to Japan.[66] Whether a graceful exit or not, however, the agreement signaled the end the era of Japan's modern telecommunications that was dependent on foreign commercial interests.

Final Efforts in the Pacific War

A little more than a year after the signing of the agreement, the Pacific War broke out. On December 8, 1941 Japan launched simultaneous attacks on the British and American forces in Southeast Asia and the Pacific. As Japan had entered war against the United States and Britain, Japanese military authorities regarded British and American companies in China as "enemy property" and placed them under military control. That morning, the Japanese Navy also seized the Great Northern facilities in Shanghai and all of GN's telegraph cables from Shanghai were severed. The GN protested that as a Danish company it should be allowed to continue operation but to no avail. A company official told the Japanese:

> The present world conditions must necessarily greatly hamper the activities of a neutral utility concern such as the GNTC, but a day will assuredly come when the commercial intercourse between the countries will resume another character than now prevailing, and in the meantime I pray that the Imperial Japanese authorities vis-à-vis the company will be guided by the same goodwill and consideration, which, I am happy to say, is traditional in the mutual relations between Denmark and Japan.

He also reminded the Japanese of the fact that "GNTC in Shanghai has been able to collaborate fully with the Japanese authorities in the latter's exercise of censorship of telegrams during recent years."[67]

---

65 Comments by a member of the Communication Association at the "Daihoku Denshin Kaisha" no sesshû o kataru (1), p. 17. Another former Japanese MOC official even described the settlement as "merciless," p. 26.

66 "Daihoku Denshin Kaisha" no sesshû o kataru (3), p. 57.

67 Scheel to Sone (December 22, 1941); Tillitse to Togo (February 3, 1942), vol. 63, Great Northern Archive, Shanghai.

Despite intervention from the Danish government, the Japanese diplomatic authorities in Shanghai refused to grant GN "non-enemy status." Suspecting heavy British involvement in GN, Japan demanded all the records from the company so as to further investigate its capital composition. Moreover, the Japanese Foreign Ministry concurred with the Japanese military authority regarding the latter's right to seize the GN facilities in theaters of war.[68]

In January 1943, GN's Far Eastern General Manager Poulsen once again visited the MOC in Tokyo. He asked for an extension of the company's landing rights in Japan as well as financial compensation since Japan's military operations had led to the cessation of cable operations ahead of the agreed date. Japanese officials were unmoved and flatly rejected both requests. They referred the latter to the Foreign Ministry since the cause was unrelated to the landing permit "whereas compensation to GNTC for loss and damage due to military necessities outside teishinsyo [sic] province." Japanese officials also resisted the company's demand of paying terminal fees, although they agreed in principle to make payments to the company against the outstanding balance. They insisted, however, that such payments not comprise the further transmission tolls due to the British and American cable companies.[69]

The meeting also represented a clash of views of GN's relationship with Japan. To further justify Japan's refusal, Japanese officials renewed the accusation about the "company's anti-Japanese attitude during the past 60 years." MOC official Kimura Yôji presented a list of charges against the company: after the outbreak of the "China Incident" in 1937, the company paid the Chinese terminal charges to the Chinese government and later refused to cooperate with the pro-Japanese government in Nanjing. Kimura then revealed that it had been reported that on December 20, 1941, a Chinese "whose name is unknown" asked H.S. Poulsen in Shanghai to forward a message from wartime Chinese capital Chungking via Vladivostok to Moscow and that Poulsen agreed to do so. In addition, at the outbreak of the "Greater East Asia War" the company had turned its two cable steamers over to the British Naval Authorities. Although Kimura doubted if these allegations were en-

68 Tillitse 16 to CPH (July 8, 1942), vol. 63, Great Northern Archive, Shanghai.

69 "Memorandum" (January 12, 1942); "Minutes of meeting at Teishinsyo Tuesday 12, 1943," and "Minutes of Meeting at Teishinsyo Friday 15, 1943," all in, vol. 64, Great Northern Archive, Shanghai. "Teishinsyo" refers to Japan's Ministry of Communications (Teishinshô).

tirely correct since they were under investigation, he considered it a pattern that had its echo in history. During the Sino-Japanese War and Russo-Japanese War, the Japanese government desired to buy the company's cables and to take over the Nagasaki station but the company refused, thereby causing much military and diplomatic damage to Japan. Kimura added that, because of its monopoly in China, the company imposed very severe and unfair conditions upon Japan in 1913 for laying the Japanese government's Nagasaki-Shanghai cable, thus showing the company's anti-Japanese attitude.[70]

To these accusations, Poulsen offered a spirited defense. He firmly denied helping the Chinese Nationalists after the outbreak of the war in China and stressed that the agreements of 1913 were "negotiated and concluded on a business basis and were not at the time considered to be unfavorable to Japan." Poulsen insisted that the company had always operated on a business basis pure and simple. He pleaded that canvassing for cable traffic was carried out with the full knowledge of the Ministry and that its one and only purpose was to acquire sufficient revenue to enable the company to maintain its cable service. He told Kimura that "I am sure that you will agree that in these circumstances the canvassing activities cannot be called anti-Japanese in any sense."[71] Finally, in a gesture that seemed to curry favor with his host and to counter the "anti-Japanese" perception, Poulsen informed the Japanese authorities in Nagasaki that he "take[s] pleasure in confirming [that the] company is prepared to donate to the Japanese government's collection of scrap iron, iron chain, two iron doors, iron manhole in Japan to replace articles in question with suitable substitutes of wood at your expense."[72]

Despite its efforts, GN only made minor progress – Japan made a one-time payment for the joint purse. It was unable to further delay its elimination from Japan even by a few months. On April 30, 1943, Japan's Director-General of Telecommunications, the top-ranking official in charge of telecommunications in the Ministry of Communications [*Teishinshô*], sent the following notice to the General Manager of the Great Northern Telegraph Company:

70 "Minutes of Meeting at Teishinsyo Wednesday 1. 20, 1943," vol. 64, Great Northern Archive, Shanghai

71 "Minutes of Meeting at Teishinsyo Wednesday 1. 20, 1943," Poulsen to DirGen Tokyo (February 10, 1942), vol. 64, Great Northern Archive, Shanghai.

72 Poulsen to T. Miyati (I.J.T.A) Nagasaki (February 18, 1942), vol. 64, Great Northern Archive, Shanghai.

> I solemnly notify you that the term of the landing right of your company cables near Nagasaki expires on this day of April 30th 1943 under provisions of the License granted by Teisinsyo [sic] and accepted by your company on May 4th 1940. I avail myself of this opportunity to thank you heartily for your kind cooperation for many years.[73]

With that, the presence of the Danish telegraph company which had operated in Japan since 1871 finally came to an end, or so it appeared to those involved at the time.

Conclusion

The relationship between Great Northern and the Japanese government has always been a complex one. For more than four decades since its arrival in Japan, from 1871 to the mid-1910s, the relationship was characterized by mutual accommodation, made possible by the asymmetric nature of that relationship: a competent global commercial telecommunications enterprise from a small Nordic country solely interested in revenue maximization and a national government with security and strategic concerns but without the necessary financial and technological capabilities.

A number of factors contributed to the relationship's gradual erosion afterwards. While GN tried to hold on to its proven business strategy, the world as well as its partner underwent great changes. Over the long-term, GN's failure to respond to the wireless challenge was a fundamental cause of its long-term decline in global telecommunications. On this issue, even an official history of GN acknowledged that "competition from the predominantly state-owned radio routes gradually became exceptionally severe for both the Northern and the other companies. This was mainly due to the fact that the State Administrations, in addition to being financially interested in favoring the radio routes, were often able to decide, or at least to influence the choice between cable and radio routes."[74] Since the 1920s, Japan's international telecommunications policy that emphasized wireless posed a major threat to GN cable operations. GN was not only threatened by the Japanese government. During the chaotic months in 1938–39, a new actor such as the

73 DirGen to Gen Man, (April 30, 1943), vol. 64, Great Northern Archives, Shanghai.

74 *The Great Northern Telegraph Company*, p. 29.

Global Wireless Company that entered the Shanghai market, also created enormous headaches for cable companies like GN. There was little doubt that the various international wireless connections now available to Japan greatly reduced the bargaining power of GN. Last but not least, starting from the early 1930s, Japan's new ambitious telecommunications policy for the East Asian region called for a cable network under Japanese control. Foreign-owned cables were not just a nuisance, let alone a financial drain, but increasingly perceived as an obstacle to achieving Japan's strategic objectives in Asia.

Several short-term factors also help explain the abruptness and almost total success of the Japanese effort: facing aggressive competition from Japanese-controlled telecommunications operations on the continent, GN may have overplayed its hand vis-à-vis the Japanese government. GN executives may have under-estimated the confidence and resolve on the part of the Japanese government due to the changing technological and political circumstances of the 1930s. In this sense, if survival of GN in Japan were the company's goal, perhaps a less confrontational stance would have prevented the formation of a strong anti-GN coalition in Japan. (If offense was deemed as the best defense, then GN should have built its own united front with other foreign cable companies).

In the realm of international telecommunications, strong government backing has always been necessary even for purely private operations, which has been amply demonstrated in GN's own history in China and Japan. With Denmark weakened by the German invasion in early 1940, GN had few diplomatic cards to play but to beat hasty retreat. These, of course, are hindsights arrived at with 20–20 vision. All in all, then, it was perhaps more remarkable that a private enterprise from a small country of four million held on to its privileges as long as it did in a rapidly expanding empire of Japan.

Previous examination of GN's end in Japan also downplayed the regional scope of their conflict as well as Japan's ambition in East Asia. The friction between Japan's promotion of wireless and GN's determination to fend off that challenge, not only took place in Japan, but also in Japan's new empire where GN met even more aggressive tactics. This was the background against which the company came up with its own tactics. Japan's policy for telecommunications hegemony in East Asia as well as its war in China were important factors, prompting GN to re-evaluate its business strategy in East Asia.

After Japan's surrender in August 1945, GN made a comeback of sorts. It again raised the issue of compensation through diplomatic channels with the

Japanese government. It had better luck this time. After negotiations, Japan agreed to pay a total of 845,000 dollars to the company. At the request of GN, telegram service through Nagasaki re-opened on the basis of the 1940 agreement. The Japanese government, and the Japan International Telegraph and Telephone Co. (commonly known as KDD, *Kokusai denshin denwa kabushiki kaisha*), a new Japanese company set up to handle international telecommunications in postwar Japan, operated the Nagasaki office. In 1954, KDD and GNTC signed a new agreement. The Nagasaki-Vladivostok cable continued to carry much of Japanese communication traffic with Europe until the late 1950s. In 1969, Japan entered into partnership with GNTC and others to build a new Japan Sea Cable (JASC), connecting Naoetsu on the Japan Sea coast with the city Nahodka in the Russian Far East, thus closing the final chapter of the Nagasaki-Vladivostok cable.[75]

Bibliography

Ahvenainen, Jorma, *The Far Eastern Telegraphs: The History of Telegraphic Communications between the Far East, Europe and America before the First World War*, Helsinki 1981.

Aoi umi o hashiru [Traversing through the blue seas], Tokyo 1969.

The Great Northern Telegraph Co. Archives, Copenhagen.

The Great Northern Telegraph Co. Archives, Shanghai.

The Great Northern Telegraph Company, *The Great Northern Telegraph Company: An Outline of the Company History, 1869--1969*, Copenhagen 1970.

Hanaoka, Kaoru, *Kaitei densen to Taiheiyô no hyakunen* [Submarine cables and the Pacific Ocean], Tokyo 1968.

Headrick, Daniel, *The Invisible Weapon: Telecommunications and International Politics 1851–1945*, New York 1992.

Ishihara, Fujio, *Kokusai tsûshin no Nihon: Shokuminchika kaishô he kutô no 99-nen* [International communication and Japan: Ninety-nine years of the bitter struggle to break away from colonization], Tokyo, 1999.

Jacobsen, Kurt, "In the Service of the Nation. Small Company, International Telecommunications and International Politics – The Great Northern Telegraph Company 1869–1945," in: Bernard Finn & Daqing Yang (eds.): *Communications under the Seas. The Evolving Cable Network and Its Implications*, Cambridge, MA 2009.

75 KDD shashi hensan iinkai, *KDD shashi*, pp. 45–53, 168–172.

—, "Wasted opportunities? The Great Northern Telegraph Company and the wireless challenge," *Business History* 52.2 (April 2010), pp. 231–250.

Japan Center for Asian Historical Resource (abbreviated as JACAR).

Japanese Ministry of Communications (MOC) Papers, Tokyo.

KDD shashi [A history of the International Telephone Co. (KDD)], Tokyo 2001.

Lebra, Joyce C. (ed.), *Japan's Greater East Asia Co-Prosperity Sphere in World War II*, Kuala Lumpur 1975.

Manshû Denshin Denwa Kabushiki Kaisha 10-nen shi [A ten-year history of the Manchuria Telegraph and Telephone Co.], Shinkyo, 1943.

Nagashima, Yôichi. "Daihoku denshin kaisha no Nihon shishutsu to sono haikei" [The entry of the Greatern Northern Telegraph Co. into Japan and its background]. *Nihon Rekishi* 576 (April 1995), pp. 77–92.

Nihon Denshin Denwa Kôsha Kaiteisen Shisetsu Jimusho, *Kaiteisen hyakunen no ayumi* [A one-hundred-year history of the submarine cable], Tokyo 1970.

Nihon musen shi [A history of the wireless in Japan], Tokyo 1951.

Sunaga, Noritake, "Senzenki Nihon no taigai tsûshin jigyô no tokushitsu to tsûshin shuken," [Characteristics of prewar Japanese international communications and communication sovereignty] *Keisei ronshû* 17 (1991), pp. 35–54.

Teishin kyôkai zasshi [Journal of the Communication Association].

Yang, Daqing, *Technology of Empire: Telecommunications and Japanese Expansion in Asia, 1883–1945*, Cambridge, MA 2011.

Zhongguo jindai yudian shi [A history of postal and telecommunications in modern China], Beijing 1984.

Progress by Technology? The Utopian Linkage of Telegraphy and the World Fairs, 1851–1880

Lars Bluma

In modern societies, discourses of technology are closely connected with social visions of public welfare, health, peace and political participation. Therefore Western societies discuss today their social, cultural and political future in terms of "information society," "globalization" and "knowledge society," which partly depend on technological changes like the use of computers and worldwide telecommunication via Internet, email etc. The beginnings of this narrative linkage between communication technologies and the self-representation of Western societies go back to the middle of the nineteenth century and were integrated within a larger context of modernist narratives about economic and political internationalism.[1]

Telegraphy as a new key technology and world fairs as a new intriguing concept of public self-representation of the industrialized economies in Europe and the United States became determining metaphors within the Western concepts of civilization and progress in the second half of the nineteenth century. Telegraphy and world fairs were linked with the idea that technology is the prime mover for social modernization.[2] To contemporaries, both were media and representations of modernity; they were symbols of a bright future of mankind.[3] However, as the first part of this article makes clear, a central premise of this discursive entanglement was that world fairs and the displayed technologies caught the visitors' imagination. The utopian potential of telecommunications and especially telegraphy was closely connected to their emotional effects.

1 Nye, *Narratives and Spaces*. Geyer, *Mechanics of Internationalism*.

2 Adas, *Machines*.

3 The term media is here used in the broadest sense as a technology or institution with the aim to communicate and to transmit information. The relation between world fairs and telegraphy is in this media-oriented perspective a complex one because both media were a mutually synchronized network of technologies and meanings. Telegraphy devices were part of the exhibitions but they were also carriers of information about these early mass events and telegraphy is itself a symbol of modernity as much as the world fairs. McLuhan's statement that the medium is the message applies in this case. McLuhan, *Medium*.

The second part of this article discusses the observation that industrialized nations in Europe as well as in the USA started to define themselves in terms of a "network society" both on a national and global level.[4] On the one hand, telegraphy was seen as a medium of spatial expansion, spreading the Western ideology all over the world. On the other hand, the world fairs of the nineteenth century were spatial media and presented the idea that the end of history as a series of wars was near and that a peaceful, technology-driven era of progress and welfare would begin soon.[5] As media of expansion and immersion, telegraphy and world fairs were complementary expressions of Western utopian visions of an imaginary future. With telegraphy and world fairs, the Western powers had mighty symbols at hand to express their idea of peaceful competition of national cultures in a new "global" world. The main actor of this vision was the bourgeoisie with its beliefs in universal values and peaceful progress. But, as the final part will show, the intensified discussions on imperial and global geopolitical leadership since the 1860s meant that such peaceful universal visions vanished into thin air and were replaced by aggressive interpretations of the competition between nations. This also infected the meaning of telegraphy and world exhibitions. At the dawn of imperialism, telegraphy and world fairs changed their meaning fundamentally from media of peace and integration to media of control and disintegration by producing connected and not connected, civilized and barbarians.[6] My purpose is to underline this shift in meaning, including the role of telegraphy and world fairs as symbols of progress and their central function in the discourse of imperialism and national competition.

Telegraphy at World Fairs: The Fascination of Technology

A close look into world fairs' catalogs from 1851 to 1880 shows that telegraph companies used the fairs as a platform to promote their latest devices. The ongoing internationalization of telecommunication markets spurred on

4 Manuel Castells uses the term, "network society," to describe the structural changes of globalization based on information networks since the 1970s. In my view, the discourse on information technology in the 19th century was a forerunner of this process. Castells, *Network Society*.

5 Sloterdijk, *Weltinnenraum*, pp. 267–268.

6 Bluma, *Globalization*.

the companies' effort to present their products at world exhibitions. The German telegraph company Siemens & Halske, for example, used the first world fair 1851 in London to open the English market for its products. Its exhibition of manifold high quality telegraph products successfully attracted the attention of English experts and Siemens received the Council Medal for his products because of this offensive marketing strategy.[7] For Siemens, the Council Medal was a stroke of luck, because in the same year, the Prussian telegraph administration broke off its business relationship with Siemens. Siemens now had to conquer foreign markets. Even though the establishment of Siemens' business especially in Russia and England was a slow process, Siemens' international reputation as a producer of high-quality devices was firmly established.[8]

Already at the first world fair in London in 1851, visitors could gaze at all different kinds of telegraph systems and their combinations which were available for all kinds of purposes. It would hence be easy to write a history of telegraph technology by using the exhibition reports. Such a history would show the ups and downs of different telegraphs like the first printing telegraphs presented in London in 1851 by Jakob Brett, the efforts to automatize the coding process which was a main theme at the second world fair in London in 1862, further developments in facsimile telegraphy and the first presentations of cables and their isolations in Paris in 1867, but also the insight at the Paris exhibition that technological progress in telegraphy had slowed down.[9]

Nevertheless, the world exhibitions of the nineteenth century were never solely sales or advertising fairs, which would have interested only a professional audience of scientists, technical experts and businessmen. Quite on the contrary, the majority of the visitors were laymen with no advanced knowledge in technology. To them, the world fairs provided insight into the miracles of the modern world and spoke to a more common desire for further education within an entertaining atmosphere. The world fairs further allowed its visitors to come into contact with myriads of people and cultures at one place; the bourgeois' "tour de monde" simply took place in the exhi-

7 Siemens, *Lebenserinnerungen*, p. 94.

8 Feldenkirchen, *Siemens*, pp. 73–81.

9 Anonym, *Amtlicher Bericht London 1851*, p. 821; Anonym, *Amtlicher Bericht London 1862*, pp. 440–445; Anonym, *Berichte Paris 1867*, p. 439; K.K. Österreichische Central-Comité, *Bericht Paris* 2. Lieferung, p. 213–221. See also Finn, *Submarine Telegraphy Technical Stagnation*.

bition halls. Rising attendance figures attest to the growing interest. More than six million people visited the first world fair 1851 in London. In 1878, the automatic counters of the entrance gates of the Paris world exhibition registered more than 16 million visitors.[10] Besides, since the mid-nineteenth century it became almost a public attitude of the bourgeoisie to show one's own knowledge about progress in technology. We hardly find any cases of technophobia.[11] As Warren Susman as well as Astrid Böger put forward, the performative character of the world fairs as multi-media events let the visitors participate in a new model of social order promoting industrialization and national unification.[12]

On the occasion of the Paris world exhibition in 1867, Hans Christian Andersen wrote the fairy tale of the Dryad in which a tree transformed itself into a young girl to attend the Paris world exhibition, thereby illustrating the great attraction world fairs had for "everybody" – even a tree.[13] To the tree-turned-girl, the exhibition building appeared as "Aladdin's magic palace of truth," and Andersen enthusiastically described the exhibition as the eighth wonder of the world.[14] One of Andersen's characters even claimed that "The miracles been built at the Mars field were even more wondrously than a Fata Morgana; since they were concrete, stone made, mundane realities and no phantasm."[15] Not only Andersen, but nearly all non-scientific, popular views used the term of the miracle if they reported on the world exhibitions. The glass architecture of the wide exhibition halls, the amassed technical apparatuses and the gathering of different people led to perceptions and experiences which were unique and fascinating. The visitors felt that they were participating in a meaningful, epoch-making event. "Every visitor was blown through a ceremonious breath," so August Oncken upon entering Crystal Palace in London in 1851, "The feeling of a big and very remarkable event overwhelmed the people."[16]

The public's great interest in electricity was nourished and spurred on by public demonstrations which had a history stretching back to the eighteenth century. Already at the beginning of the eighteenth century, electricity was

10 Beutler, *Weltausstellungen*, Appendix, no page.
11 Binder, *Wunderbarer Traum*, p. 40.
12 Susman, *Ritual Fairs*. Böger, *Envisioning*.
13 Andersen, *Dryade*.
14 Ibid., pp. 16–17; 21.
15 Ibid, p. 17.
16 Oncken, *Wiener Welt-Ausstellung*, p. 8.

not only an interesting field of investigation for some specialized scientists, but had also attracted the attention of the general public through impressive experiments which celebrated electricity as a device of entertainment and fascination.[17] The mysterious effects of electricity often contradicted current physical theories and offered hopes that this natural force was a kind of universal key for all as yet explained physical phenomena.[18] The rational discourse of the Enlightenment changed this appreciation only marginally and despite theorization, electricity remained for most people an inexplicable force of nature. Probably quite similar to first responses to electricity in the eighteenth century, visitors of the world fairs marveled about the telegraph. The telegraph devices were displayed as "miracles" or they were just a "mystery."[19]

As late as 1867, almost three decades after the invention of the telegraph, telecommunications had lost nothing of their appeal and still caused visitors to rush to the respective stands of the exhibitors.[20] The public was particularly fascinated with attempts to connect the European with the American continent by telegraph. The first successful transatlantic connection in 1858, which was intact only for one month, led to worldwide festivities. In the United States, coastal towns as well as big cities such as New York celebrated the event with services and fireworks; resourceful dealers such as Tiffany's sold parts of the defective line as souvenirs.[21] However, rational interpretations could not make everything that people produced for the supposed "progress of civilization" and social advance completely comprehensible. The strangeness of technology, whether expressed as an affirmative fascination or negative techno-phobia, cannot be ignored or repressed completely. Anthropomorphic analogies between technology and the human body,[22] for example the equation of telegraph lines with nerve cords,[23] embedded the principal strangeness into the existent knowledge and experience of the human body.

17 Meya, *Element.*

18 Ibid., p. 61.

19 K.K. Österreichische Central-Comité, (ed.), *Bericht Paris 1. Lieferung*, p. 139.

20 *Deutsche Ausstellungs-Zeitung*, August 27, 1867, column 6.

21 Finn, *Submarine Telegraphy Grand Victorian*, p. 21.

22 On the interrelation between technology and the body see also Martin Doll's chapter in this book.

23 For an early example, see Sömmering, *Telegraphen*. See also Siegert, *Kant*. Breidbach, *Materialisierung*, pp. 160–165.

The meaning of telegraphy changed, however, over the years. Already in the 1860s, it had ceased to be a fascinating innovation and transformed into an every-day technology. With the propagation of the telegraph, its use in business life, politics and above all, for the private information transmission of an elite, the fascination for this technology decreased successively. Nevertheless, the operating mode of electric communication remained foreign to most users and telegraph devices were still "black boxes," though they no longer produced surprise and fascination. However, the latest telegraph devices were still displayed at world fairs. Now everybody could use the technology in the specially furnished telegraph offices of the exhibitions.[24] The main focus of the presentation of electricity at the world exhibitions shifted to more and more spectacular lightshows and thus the invention of electric lightning. Furthermore, a new communication device, the telephone, caught visitors' attention from the 1870s onwards. Patented by Alexander Graham Bell in February 1876, the telephone was presented with great success at the world exhibition of Philadelphia that very year. In presence of the emperor of Brazil, Pedro II, the jury members of the world exhibition tested the new invention and showed themselves positively surprised by the transmission of the human voice.[25] At the following exhibitions, the telephone quickly became a sensation of the first rank. Over and over again, music transmission through the telephone enthused the listeners on exhibitions. The possibility to take part in an opera without being physically present held more charm then a dialogue with other people.[26] Telephony and lightshows stimulated the human senses in a new way and thus allowed visitors new bodily experiences. This was an important condition in order for this specific technology to become a fascination and replace telegraphy as a public enthrallment.[27] But telegraphy still remained a mighty symbol of social progress and imperialism at that time.

Telegraphy and World Fairs in the Discourse of Progress

One has to bear in mind the limited generalizability of official sources like the exhibition catalogues as a source of historical investigation. Written by

24 Hesse-Wartegg, *Pariser Weltausstellung*, p. 52.
25 Bruce, *Bell*, pp. 196–197.
26 Binder, *Wunderbarer Traum*, pp. 33–34.
27 Osietzki, *Wirklichkeitsallegorien*, pp. 57–58.

state officers, entrepreneurs or scientists, who took over the function of public representatives and jury members etc., these reports show only the perspective of a particular group of actors, which we can identify as the new cosmopolitans or "Weltbürger" of the nineteenth century.[28] They published mostly positive statements about the world exhibitions that they visited and progress that was supposedly driven by technology.[29]

The defeat of nature and its utilization were the main ideological functions of technology in the nineteenth century. In this easy oppositional thinking, telegraphy fought on the side of technology created by the human ingenious mind, penetrating the other side, nature. With the installation of the first telegraph connection between Europe and America in 1858, it seemed that technology was able "to defeat the separation of nations by the ocean" and telegraphy was seen as "the last stage of the human mind to control economy."[30] Where nature appears only as a "disqualified nature," the supporters of progress and development saw the light of Enlightenment shining most brightly.[31] Seen as a result of rational, systematic and intentional actions of people, progress could be equated with technical change and a "technological sublime" that was man-made.[32] Therefore, telegraphy and world fairs functioned as important media to connect people and to enable social interactions; they thus became the most important expressions of this sublime. Sartorius for example described the Crystal Palace of the Great Exhibition in London in 1851 as a "sanctuary," built to praise the "creative and ingenious

28 I use the term, bourgeoisie, not as a classification for a clearly defined social class. The bourgeoisie includes different social groups (like intellectuals, merchants, engineers, entrepreneurs etc.) which share the same social habit or lifestyle. Most bourgeois ideas on peaceful progress, world society and universal equality in the second half of the nineteenth century had been developed by the cosmopolitans of the eighteenth century. In Germany, the term "Weltbürgertum" is used for this special bourgeois network of the Enlightenment. Although the "Weltbürger" of the eighteenth century was an international literary elite of the Enlightenment, it is worth considering using this term for those civic actors of the nineteenth century who were likewise part of an international network and who think in "global" dimensions. Thus, internationally active engineers, merchants, intellectuals and entrepreneurs, who dominated the discourse on world fairs and telegraphy, would become the successors of the eighteenth-century "Weltbürger." Schäfer, *Geschichte des Bürgertums*, p. 38–43; Reitmayer, *Bourgeoise Lebensführung*; Kocka, *Bürgertum*; Bolz, *Weltbürgertum*.

29 Sieferle, *Fortschrittsfeinde*.

30 K.K. Österreichische Central-Comité, Bericht Paris *16./17. Lieferung*, p. 22.

31 For example: Mohl, *Wunder*, p. 333.

32 Nye, *Technological Sublime*, p. 62–63. About the sublime of technology see also Noble, *Technology and Religion*; Kang, *Sublime Dreams*.

mind as a divine principle of mankind."[33] Even the architecture of the world fairs seemed to show signs of holy places. The buildings were often characterized as cathedrals of modernity and for correspondent Lothar Bucher the summerhouse at Hyde Park, one of the buildings at the world exhibition in 1862, was part of the temple of the Holy Grail.[34]

As one of the advanced technologies of the second half of the nineteenth century, telegraphy too was interpreted as a realization of the divine will: "In spite of these gaps [in the global telegraph network, L.B.] the word of the Psalmist (Ps. 19, 4.5.) has now been applied rightly to telegraphy in its unstoppable spreading all over the globe: 'There is no speech nor language, where its voice is not heard. Its line has gone out round about the entire earth, and its words to the end of the world.'"[35] Fischer's quote shows clearly that the spread of religion had entered the authority of a technical communication device, telegraphy. Basically, all commentators agreed that the industrial exhibitions were "milestones of progress."[36] If technology could not only guarantee the satisfaction of one's need, but was mainly interpreted as an expression of certain moral attitudes, this had corresponding consequences for the notion of world fairs. With their accumulation of all kinds of technical products, they presented the refinement of human mastery and skill that was supposed to lead to an ideal society. The bourgeoisie was the main advocate of this utopian vision of technology. They prepared in the halls of the exhibitions an artificial world of technologically-driven utopia based on harmony and social consensus without class conflicts. Bourgeois voices also claimed that technological change and social progress would end in a free and democratic society.[37] The exaggerated hope that the very existence of world exhibitions could solve all social problems demonstrates the overestimation of Western self-confidence. The bourgeoisie's discourse on world fairs and technology is an impressive confirmation of the mental and cultural shift of the bourgeoisie's self-image in the first half of the nineteenth century. With industrialization, the idealistic universalism of the bourgeoisie was now transformed into a more pragmatic, technologically oriented universalism. The rationality of railways, steamships and telegraphs and the dy-

33 Sartorius, *Industrieausstellung*, p. 2. See also the passage in Jonas Harvard's chapter in this book on how technology takes on the role of religion.

34 Bucher, *Bilder*, p. 2.

35 Fischer, *Post*, p. 92. In fact, Fischer means Ps 19, 3.4.

36 Hamm, *Illustrirter Katalog*, p. 3.

37 Chevalier, *Industrie*, p. 20.

namics of industrialization were seen as the prime movers for the realization of a civil-liberal society.[38]

Like the world exhibitions, telegraphy acted as a metaphor for progress and was included into the discursive network of politics, economy and technology. Due to the important role of telegraphy for economic and political needs, this insight is no surprise. As the fastest technology to transport information in the nineteenth century, contemporaries assessed telegraphy as one of the most advanced technologies. Franz Xaver Neumann ascertained in his cultural-historical considerations, which were part of his report on the Vienna exhibition in 1867, that: "Above all the special applications of electromagnetism, which foster the commerce, telegraphy, outshone all other inventions in its cultural-historical importance."[39]

The rapid expansion of telegraph lines and the huge growth rate of telegram transmission were a clear evidence for the rapid and far-reaching technological change in general.[40] That the application of telegraphy was a crucial factor for further economic growth, advanced industry and social progress was commonly assumed.[41]

Unlike world exhibitions which concentrated goods and trade like a big market on a certain geographic place and for a fixed period, telegraphy seemed to neglect the dimensions of space and time.[42] Anthropomorphized as a "nerve tissue," which expanded into the whole world, telegraphy was supposed to enforce capitalism and commerce on a global level.[43] Every place that was reachable by means of transportation, whether by shipping, railroad or telegraphy, counted as a potential market.[44] According to Franz Reuleaux, the famous German rapporteur at several world fairs, telegraphy and postal services became the "elite troops of global trade."[45]

Whereas the bourgeoisie used the world fairs to demonstrate their economic and cultural power, telegraphy seemed to be the ideal device to spread bourgeois ideas and ideology into the last corners of the world untouched

---

38 Glaser, *Bildungsbürgertum*, pp. 59–73.

39 K.K. Österreichisches Central-Comité, *16./17. Lieferung*, p. 153.

40 See Jonas Harvard's article in this volume.

41 Michael Chevalier named three central pillars of progress: communication devices, credit banks, industrial education. Chevalier, *Industrie*, p. 54.

42 Wenzlhuemer, Less Than No Time. See also Florian Sprengler's article in this volume.

43 Reuleaux, *Weltverkehr*, p. 20.

44 On the correlation of bourgeois power and spatial order by cartography, traffic routes etc., see Schlögel, *Im Raume*, pp. 211–219.

45 Reuleaux, *Weltverkehr*, p. 508.

by civilization. Therefore, contemporaries also regarded telegraphy as the medium of modern culture.[46] Europe saw itself as a starting point and avant-garde of this world occurrence of Western civilization which started its "triumphal procession with the discovery of America by Columb[us] in 1492," to use the words of August Oncken.[47] Later, telegraph, railroad and steamships took over the role of the early-modern sailing-ships for the domestication of the world as a "walk of the civilization around the earth."[48] In fact, Euro-Americans equated civilization and telegraphy with each other. Where telegraph wires stretched through the landscape, civilization seemed to have been successfully established: "[...] civilization started primarily with the telegraph wire."[49]

The repression of the "primitive" nations and races within the frame of the universal approach of the bourgeois project seems to be a necessity. It is obvious that this discourse of modernity, connected with Western universalism, was a strategic element to legitimize European imperialism as a "white man's burden" to rule underdeveloped nations for the benefit of those people.[50] Additionally, agents of imperialism transformed Darwin's theory of evolution into a concept to explain the powerful dominance of Western civilization. Within this line of argument named Social Darwinism, every repression of other people seemed justified because the winner's culture demonstrated its supremacy automatically with its technical abilities and scientific knowledge .[51] World fairs and technologies like the telegraph served as proof for this concept.[52] In fact the bourgeois concept of a uniform world in which all people were equal and had the same rights obviously broke down. Now the dividing line ran exactly between the civilized cultures and the "primitive natives." This corresponded to the basic common opposition of culture and nature up to the end of the nineteenth century.

---

46 Fischer, *Post*, p. 47.

47 Oncken, *Wiener Welt-Ausstellung*, pp. 16–17. Ironically, more than 130 years later, Pulitzer-Prize winning author Thomas L. Friedman starts the history of globalization again with Columbus' travel to America. Friedman, *World is Flat*.

48 Oncken, *Wiener Welt-Ausstellung*, pp. 16–17.

49 Anonym, Elektrotechnik, p. 239.

50 Hawley, *Encyclopedia*, p. 445. The phrase "white man's burden" traces back to a poem by Rudyard Kipling, published in 1899. It is an apology for the imperial mission of Western powers to overcome the well-established colonial politics of Spain etc. Kipling, White Man`s.

51 Adas, *Machines*, p. 203.

52 See for example K.K. Österreichische Central-Comité, Bericht Paris *16./17. Lieferung*, pp. 19–20.

Africa especially was seen as a dark continent that had to be brought into the light of Western civilization. Imperial penetration with telegraph lines became a primary mission of the Western powers not only for stabilizing their political regime, but also to enable Africa to be part of the civilized world.[53] Africa was supposed to be integrated into the "große Verkehrsgemeinschaft" (huge community of commerce).[54] Therefore, building telegraph lines was the major "burden of the white man" in Africa. The destruction of telegraph lines by "natives" as a form of resistance was consequently seen as an attack against the values of Western civilization. Franz Xaver Neumann, an Austrian economist, described such a demolition of a telegraph line in 1865 in his exhibition report. The draconian punishment of the emperor of Morocco against the Kabyle people who were responsible for the attack consisted in putting the beaten heads of the leaders ostentatiously on the telegraph poles. Neumann commented on this cruel punishment: "Admittedly, civilization has to fight many battles before it will be accepted."[55] Additionally, he expressed the hope that the telegraph might defeat the African continent rather than the Christian missionaries.

There is also an interesting correlation between world fairs and imperial telegraphy in the construction of difference/equality and center/periphery as fundamental concepts of colonial discourse.[56] Racial otherness and the assumed Western superiority were displayed at world fairs by "living exhibits," presenting war dances, rituals and the supposed everyday life of colonized people, who were degraded as semi-naked primitives. This construction of racial difference/alternity was enhanced by the fact that the "natives" were shifted from the colonial peripheries to the centers of Western civilizations.[57] Whereas the exhibited "natives" were shipped to the Western centers of assumed cultural and political predominance, telegraphy was supposed to connect these centers with its periphery. Telegraphy was in this perspective both a technical infrastructure to gain control over the colonized people and a symbol or metaphor in the imperial discourse of modernity to justify and stabilize imperial ideology.[58] The spatial organization of imperial infrastruc-

53 Bluma, Globalization.

54 Veredarius, *Buch*, p. 359.

55 K.K. Österreichische Central-Comité, Bericht Paris *2. Lieferung*, p. 29

56 Said, *Orientalism*. Hawley, *Encyclopedia*, 85.

57 All world fairs of the nineteenth century except the Australian exhibitions in 1879 and 1880 were staged in Europe and the USA, mostly in metropolises like Paris, London etc.

58 For the role oft telecommunications as a technology of imperialism and geopolitics see Hugill, *Global* Communication; Headrick, *Invisible Weapon*; Headrick, Double-Edged Sword; Yang, Telecommunication; Quevedo, Telecommunications.

ture symbolized the endeavor of the imperial powers to aggrandize themselves in the center by treating all else as peripheral.

Telegraphy and World Fairs in the Discourse of National Competition

At least since industrialization, the different institutions, rituals, procedures and performance measurements of the Western economic system established a discourse of competition. Within this discourse of competition, the world exhibitions became the place to celebrate and to represent the economic spirit of capitalism. According to contemporaries, the basic function of world fairs lay in the concentration of international entrepreneurs and companies competing with each other.[59] This gathering of international suppliers and their products of all branches constituted in fact a technological and "economic battlefield."[60]

Besides the quality inspection by potential buyers, the institutionalized assessments were also important modes of this discourse of competition. Firstly, commissions at the exhibitions awarded certifications of excellence to particular products. The exhibition reports also took care of a comparative description and review of the exhibited goods. The distinctive competition situation at the world exhibitions incorporated the products of telegraphy as well. Nevertheless, it did not come to such spectacular contests as for example in the steel industry. One of these spectacular dramatizations of this fierce competition was the so called "Gußstahlstreit" (cast-steel-fight) between the German companies Krupp and Bochumer Verein at the Paris world fair 1855.[61] With the presentation of increased large and heavy steel ingots, both companies tried to trump not only their respective national business rival, but also the entire English steel industry.[62] These imposing shows portrayed the large size of materiality and operated according to the principle "size matters;" they were foreign to the demonstration of telegraph technology, because viewers could not see how it functioned. Neither size

59 Blenck, Pariser Welt-Industrie-Ausstellung, p. 174.

60 Goldschmidt, *Weltausstellung*, p. 13.

61 Kroker, *Weltausstellungen*, p. 62–63.

62 Engel, *Biographisches Lexikon*, p. 59.

nor weight stood in the foreground, but, first and foremost, the speed of the telegraph apparatus, its security and user-friendliness.

The discourse of competition as a necessary part of capitalism could be found in other social relevant discourses and behavior patterns. Most obviously, the topos of competition trickled into political discourse, so that the comparison of products was used to reassert national identities.[63] But such a pronounced aggressive idea of competition between national states, which can be found in the late nineteenth century, was not yet foreseeable in the 1850s and 1860s. Rather the first world exhibitions operated under the aegis of the bourgeoisie's idea of a peaceful competition of nations. Commentators interpreted the world exhibition in London in 1851 as a "union unique in the history of mankind of the people of the earth."[64] Competition was applied to the concept that the world as a whole would benefit in the course of progress and not just a single nation: "The educational degree of single nations is unsteady, but the progress of the whole humanity is always steadily."[65] Within this peaceful concept of competition between nations, symbols and metaphors dominated which emphasized international cooperation. Nevertheless, cooperation was not seen as the opposite of competition, but as a result of peaceful advancement. The journalist Saphir expressed this idea in passionate but typical words:

> Well then, on the 15th May the peaceful tournament will begin in the industry palace, all people will take her art and industry lances, it will be a fight, but no bloody one, the defeated people will sustain no wounds, the winners did not fight for themselves, but they will have won for their nation, for the universal, for the whole humankind. I say for humankind, because now industry has become a matter of humanity.[66]

In this view, the cathedrals of industry were educational institutions for the nations, which acted like students competing with each other for further achievements. Based on the ideas of humanistic philosophy and the bourgeois approach of universalism, a concept of education emerged, which saw technological knowledge and economic education as a precondition for a

63 Astrid Böger shows the importance of the American world exhibitions for the formation of the American nation, but this was also significant for the European nations and their representation on world fairs. Böger, *Envisioning*.

64 Sartorius, *Industrieausstellung*, p. 1.

65 Reuleaux, *Weltverkehr*, p. 475.

66 Saphir, *Pariser Briefe*, pp. 48–49.

peaceful world.[67] Within this cooperative discourse of competition, telegraphy earned a dominant position as a concrete device for communication and as a metaphor, because telegraphy served as one of the most modern devices of transporting human ideas and knowledge all around the world. The peaceful aspect of the world exhibitions was for Saphir "a thought of world belt (*Weltgürtelgedanke*), it flies around the globe, it navigates through seas, it runs down the poles, it fraternizes nations and combines the genius of the industry."[68] The idea that the global telegraph network is something like a belt or band embracing and connecting the whole world was emphasized over and over again and this perception corresponded with the meaning of world fairs bringing different nations and cultures together. Let me cite Ferdinand Lesseps, the famous builder of the Suez Canal, who explicated this idea: "All these enterprises of universally interest – some already completed, others under construction or projected – have an identical goal: drawing peoples together, and thereby bringing about an era in which men, by knowing one another, wants to finally cease fighting."[69] In order to guard against misunderstandings, indigenous people were not included in this concept. "Drawing peoples together" actually meant drawing peoples of Western nations together.

Because telegraphy brought people in contact with each other – as well as the world exhibitions – it had in the view of the nineteenth century bourgeoisie essential influence on the exchange of ideas, meanings and views: "Telegraphy is the carrier of the modern cultural life."[70] Therefore, international cross-border communication by telegraphy was one of the important foundations for activists of internationalism and the peace movement who used the world fairs also for their congresses and meetings. The founding of the International Telegraph Union (ITU) in 1865 was an attempt to institutionalize this idea of an international community.[71] The ITU and its successors established international rules for telegraphy, first on a European and later on a global level. It defined the use of standardized technical equipment, unified the operation processes and regulated the international tariff system. Through coordination of technical studies, testing and measurement, the ITU was responsible for international sharing of knowledge and

67 Kraul, Bildung; Giesen, Mythos; Wehler, *Zielutopie*.

68 Saphir, *Pariser Briefe*, 66.

69 Letter to Maxime Hélène, 6 November, 1882, in: Hélène, *travaux publics*, p. 7.

70 Fischer, *Post*, p. 47.

71 Lyall, International Communications. Bluma, *Globalization*, pp. 250–252.

thereby shaped a new group of internationally-oriented technical experts in the field of telegraphy.[72] Commerce and communication was now detached from national contexts and embedded in an "international and cosmopolitan mission" of free exchange of ideas.[73] The implementation of the Morse alphabet as an international standard in telegraphy was a confirmation of this mission: "Telegraphy has thus achieved the goal of an unified alphabet for all nations of the world, what still will remain a wishful thinking for the usual writing and the letterpress."[74] No wonder that the new "Weltbürgertum" was so fascinated by telegraphy because it promised a reconciliation of literary culture ("unified alphabet," "writing and the letterpress") and new technologies of communication ("telegraphy").

The competitive exhibitions of the nineteenth century, actually an institution to compare products and technologies of single companies, were now conventionalized as a competition of nations, which used the world exhibitions more and more for self-representation. Even at the first world exhibition in London in 1851, national arguments appealing to the patriotism of the exhibitors appeared: "To be represented in a worthy manner with products is as well necessary because of commercial interests, as a case of the honor and the fame of the national industry."[75] Even more than with the guest's nations, the presentation of national dignity and identity was a central concern of the respective host nation. This is best represented by the opening ceremonies with their exorbitant splendor and pomp.[76] Also this ritual practice had the function to envision a collective identity called nation and to deny existing social divergences and conflicts.

The exhibition catalog of the Brazilian empire shows the importance of telegraphy for this kind of national representation. The catalog documented in detail the own telegraph history.[77] It stressed not only the important role of telegraphy for the industrial development and improvement of Brazil, but also the nationwide solidarity and identity of the Brazilians affected by telecommunication. Telegraphy also laid the groundwork for national myths: the Brazilian catalog describes in detail, for example, the military role of

72 Ahvenainen, *International Telegraph Union*. See also Leonard Laborie's article in this volume and Simone Müller about the formation of transnational networks of telegraph engineers: Müller, Transatlantic Telegraphs.

73 K.K. Österreichische Central-Comité, *Bericht Paris 16./17. Lieferung*, pp. 255–257.

74 Geistbeck, *Weltverkehr*, p. 472.

75 Anonym, *Amtlicher Bericht London 1851*, pp. 91–92.

76 Haltern, *Londoner Weltausstellung*, pp. 160–161.

77 Anonym, *Kaiserreich Brasilien*, pp. 326–327.

telegraphy for the defense of the bay of Rio de Janeiro in the Anglo-Brazilian war 1863.[78] The presentations of the German-speaking countries on the world fairs before the foundation of the German empire in 1871 turned out to be especially problematic. They appeared together under the name "Zollverein" (Customs Union), however, the single German states retained their sovereignty in all exhibition questions. This included the equipment in each case of the different showrooms as well as the right of every single country to negotiate with the exhibition organizers independently.[79] The "Zollverein" as an economic coalition of the different German states barely offered a national identity, though the commissioners of the union organized the publication of a joint exhibition catalog.[80] In spite of this, Schubarth and Viebahn, the commissioners of the Customs Union at the world exhibition in 1855 in Paris, noticed a sentiment of unity encompassing all involved parties in spite of the lack of political unity in reality.[81] However, there was a certain national inferiority complex on the part of German officials in comparison to England and France prior to 1871.[82]

Modern technology had the function to compensate for the national disunity of Germany over and above its use as a practical tool.[83] German reports repeatedly pointed to the necessity to draw level with England and France, and they highlighted the technical fields in which the Germans felt that they were leading. One of these innovative fields was, at least from Prussia's point of view, electrical engineering or the physical instrument making under which telegraphy was also categorized for a long time in the world exhibitions.[84] Prussia thus prioritized the exhibition of the technical leadership sectors as particularly prestigious. As a result, the Prussian commission of the Paris exhibition of 1855 focused on presenting products which would prove successful in comparison to foreign products.[85] This competition at an international level created clichés of the different nations that have partially remained to this day. The "elegance of French products," the "solidity of English manufactured products" and the "working enthusiasm" of the American people were such common stereotypes as well as the conviction

78 Ibid., p. 325.
79 Kroker, *Weltausstellungen*, p. 31–32.
80 Ibid., p. 34.
81 Viebahn, *Amtlicher Bericht*, p. VI.
82 Schirges, *Weltausstellung*, p. VII.
83 Radkau, *Technik in Deutschland*, p. 149.
84 Reuleaux, *Weltverkehr*, p. 343.
85 Anonym, Zirkular, p. 17.

that "the Germans were on the first position under all nations of the world if looking on education and knowledge."[86] Also the origin of so-called "rankings" had its cause in the use of technology as a comparative instrument of nations. Depending on the technical field, the performances were compared with each other – England, France, the United States and the German states (mostly denoted as an imaginary unit) were competing with each other for the top spot on the ranking lists.

This was no different for presentations of telegraphy organized by states at the world exhibitions. From the Paris world fair in 1867 onwards, the governmental telegraph administrations, at first Spain, France, and Prussia, acted as exhibitors but only showed a limited range of samples, especially individual pieces of telegraphs and cables as well as models of the national telegraph lines.[87] Because they had no interest in selling products, telegraph administrations focused on the representation of the national history of telegraphy. While qualitative categories like solidity and ease of use stood in the foreground, the state exhibitors emphasized quantifiable categories, such as the length of the national network and the number of telegrams. They favored presenting their telegraph networks on maps: "Let us stay a moment in front of a large special map of France on which the telegraph network is charted. Many thousand wires are presented by elegant small silk threads which not only pass the most remotest coasts of the empire, but also range from the Mediterranean Sea to Italy and Algeria, and to the Canal to England and Scotland. In Paris all threads were united in the Ministry of Interior as a nerve center of communication [...]. A more suitable and striking image of the French Centralization can hardly be imagined."[88] Ebeling concluded in his comment that the organization of the telegraph network symbolized the political state of France and he was certainly right with his assessment.

After unification in 1871, the German empire demonstrated its new national power with an impressive exhibition on the history of state telegraphy in Germany at the Vienna world fair in 1873.[89] The postal service and the integrated telegraph administration were, with the exception of Bavaria and Wuerttemberg, which still had their own postal systems, one of the few centralized political services of the German empire and therefore telegraphy was an ideal object of public representation. This was the starting point for the

86 Mohl, *Wunder*, p. 233.
87 Anonym, *Berichte Paris 1867*, p. 424–425.
88 Ebeling, *Wunder*, p. 160–161.
89 Anonym, *Amtlicher Katalog*, p. 508.

increase in states' activities at subsequent world fairs to demonstrate their national prowess, whereas telegraph companies lost interest in participating in these expensive events. To motivate German companies to play an active part at the Philadelphia fair in 1876, the state had to cover travel costs[90] and even Werner Siemens, one of the main promoters of the big historical telegraph exhibition at Vienna, was not very enthusiastic about acting as an agent of national interests.[91] All in all, commentators noticed a common world fair tiredness (*Ausstellungsmüdigkeit*) on the part of businesses, which complained about the transformation of world fairs into touristic events and national self-representations.[92] In this context, it is evident that the Philadelphia exhibits by the German telegraph administration, complemented by telegraph devices at the Siemens & Halske presentation, were the cornerstone of the first German museum dealing with the history of technology, the Deutsche Reichspostmuseum.[93] Telegraphy was now an object of museums and historical exhibitions to represent national power and no longer an innovative technology presented by companies at world exhibitions.

One main effect of the state activities was that quantitative measures of performance like tables, graphics, charts etc. were increasingly integrated in the discourse of competition, because they produced, according to Joachim Radkau, the ideal of "faster, stronger, higher" as an individual and collective obsession of the late nineteenth century.[94] From the Viennese world exhibition in 1873 onwards, statistical representations of all kind were compulsory for every available class of goods.[95] Statistical evaluation documented the national expansion and growth of the telegraph network and had a symbolic character. Telegraph wires on the maps of the national telegraph authorities represented the range of Western civilization. The blind spots on the maps corresponded to the uncivilized, who were not integrated into the communication system of industrialized societies.

The exhibitors expected further quantification of the technological and economic advancement by comparative assessments, as shown by the many attempts for standardization of different measurement and numerical units

90 Kroker, *Weltausstellungen*, p. 38.
91 Matschoß, *Siemens*, p. 477.
92 Huber, *Ausstellungen*, p. 16.
93 Museum für Post und Kommunikation Frankfurt am Main, Archive: I CC 10, I CC 15, I CC 24, I CC 33. See also Hübner, Leidenschaftlicher Sammler.
94 Radkau, Technik im Temporausch, p. 65.
95 Oncken, *Wiener Welt-Ausstellung*, p. 55.

on special congresses at the world exhibitions.[96] Furthermore, at the end of the nineteenth century, the discourse of competition constituted not only the collective perception of the individuals as a nation, but also highlighted them as individuals by elaborated performance criteria which penetrated the whole social field, obviously demonstrated by the rising interest in sport meetings, which were oriented by the "ideal of faster, stronger, higher."[97] The ubiquitous discourse of competition structured the process of nation building and therefore offered a national identity for all relevant social groups. At first, this competition of nations was a peaceful project to foster the common progress of all nations, but in the late nineteenth century national competition perverted into an aggressive and militant race of nations. Imperialism and its run for colonies and global power was the political manifestation of this cultural process. Likewise, we can detect a shift from free trade to protective trade policy from the 1870s onwards, especially through the implementation of prohibitive tariffs.[98]

Conclusion

A consolidated view on the history of world exhibitions linked with the history of telegraphy indicates that both were embedded in a complex symbolic order which provided identities for individuals as well as social groups like nations or the global community. This article aimed to reconstruct the shifting meaning of telegraphy and world fairs through the broad framework of utopian thinking between 1851 and 1880. This thinking took for granted the idea of technologically-driven progress. World fair visitors' amazement expressed the sublime of technology and paved the way for the imaginary linkage between telegraphy and social progress. This was initially embedded in the bourgeois project of liberal trade, free exchange of information and peaceful contest to overcome the clash of nations and to raise the cultural level of all civilizations for the good of humanity.

96 For the process of standardization in the second half of the nineteenth century see Geyer, One Language.

97 Eichberg, Schneller, Höher, Stärker. Pierre de Coubertin suggested this as the motto for the Olympic games. In fact, the Olympic games of 1900 and 1904 took place under the aegis of the respective world exhibitions in Paris and St. Louis.

98 Polland, Free Trade; Aldenhoff-Hübinger, *Agrarpolitik*.

But from the 1860s on, this concept of peaceful contest was replaced in the examined world flair's reports by more and more aggressive narratives that underlined the clash of Western nations and paralleled the implementation of Social Darwinist ideas about the struggle of races. Telegraphy and world fairs were used in both contexts as important symbols for an imaginary future. The double function of telegraphy and world fairs as social practices and as symbolic figures make them particularly significant for analyzing politics and culture in the second half of the nineteenth century.

Bibliography

Adas, Michael, *Machines as the Measure of Men. Science, Technology, and Ideologies of Western Dominance*, Ithaca, NY/London 1989.

Ahvenainen, Jorma, "The International Telegraph Union: The Cable Companies and the Governments," in: Finn, Bernard/Yang, Daqing (eds.), *Communication Under the Seas. The Evolving Cable Network and Its Implications,* Cambridge, MA/London 2009, pp. 61–79.

Aldenhoff-Hübingern, Rita, *Agrarpolitik und Protektionismus. Deutschland und Frankreich im Vergleich, 1879–1914*, Göttingen 2002.

Andersen, Hans Christian, *Dryade. Von der Weltausstellung 1867. Ein Abenteuer*, Bremen 1869.

Anonym, *Amtlicher Bericht über die Industrie-Ausstellung aller Völker zu London im Jahre 1851, von der Berichterstattungs-Kommission der Deutschen Zollvereins-Regierungen,* vol. 1. Berlin 1852.

Anonym, "Zirkular der Central-Kommission für die Pariser Ausstellung an die Bezirks-Kommissionen Preußens vom 2. Okt. 1854," in: Viebahn, Georg W., *Amtlicher Bericht über die Allgemeine Pariser Ausstellung von Erzeugnissen der Landwirthschaft, des Gewerbefleißes und der schönen Kunst im Jahre 1855*, Berlin 1855, pp. 17–21.

Anonym, *Amtlicher Bericht über die Industrie- und Kunst-Ausstellung zu London im Jahre 1862, erstattet nach Beschluß der Kommissare der Deutschen Zollvereins-Regierungen,* vol. 1. Berlin 1865.

Anonym, *Berichte über die Allgemeine Ausstellung zu Paris im Jahre 1867, erstattet von den für Preußen und die Norddeutschen Staaten ernannten Mitgliedern der internationalen Jury, VI. Heft.*, Berlin 1868.

Anonym, *Amtlicher Katalog der Ausstellung des Deutschen Reiches, Wiener Weltausstellung*, Berlin 1873.

Anonym, *Das Kaiserreich Brasilien auf der Weltausstellung von 1876 in Philadelphia,* Rio de Janeiro 1876.

Anonym, "Die Elektrotechnik im 20. Jahrhundert," *Das Neue Universum*, vol. 22 (1901), pp. 236–239.

Beutler, Christian, *Weltausstellungen im 19. Jahrhundert.* München 1973.

Binder, Beate "'...es ist uns, als ob ein wunderbarer Traum unsere Sinne umgaukle.' Die Inszenierung einer elektrischen Welt auf der Frankfurter 'Internationalen Elektrotechnischen Ausstellung'," in: Becker, Jörg (ed.), *Telefonieren*, Marburg 1989, pp. 31–44.

Blenck, "Die Pariser Welt-Industrie-Ausstellung im Jahre 1867," *Zeitschrift des Königl. Preussischen Statistischen Bureaus*, no. 7 (1865), pp. 174–179.

Bluma, Lars, "Globalization and the Rise of Telecommunication Networks," in: Anderl, Rainer/Arich-Gerz, Bruno/Schmiede, Rudi (eds.), *Technologies of Globalization*, Darmstadt 2008, pp. 249–261.

Böger, Astrid, *Envisioning the Nation. The Early American World Fairs and the Formation of Culture,* Frankfurt am Main/New York 2010.

Bolz, Norbert/Kittler, Friedrich/Zons, Raimar (eds.), *Weltbürgertum und Globalisierung,* München 2000.

Breidbach, Olaf, *Die Materialisierung des Ichs. Zur Geschichte der Hirnforschung im 19. und 20. Jahrhundert*, Frankfurt am Main, 1997.

Bucher, Lothar, *Bilder aus der Fremde, 2. Band: Die Industrieausstellung von 1862,* Berlin 1863.

Bruce, Robert V., *Bell. Alexander Graham Bell and the Conquest of Solitude*, London 1973.

Castells, Manuel, *The Rise of the Network Society, The Information Age: Economy, Society and Culture,* vol. I., Malden/Oxford 1996.

Chevalier, Michael, *Die heutige Industrie, ihre Fortschritte und die Voraussetzungen ihrer Stärke. Weltausstellung von 1862*, Berlin 1863.

Deutsche Ausstellungs-Zeitung, August 27, 1867, column 6.

Ebeling, Adolf, *Die Wunder der Pariser Welt-Ausstellung 1867*, Köln 1876.

Eichberg, Henning, "'Schneller, höher, stärker.' Der Umbruch in der deutschen Körperkultur um 1900 als Signal gesellschaftlichen Wandels," in: Mann, Gunter/Winan, Rolf (eds.), *Medizin, Naturwissenschaft, Technik und das Zweite Kaiserreich*, Göttingen 1977, pp. 259–283.

Engel, Moritz (ed.), *Biographisches Lexikon der Wiener Weltausstellung,* Wien 1873.

Feldenkirchen, Werner, *Werner von Siemens. Erfinder und internationaler Unternehmer*, München/Zürich 1996.

Finn, Bernard S., *Submarine Telegraphy. The Grand Victorian Technology*, London 1973.

— "Submarine Telegraphy: A Study in Technical Stagnation," in: Finn, Bernard/Yang, Daqing (eds.), *Communication under the Seas. The Evolving Cable Network and its Implications*, Cambridge, MA/London 2009, p. 9–24.

Fischer, Paul David, *Post und Telegraphie im Weltverkehr*, Berlin 1879.

Friedman, Thomas L., *The World is Flat. A Brief History of the Globalized World in the Twenty-first Century*, New York 2005.

Geistbeck, Michael, *Der Weltverkehr. Seeschiffahrt und Eisenbahnen, Post und Telegraphie in ihrer Entwicklung dargestellt*, Freiburg im Breisgau 1895.

Geyer, Martin H., "One Language for the World: The Metric System, International Coinage, Gold Standard, and the Rise of Internationalism, 1850–1900," in: Geyer, Martin H./Paulmann, Johannes (eds.), *The Mechanics of Internationalism. Culture, Society, and Politics from the 1840s to the First World War*, New York 2001, p. 27–53.

Geyer, Martin H./Paulmann, Johannes (eds.), *The Mechanics of Internationalism. Culture, Society, and Politics from the 1840s to the First World War*, Oxford 2001.

Giesen, Bernhard/Junge, Kay, "Der Mythos des Universalismus," in: Berding, Helmut (ed.), *Mythos und Nation. Studien zur Entwicklung des kollektiven Bewußtseins in der Neuzeit*, vol. 3, Frankfurt a.M. 1996, pp. 34–64.

Glaser, Hermann, *Bildungsbürgertum und Nationalismus. Politik und Kultur im Wilhelminischen Deutschland*, München 1993.

Goldschmidt, Friedrich, *Die Weltausstellung in Philadelphia und die deutsche Industrie*. Berlin 1877.

Haltern, Utz, *Die Londoner Weltausstellung von 1851. Ein Beitrag zur Geschichte der bürgerlich-industriellen Gesellschaft im 19. Jahrhundert*, Münster 1971.

Hamm, Wilhelm, *Illustrirter Katalog der Londoner Industrie-Ausstellung von 1862*, Leipzig 1863.

Hawley, John C. (ed.), *Encyclopedia of Postcolonial Studies*, Westpoint/London 2001.

Headrick, Daniel R., *The Invisble Weapon. Telecommunications and International Politics 1851–1945*, Oxford 1991.

—, "A Double-Edged Sword: Communications and Imperial Control in British India," *Historical Social Research*, vol. 35, no. 1 (2010), pp. 51–65.

Hélène, Maxime, *Les travaux publics au XIXe siècle. Les nouvelles routes du globes*, Paris 1883.

Hesse-Wartegg, Ernst von, *Die Pariser Weltausstellung 1878*, Wien 1878.

Huber, F. C., *Die Ausstellungen und unsere Exportindustrie*, Stuttgart 1886.

Hübner, Hans, "Ein leidenschaftlicher Sammler. Von der Raritätenkammer zum Reichspostmuseum," in: Beyrer, Klaus (ed.), *Kommunikation im Kaiserreich. Der Generalpostmeister Heinrich von Stephan*, Heidelberg 1997, pp. 173–179.

Hugill, Peter J., *Global Communications Since 1844. Geopolitics and* Technology, Baltimore, MD 1999.

Kang, Minsoo, *Sublime Dreams of Living Machines. The Automaton in the European Imagination*, Cambridge, MA 2011.

Kipling, Rudyard, "The White Man's Burden," *McClure's Magazine*, vol. 12 (Feb. 1899).

K.K. Österreichische Central-Comité (ed.), *Bericht über die Welt-Ausstellung zu Paris im Jahre 1867. Officieller Ausstellungsbericht, 1. Lieferung: Instrumente für Kunst und Wissenschaft*, Wien 1868.

K.K. Österreichische Central-Comité (ed.), *Bericht über die Welt-Ausstellung zu Paris im Jahre 1867. Officieller Ausstellungsbericht, 2. Lieferung: Verkehrsmittel*, Wien 1869.

K.K. Österreichische Central-Comité (ed.), *Bericht über die Welt-Ausstellung zu Paris im Jahre 1867. Officieller Ausstellungsbericht, 16./17. Lieferung: Einleitung. Die Civilisation und der wirthschaftliche Fortschritt von F. Xav. Neumann*, Wien 1869.

Kocka, Jürgen (ed.), *Bürgertum im 19. Jahrhundert*, Göttingen 1995.

Kraul, Margret, "Bildung und Bürgerlichkeit," in: Kocka, Jürgen (ed.), *Bürgertum im 19. Jahrhundert. Deutschland im Europäischen Vergleich*, vol. 3, München 1988, pp. 46–73.

Kroker, Evelyn, *Die Weltausstellungen im 19. Jahrhundert. Industrieller Leistungsnachweis, Konkurrenzverhalten und Kommunikationsfunktion unter Berücksichtigung der Montanindustrie des Ruhrgebietes zwischen 1851 und 1880*, Göttingen 1975.

Lyall, Francis, *International Communications. The International Telecommunication Union and the Universal Postal Union*, Burlington, VT 2011.

Matschoß, Conrad (ed.), *Werner Siemens. Ein kurzgefaßtes Lebensbild nebst einer Auswahl seiner Briefe*, vol. 2. Berlin 1916.

McLuhan, Marshall/Fiore, Quentin, *The Medium is the Message: An Inventory of Effects*, New York 1967.

Meya, Jörg/Sibum, Heinz Otto, *Das fünfte Element. Wirkungen und Deutungen der Elektrizität*. Reinbek 1987.

Mohl, Oswald M., *Die Wunder der Pariser Weltausstellung im Jahre 1867*. Leipzig 1869.

Müller, Simone, "The Transatlantic Telegraphs and the Class of 1866 – the Formative Years of Transnational Networks in Telegraphic Space, 1858–1884/89," *Historical Social Research*, vol. 35, no. 1 (2010), pp. 237–259.

Noble, David F., *The Religion of Technology: The Divinity of Man and the Spirit of Invention*, New York 1999.

Nye, David E., *American Technological Sublime*, Cambridge, MA/London 1994.

—, *Narratives and Spaces. Technology and the Construction of American Culture*, Exeter 1997.

Oncken, August, *Die Wiener Welt-Ausstellung 1873*, Berlin 1873.

Osietzki, Maria, "Wirklichkeitsallegorien der Elektrizität als 'Wunschmaschinen'," *Technikgeschichte*, vol. 63, no. 1 (1996), pp. 47–70.

Polland, Sidney, "Free Trade, Protectionism, and the World Economy," in: Geyer, Martin H./Paulmann, Johannes (eds.), *The Mechanics of Internationalism. Culture, Society, and Politics from the 1840s to the First World War*, New York 2001, pp. 27–53.

Quevedo, Javier Márquez, "Telecommunications and Colonial Rivalry: European Telegraph Cables to the Canary Islands and Northwest Africa, 1883–1914," *Historical Social Research*, vol. 35, no. 1 (2010), pp. 108–124.

Radkau, Joachim, *Technik in Deutschland. Vom 18. Jahrhundert bis zur Gegenwart*, Frankfurt a.M. 1989.

—, "Technik im Temporausch der Jahrhundertwende," in: Salewski, Michael/Stölken-Fitschgen (eds.), *Moderne Zeiten. Technik und Zeitgeist im 19. und 20. Jahrhundert*, Stuttgart 1994, pp. 61–76.

Reitmayer, Marten, "Bourgeoise Lebensführung im ersten Drittel des 20. Jahrhunderts," in: Plumpe, Werner/Lesczenski, Jörg (eds.), *Bürgertum und Bürgerlichkeit zwischen Kaiserreich und Nationalsozialismus*, Mainz 2009, pp. 59–69.

Reuleaux, Franz, *Der Weltverkehr und seine Mittel*, vol. 2: *Rundschau über Weltwirtschaft und Industrieverkehr*, Leipzig 1889.

Said, Edward W., *Orientalism*, London 1978.

Saphir, Moritz Gottlieb, *Pariser Briefe über Leben, Kunst, Gesellschaft und Industrie zur Zeit der Weltausstellung im Jahre 1855*, Pest/Wien/Leipzig 1856.

Sartorius, Carl, *Die Industrieausstellung in London*, Darmstadt 1851.

Schäfer, Michael, *Geschichte des Bürgertums*, Köln 2009.

Schirges, Georg. *Die zweite Weltausstellung mit besonderer Berücksichtigung der Deutschen Industrie. Briefe aus Paris*, Frankfurt a.M. 1855.

Schlögel, Karl, *Im Raume lesen wir die Zeit. Über Zivilisationsgeschichte und Geopolitik*, München/Wien 2003.

Sieferle, Rolf Peter, *Fortschrittsfeinde? Opposition gegen Technik und Industrie von der Romantik bis zur Gegenwart*, München 1984.

Siegert, Bernhard, "Kant, Sömmering und die Ästhetik Des Dings. Paralogische Anfangsgründe der Signaltechnologie," *Weimarer* Beiträge, vol. 43, no. 1 (1997), pp. 89–98.

Siemens, Werner von, *Lebenserinnerungen*, Berlin 1893.

Sloterdijk, Peter. *Im Weltinnenraum des Kapitals. Für eine philosophische Theorie der Globalisierung*, Frankfurt a.M. 2006.

Sömmering, Samuel Thomas, "Über einen electrischen Telegraphen," *Denkschriften der Königlichen Akademie der Wissenschaften zu München 1809/10*, München 1811, pp. 401–414.

Susman, Warren. "Ritual Fairs," *Chicago* History, no. 3 (1983), pp. 4–9.

Veredarius, O., *Das Buch von der Weltpost. Entwickelung und Wirken der Post und Telegraphie im Weltverkehr*, Berlin 1885.

Viebahn, Georg W., *Amtlicher Bericht uber die Allgemeine Pariser Ausstellung von Erzeugnissen der Landwirthschaft, des Gewerbefleißes und der schönen Kunst im Jahre 1855*, Berlin 1855.

Wehler, Hans-Ulrich, "Die Zielutopie der 'Bürgerlichen Gesellschaft' und die 'Zivilgesellschaft' heute," in: Lundgreen, Peter (ed.), *Sozial- und Kulturgeschichte des Bürgertums. Eine Bilanz des Bielefelder Sonderforschungsbereich (1986–1997)*, Göttingen 2000, pp. 85–110.

Wenzlhuemer, Roland, "Globalization, Communication and the Concept of Space in Global History," *Historical Social Research*, vol. 35, no. 1 (2010), pp. 19–47.

—, "Less Than No Time: Zum Verhältnis von Telegrafie und Zeit," Geschichte und Gesellschaft, vol. 37, no. 4 (2011), pp. 592–613.

Yang, Daqing, "Telecommunication and the Japanese Empire: A Preleminary Analysis of Telegraphic Traffic," *Historical Social Research*, vol. 35, no. 1 (2010), pp. 66–89

The "Manly Telegrapher": The Fashioning of a Gendered Company Culture in the Eastern and Associated Telegraph Companies

Wendy Gagen

In 1886, as part of an editorial of one of the earliest cable company in-house magazines, *The Monthly Correspondent of the Brazilian Submarine Telegraph Company*, we catch a glimpse of the identity of the "manly telegrapher:"

> [...] a tribute of admiration to the brave men whom the cable companies are proud to claim as their servants all over the world [...] Englishmen have peopled and almost subdued the world by their physical capacity for adapting themselves to every climate; but none the less are those heroic qualities deserving of the highest praise which keep men at their posts under such trying circumstances, as almost everywhere prevail where cables are landed on distant and foreign shores.[1]

This chapter sets out to explore the fashioning of this British identity in the late nineteenth and early twentieth centuries as the numbers of telegraphers grew alongside the cables of the Eastern and Associated Telegraph Companies (hereafter cited as E&ATC) that snaked around the world. [2] Whilst this identity may have been transnational in terms of geographical positioning, it was somewhat national in character, much like E&ATC. By considering the creation of a gendered telegrapher identity within a group of interrelated companies, a more complex understanding of the role of masculinities within colonialism can be uncovered. Rather than seeing imperial masculinity as solely a tool that aided expansion and domination of a particular nation, in the case of the "manly telegrapher," it was fundamentally about the cohesion and success of a global company.

Formed in 1902, the E&ATC was undoubtedly one of the largest and most successful groups of British submarine telecommunication companies

1 *The Monthly Correspondent of the Brazilian Submarine Telegraph Company,* 1884–1887, vol. 3, September 1886, p. 5. PUB//TCOM/48-1-4.

2 Here I am using the term telegrapher to encapsulate a number of land based employments that included electricians, key operators, superintendents and the like. In short, all key staff who were employed to run a telegraph office overseas.

in the world. Of course, the E&ATC's ownership of so many cables that reached vast geographical regions was one of the reasons it held such a prominent position in the world of telegraphy. Importantly, its cables stretched beyond British territorial lands and waters around the globe. Nevertheless, it would be foolish to think that cables alone made such a business so profitable. One key factor in any business is its workforce, and telecommunications was no different. Being truly global meant that its employees were spread across the globe ready to lay, repair, mend and work the cables. Without a team of men to service the cables, the cables would have been redundant, and without a coherent, strong and pervasive company culture, the employees of the E&ATC would not have been a world leader in telecommunications.

This chapter will explore company culture and in particular the fashioning of a telegrapher identity within the E&ATC to highlight the importance of the "manly telegrapher" in creating a successful global company. By looking at the lifestyle of telegraphists, how they were supported and cared for by the company, and how men expressed their life experiences, the shape of the company culture and the routes it took can be studied. This is a complex picture to unpick and it is clear that the gendered E&ATC company culture was multi-stranded, and was both a top down and bottom up creation. It was maintained, created and built upon via company rules, exams, hierarchies, geographical location and architecture and also by sports, leisure activities, telegraphic dynasties, friendships, letters, diaries, the in-house magazine, and most importantly, through shared experiences. However, it would be a rather two-dimensional picture without considering wider factors. These were wrapped up into two inter-linking ideas: the British Empire and dominant notions of masculinity.[3]

It is clear that the pervasive reality and popular notions surrounding the idea of the British Empire and imperialism influenced the way telegraphers saw themselves as men and as workers within "foreign" lands. "Englishness" was a key component in creating the telegraphic man as something that created a bond to the central companies' headquarters in London, but also in relation to the idea of "civilized" behaviour encapsulated in imperial masculinity. This was accompanied by the embracing of the populist ideal of the

3 It would be interesting to consider whether a gendered company culture aided the success of other companies (especially telecommunication companies) that traversed the globe and if they were affected by contemporary, dominant forms of masculinity but such comparisons are beyond the purview of this study.

heroic imperial adventurer and pioneer that was expressed through various avenues of popular culture such as literature, art and education.[4] Such an image and identity was something to aspire to and emulate, but like most ideals it was something that was fundamentally unobtainable outside of the pages of a novel or a fantasy. This gendered fantasy was closely allied with dominant notions of the contemporary manly ideal, in a mirage that became the basis of their identity creation and, which was played out and expressed in their work, their leisure time, and how they portrayed themselves. Whilst company culture, and company and personal identity may have been articulated through such common tropes as empire and hegemonic masculinity, they were not necessarily thoroughly claimed by them. The "manly telegrapher" was able to stand apart from others and recognise a fellow telegrapher from the crowd. This ability was paramount for both the success of the companies and the men as they moved around the globe. Thus, they needed to create an identity that was essentially British or rather "English" that represented the home nation of the companies and one that did not have to necessarily engage with the host nation nor with British empire-building. That is not to suggest that the companies were not part of supporting and maintaining the British Empire, but rather that this gendered identity was not focused on this goal. In this case the "manly telegrapher" and the E&ATC were engaged with the ideas of the imperial adventurer and hegemonic masculinity, whether living within British territories or not, to create pockets of "Englishness" around the globe that would provide company cohesion and made them a force to be reckoned with.

A note of caution needs to be sounded as this company identity was not fully adopted by everyone and there were those who embraced some elements more enthusiastically than others. There were also moments when this identity broke down, and the last section of this chapter will consider the errant, "manly telegrapher." However, even the "misfit" can shed light on the dynamism of this highly successful company identity. There were some who were unable to fit into the mould and were soon dismissed. Others shaped by their experiences of being a telegrapher overseas, took this ideal further. What becomes clear for these men is that they desired a more visceral experience than merely working the cables could deliver. Perhaps the inability to fulfil the colonial mission when the "manly telegraphers" identity was steeped in them lay at the heart of such feelings of dissatisfaction. Our suc-

4 "Englishness" is used here as it becomes much more closely connected with ideas of imperialism and identity than the term British did.

cessful E&ATC employees had to embrace the imperial adventurer fantasy, but just enough. In short, to be a "manly telegrapher" was to be a bit of an action hero, to be a bit of an emblem of imperial Englishness, but first and foremostly, to be a skilled worker and representative of the E&ATC.

The Companies

The E&ATC was formed in 1902, but they had a complex history that led back to the early attempts at laying a transatlantic cable from 1854 on. It was not until 1866 that a successful attempt was made under the Anglo-American Telegraph Company and it is here that John Pender took a more central role. Pender, originally concerned with supporting communications between Britain and India due to his business interests in cotton, was never shy in becoming involved in new enterprises. Thus, he helped to create and form a range of companies connected to submarine telegraphy. The most well-known was the Eastern Telegraph Company. The make-up of the various telegraph companies under the Pender umbrella was extremely complex.[5] Not until the 1870s and the completion of cables to India, Australia and Southeast Asia and the formation of the Eastern Telegraph Company did the Pender empire really take off.

In reality, many of the companies that came to make up the E&ATC were already part of the Pender empire and this amalgamation was a formality.[6] Before the formal merger many of the employees and the company structure either embraced more than one company or could easily slip between companies. The Pender umbrella also included connections to cable manufacturing (Telegraph Construction and Maintenance Company), financial support (Sub-

5 To get a better understanding of just how many companies and cables were laid by the Pender empire the website run by Bill Burns is most useful http://www.atlantic-cable.com/. That said, there is still work to be done locating just how far this empire stretched as family members became involved.

6 Formed in 1872, The Eastern Telegraph Company was created from the Anglo-Mediterranean Telegraph Company, The British-Indian Submarine Telegraph Company, The Falmouth, Gibraltar and Malta Telegraph Company and The Marseilles, Algiers and Malta Telegraph Company. In 1902 the Eastern Telegraph Company merged with The Eastern Extension, Australasia and China Telegraph Company (itself made up of three companies) and a further ten companies to form the E&ATC. History of the Atlantic Cable and Undersea Communications, accessed December 12 2010, http://www.atlantic-cable.com/.

marine Cables Trust), various utility and transport companies and also links to marine salvage and docks. Over time, such involvement may have waxed and waned but it does show the breadth of the Pender family's involvement.[7]

It is a fascinating job trying to tease out such interconnections, which is often hindered by each company having both individual and shared documents and board members. Thus, rules and regulations may pertain to several companies under the E&ATC umbrella, and staff books often remained separate. Likewise Sir James Anderson, initially involved in the first transatlantic attempts, was director of all four companies that came to make up The Eastern Telegraph Company, then Managing Director of that Company, Director of The Brazilian Submarine Telegraph Company, The British, Australian Telegraph Company, The West India and Panama Telegraph Company and the Submarine Cables Trust (there may have been more). In most cases the companies did share many similarities and identities and also a headquarters in London, but a particular company will be highlighted where appropriate. This was also true of the documents created by the men. Many men were employed at various times in different companies under the Pender umbrella, and they remained friends over time. Stories, letters and time were shared between differing company employees and *The Zodiac* (the in-house company magazine), although based within The Eastern Telegraph Company, was written and read by such men.

Creating an "English" Based Company Culture and Identity That Could Traverse the Globe

Remember that Englishmen are known all over the world, and that you will have in your safe keeping the good name of England. I am sure you will show an example to those foreigners in the different countries you are about to visit; that you will let them see what Englishmen are made of; […][8]

7 Various generations of Penders and Denison-Penders played central parts in running the companies. Indeed there was also much cross over between board members in the various companies too. The extent can be seen in the various Boards' Minute Books of the various companies that made up the E&ATC held at the Cable and Wireless Archive such as the Board Minute Books for The Eastern Telegraph Company, vols. 1–16, 1870–1916. DOC/ETC/1/11-26.

8 "The Eastern Telegraph School's Athletic Club Dinner," p. 294.

This speech was made at the annual Eastern Telegraph School's Athletic Club Dinner by Major A.G. Broderick, who was Managing Director of the African Direct Telegraph Company in 1909. Imperialistic in tone, the speech expressed what the company expected from their employees, namely to show "foreigners" the power and "civilized" nature of good "Englishmen." The companies wanted to make sure they employed a particular sort of "English" man.[9] They chose young middle-class men to train into the world of the E&ATC. Some had fathers, brothers or uncles in the service and thus some telegraphic dynasties were ready to pass ideas and identities down through generations.[10] Other young men were taken on after recommendations by senior members of staff. All men went through technical training and they were also medically examined to see whether they were fit to enter service:

> He must be in good health and free from any physical defect of body, and also from any predisposition to constitutional or hereditary disease, epileptic or other fits, or weakness of any kind, and must have been vaccinated within a comparatively recent period.[11]

The need for men to be medically and physically sound as well as technically proficient was of great importance for a company interested in keeping the cables working. Geographically, by the early twentieth century, telegraphers certainly spanned the globe. Some were stationed in bustling metropolises but most were stationed on quiet islands or far-away places. Seemingly, certainly from a British viewpoint, exotic locals such as Jamaica, Rangoon, Accra and Tangier would become home for several years with only infrequent trips back to Britain. Setting up these stations could be nerve-wracking work and men needed to have a healthy constitution to cope with it as Mr Simpson, who was Director-in-Chief of the Indo-European Telegraph Company, recalled:

> In such circumstances it is the case of one Englishman alone with a handful of Persian labourers, drinking brackish water till he can hardly swallow, eating anything that comes his way and sleeping on the ground.[12]

9 Many other men were employed with different ethnic backgrounds, but all the focus of the company culture and telegrapher identity was towards the notion of "Englishness."

10 The companies did employ some non-British telegraphers to train, but it appears that the majority in positions of importance and trust remained under British control. The position that most non-Britishers found themselves employed in was that of servant or laborer.

11 Rules and Regulations. DOC/CW/5/186.

12 "History of the Indo-European Telegraph Department," pp. 243–244.

Just getting to the stations could be straight out of an adventure story. Writing about his time in South America, James Golding wrote in his diary of his arrival in Chile in 1881. Arriving by ship into Valparaiso on 22nd June, he noted:

> Early morning a small steam launch came off for mails. And our ship being in quarantine the men were not allowed on Board. The bags were thrown onto the launch where one man was standing ready with a can full of disinfecting fluid which was poured on everything handed from our ship. [...] The natives are black and copper coloured and appear very good natured and amiable.[13]

In both of the above quotations, the hardships and physical problems soon become apparent. Whether it was infection or difficult surroundings, the employees had to be constitutionally strong. Coupled with Broderick's depiction of the type of man, the companies pushed the idea of imperial manliness to this fore. Susan Kingley Kent outlines some of the characteristics that were attached to this British hegemonic masculine ideal: physicality and action.[14] George Mosse explains this further and considers its origins through a Nietzschean strength of will or "will to power" that raised courage and the ability to face danger and pain as central components of manliness.[15] Whilst these characteristics were important aspects of a hegemonic masculinity that aided colonial expansion, they were also characteristics that the cable companies were looking for when they sought employees to send around the world. This ideal of manliness was common currency in British popular culture of the late nineteenth and early twentieth centuries. Literature, art, education, music, theatre, volunteer movements and newspaper reporting helped to create what John Mackenzie would term popular imperialism which, was more interested in generalised ideas and ideals than just imperialistic events, theories or political entanglements.[16] Here men were hypermasculinised. As the work of Mosher and Serkin have shown, one of the key tenets of hypermasculinity is the connection between danger and excitement, and the work of writers such as H. Rider Haggard, that was steeped in risky thrill seeking, helped to create an image of manliness that longed for heroic adventure in far-off lands.[17]

---

13 Personal Diary of James Golding, DOC/WCATC/5/3.

14 Kingsley-Kent, *Gender and Power in Britain*, p. 226.

15 Mosse, *The Image of Man*, p. 100.

16 MacKenzie, Introduction, pp. 1–3.

17 Mosher and Serkin, Measuring a Macho Personality, pp. 150–152. H. Rider Haggard's two most interesting works would be *She* and *King Solomon's Mines*.

The adventurer style emerges in the writing of many telegraphers whether for private perusal (diaries and letters), or for the consumption of other telegraphers (in-house magazines such as *The Zodiac*). In many ways, this follows Kathleen Canning's work on discourse and experience. She suggests that an experience which becomes a linguistic, or in this case a literary event, is based on established meanings.[18] Here a new personal experience becomes fashioned into a pre-existing mould. In the case of our "manly telegraphers," experience was expressed through the imperial adventurer narrative that was based in popular culture. Often this connection began with new experiences as seen in Golding's thoughts in 1881 of River Plate:

> St Antonia a neighbouring island some 3 or 4 miles distant nearly all tropical fruit can be obtained in great abundance and it is from there the natives obtain their supply of oranges, bananas, coker nuts, and all kinds of vegetation. The scenery is grand in the extreme [...][19]

If we contrast this to a telegrapher writing, "Notes from San Thiango" in 1886, there is less wonder at the exotica of the spectacle and he sees more irritation over cockroaches, mosquitoes and the like, which generally made life uncomfortable and somewhat disease-ridden.[20] Whilst bugs may have made life uncomfortable, they were still markers of something different than everyday life in Britain. As Anne McClintock suggests in her work on H. Rider Haggard, whose existence shifted from something he felt to be mediocre when living in Britain and more extraordinary when overseas, the ability to explore a foreign land meant that men had the chance to be something extraordinary.[21] What is interesting about these narratives is the lack of tension about what constituted an ideal. Roper and Tosh have argued that "masculinity is a site of struggle" and something to be perpetually achieved. It is clear that the fantasy of the "manly telegrapher" was certainly that, but what comes across in these narratives is a very *solid* image and experience of masculinity.[22] This aided its hegemonic status, but it also hints at an interesting quality too. This form of masculinity was both comforting in its clear sighted construction, but fear inducing in its lack of attainability. Working in another country may have been a way to experience reinvention, like it

18 Canning, Feminist History, p. 369.

19 Personal Diary of James Golding, 1880–1881. DOC/WCATC/5/3.

20 "Notes from San Thiango," *The Monthly Correspondent of the Brazilian Submarine Telegraph Company*, 1884–1887. PUB//TCOM/48-1-4, vol. 36, 1886.

21 McClintock, *Imperial Leather*, p. 239.

22 Tosh and Roper, *Manful Assertions*, p. 16.

was for Haggard, but becoming a day-to-day, living embodiment of a story-book imperial adventurer was unlikely.

For our telegraphers, experiencing both the positive and negative sides of living in a new land meant that they could at least attempt to make themselves over in a new image or fantasy even if they were never to be entirely successful. This follows Graham Dawson's assertion of fantasy identities that are lived out or experience rationalised to fit a powerful narrative. As he suggests, "masculinity is lived out in the flesh but fashioned in the imagination."[23] The telegrapher identity was certainly based on the already powerful popular culture surrounding manliness and Empire, but then honed and expressed via narratives shared through the in-house company magazine, *The Zodiac*. In short, this identity was created in relation to other telegraphers. As an imperial fantasy it was also created in relation to its colonial subjects.

R. W. Connell and James W. Messerschmidt, in discussing hegemonic masculinity, cite the importance of hierarchies in creating a dominant form of masculinity.[24] In relation to British imperial masculinity, one way to support white male supremacy was via racial difference. In this period the black "native" was depicted negatively in relation to the white imperialist adventurer. Seen as often lazy, untrustworthy and often bestial, the "conquering west" could justify its actions through the belief in its "civilising presence."[25] We get a hint of that in Broderick's speech showing "foreigners" how to behave and also by Golding's depiction of the locals on his arrival into Valparaiso as being "good natured." Tension is created as Golding adds the term "appear" to this epithet. Can we really trust them (the "natives")? L. E. Caine expresses the same sentiment more extremely when writing of an elephant hunt in Africa in 1909:

> The natives had their Christmas feast — digging into the carcase [of the elephant Caine had shot] with their knives, a hundred of them cutting off chunks wherever they could distinguish a part of the animal from the seething mass of black, naked humanity giving full vent to black savagery, fighting and cutting and pulling at the steaming, bloody carcase, and smeared and dripping with blood themselves.[26]

23 Dawson, The Blond Bedouin, p. 114.

24 Connell and Messerschmidt, Hegemonic Masculinity, p. 831.

25 Said encapsulates this concept best when discussing "a mock chain of being" in *Orientalism* that saw the West at the top and thus a justification of Western control over its colonies. Said, p. 45.

26 Caine, An Elephant Shoot, p. 6727.

Caine was disgusted by the behaviour of the "natives" in contrast to his cool, "civilized," "English" way of killing with a gun. However, these commonly held views could be complicated through experience. Written by a telegrapher stationed in St Jago, calling himself Fluta, he shows us a different angle:

> Blacks here spit on the ground after we pass. This teaches us that a white can only be as good as a black if he behaves himself.[27]

Whilst this may have been written with a touch of wry humour, it nevertheless highlights the reality that many telegraphers were not necessarily revered as a superior race. Nor could they always expect to be stationed within the British colonies and thus part of a politically dominant elite. Such a position could have caused tensions in the creation of the identity of the "manly telegrapher" and subtly alter how he acted in relation to geographical positioning, which in turn would have potentially affected the companies' position. However, they could fall back on the notion of the British imperial adventurer as depicting the highest point of "the chain of being," which placed him above all others. In this sense, the strength of the "manly telegrapher" lay in its engagement with the idea of the imperial adventurer and hegemonic masculinity rather than being part of a concerted effort at British empire-building. These were experiences of an imperial adventurer. They were something new, whether positive or negative. Such experiences were heightened by natural disasters such as earthquakes, dealing with local political squabbles and most keenly solidified by the segregation faced by being a small group of telegraphers placed in an isolated spot. Of course, this was not necessarily the experience of all telegraphers all the time, but the fact that they would be posted to more than one station meant that most of them would have experienced this imperial adventure at some time in their careers, all in the knowledge that they had to live up to the ideals of being a superior and "civilized Englishman."

Whilst being stationed in a new country and facing new people could be an adventure, the facilities that the companies offered were decidedly a home from home. Writing of his time at Midway, A. Beach Comber described what he found:

> There are now four buildings; operating room, superintendent's office, and battery room occupy one building; superintendent's living quarters take another; the balance of the staff live in the third; and the fourth contains mess room, billiard room,

27 *The Monthly Correspondent of the Brazilian Submarine Telegraph Company,* 1884–1887. PUB//TCOM/48-1-4, vol. 47, 8, October 1887.PUB//TCOM/.

library, and kitchen and on the upper, and engineer's workshop, laundry, and cold storage plant on the lower floor.[28]

This particular station had nineteen inhabitants: a superintendent, four operators, a physician, an engineer, a batteryman, six Chinese servants and five Japanese labourers.[29] Different stations had different layouts and varying numbers of staff, but all seemed to hold sleeping quarters, recreational spaces and rooms given over to telegraphy. The sameness of the stations helped to provide stability as they moved from station to station and something familiar for men far away from home. Home comforts were provided in the form of food, leisure activities (library and billiards table) and furniture.[30] This also meant that those living at the station worked and played alongside their fellow telegraphers creating shared moments that aided cohesion and added in, some ways, to their isolation. R.W. Connell's work on masculinities has shown the importance of forms of collectivism and shared experience in supporting hegemonic masculinity. Such homosocial practices would not have been unknown to middle-class men who had experienced all-boy schools. [31]This segregation was not unnoticed by the men and they named themselves "The Exiles."[32]

Some stations were particularly isolated from the rest of the world. Midway, in the Pacific Ocean, held little more than the cable station buildings. This isolation and the realities of it come forward in the very title of an article in *The Zodiac*: "Robinson Crusoes of the Cable: Life on the White Coral Sands of Lonesome Midway Island: Cablemen as Settlers."[33] Abandonment and loneliness comes through in this title, as does the more pioneering or frontiersman ethic of the settler. The men at Midway did attempt to fight against nature to become self-sufficient in vegetables and it is clear that they also brought over animals to keep them stocked in eggs, fresh meat,

28 Beach Comber, Robinson Crusoes, p. 410.

29 Beach Comber, Robinson Crusoes, p. 415.

30 Furniture was often provided or at least a list of requirements were given by the company. List of Furniture and Outfits for Quarters of Superintendent and Assistant Superintendent and Electricians, 1910. DOC/ETC/9/1.

31 Connell, *Masculinities,* p. 106–109

32 Whilst I am unaware of the origin of this term, it is clear they embraced it wholeheartedly through their names of the sports teams and in *The Zodiac*.

33 Beach Comber, Robinson Crusoes, p. 408. There was no mention of any indigenous peoples here by Beach Comber, although the telegraphers did share the island with the American military.

and milk.[34] The extent to which this was an authentic frontiersman experience is not necessarily relevant here. What is important is that the men either believed that this was the case or wanted to depict themselves in this fashion to other telegraphers. This settler spirit was something that the companies wanted to foster and *The Zodiac* was keen to push membership of the Legion of Frontiersmen Club where they would learn skills to protect their cables and routes of communication in difficult times.[35] This pioneering attitude was also useful in times when the cables and telegraphic machinery needed mending and when supplies were low. This spirit of the frontier was very much part of an imperial adventurers' narrative as was its link to the military.

There is some suggestion that stations were run on militaristic lines and with "quasi- regimental mess rules," with senior and junior common rooms that were equivalent to separating the officers from the other ranks.[36] Van Hoven and Horschelmann have suggested that space is an important way to understand both the ideals of masculinity and its performance.[37] This is certainly true in terms of both geographical location in relation to imperialistic dreams and the layout of the stations. Separating spaces into work and leisure, and further into hierarchical and even racial lines assured the structure of the company. Servants and local telegraphists often lived at home, and the lowly telegraphers were kept from the superintendents. Whilst the spaces inside the stations already hinted at the hierarchy between the inhabitants, the companies made this crystal clear through the rules and regulations they set down. These included details about working patterns, medical attention, and appropriate behaviour. Secrecy was paramount, insobriety was problematic and the company had to grant permission to get married, although none of these were unusual edicts in this period. The company was keen to keep control of the behaviour of their workers to secure their reputation and the smooth running of their service, but just as happy to reward them for good behaviour, excellence and innovation through pay, promotion, and support when they needed it.[38] In short, if men were loyal to the company the patriarchal board promised to care for them in their hour of need.

34 Beach Comber, Robinson Crusoes, p. 411.

35 Editorial, Legion of Frontiersmen, p. 290.

36 Personal Papers of James Anderson, 1866 onwards. DOC/ETC/5/58.

37 van Hoven and Horschelmann, Introduction, pp. 2–7.

38 Rules and Regulations. DOC/CW/5/186. Cable and Wireless Archive. The companies often provided extra care and pensions for those who had worked for the companies for some time and even went so far as to support dependents.

It is clear that the station as a home was meant to represent a little bit of England by providing home comforts, but it also epitomised the hierarchy that most had experienced in their family homes or in their schooling. Routine helped to keep a balance and the cables running smoothly with duties being generally arranged in various six hour shifts (6am-12pm and 12pm -6pm etc.).[39] However, these hours, coupled with the original five-year service engagement and leave only to be granted after the first four years of uninterrupted service, meant time to be bored. Isolation and men who did not keep their eye on their work could, as the companies knew, be dangerous for business.[40]

Leisure

The arts were one way to alleviate boredom. Amateur dramatics and musicals were commonplace and the Cable and Wireless Archive at the Porthcurno Telegraph Museum in Cornwall holds many images and programmes that attest to their obsession with entertainment. George Spratt, Assistant Superintendent at Porthcurno, was heavily involved in supporting amateur dramatics. In his diary written during his time in Cornwall between 1871–1909, he wrote often of his experiences with organising theatricals with the telegraphers. On such entry was dated 9th April 1885:

> Went to Penz[ance]; with Maunsell to see Carne and others about arrangements for forthcoming dramatic performance at the Central Hall. Hear the Russians have forced matters in Afghanistan by attacking & defeating the Afghans with heavy loss.[41]

This was performed in a fully functioning theatre created for the use of the men.[42] Telegraphers threw themselves wholeheartedly into this form of entertainment and the archive holds many images of them dressed in outland-

39 Report of work on the Telegraph System in Karachi by Charles Tooley, 1916–1918. DOC/IETC/12/3. It is also clear that their odd working hours helped to keep them isolated from other groups of people.

40 The Halifax and Bermuda Cable Company and Direct West India Company: Rules and Regulations no. 89, 1900. DOC//6/39-350, and Rules and Regulations to be Observed by the Employee of The Eastern Extension Australia and China Telegraph Company, 1906. DOC/CW/5/186.

41 The Spratt Diaries can be viewed at http://www.porthcurno.org.uk/nerve-centre/digital-archive/george-spratt-diaries. Spratt, 9th April 1885, vol. 3, p.4.

42 It is now the Cable Station Inn.

ish costumes. Music also formed part of their many dinners. Annual dinners held in Britain were black tie events and were important moments of cohesion. There were dinners for all the cables companies, sport, anniversaries and conferences. There were lavish menus and company silverware that attest to their love of dining. This was also true of overseas stations where food and drink served as moments to create unity through shared enjoyment and full stomachs. The drink often flowed well into the small hours and it seems that many stations needed little excuse to hold a party whether formal or informal. John Norman, writing in the early twentieth century from St. Vincent, detailed a picnic for 35 that boasted ham and tongue, cold duck, tinned trout, cheese and butter, cocktails, ice-cream and beer. As part of the day they played "rugger" swam in the sea and threw seaweed at each other.[43] This event had all the hallmarks of a school boy's outing, but it was certainly a highlight for Norman.

Another of their main activities was sport. Each station and training school had quite a range of sports on offer: rugby, football, rifle shooting, swimming, golf, etc. Sport could be either formal or informal. Many activities such as swimming, tennis or rowing could be undertaken whenever men were free. The companies also had a large number of teams that played against other employees or local teams (often from ships docked nearby, local police or the armed forces). The training schools helped to foster such interest, and both London and Porthcurno schools had many teams and opportunities to play sport. By 1919, the London branch had a separate sports club for the "Exiles." Alongside paying for outings and other leisure activities, the companies also paid for silver trophies (some were also bought by the men and their families) to be commissioned as well as paying for sports equipment such as sailing boats. Why were the boards so keen to support and even pay for sporting activities?

There were a variety of reasons why sport was a high priority. Firstly, sport was believed to promote sound health. By 1890, 115 gymnasiums existed in Britain, and by 1906, physical culture was booming.[44] One of the most famous men to offer advice on training one's body, whilst showing his own off to the world, was Eugen Sandow. Like many others, *The Zodiac* also offered access to body-building techniques propounded by Sandow through a special correspondence course. These exercises offered an image of the body

43 Handwritten Diary of John Norman, 1919–1930, 25th April 1919. DOC/ETC/5/24/1.
44 Radley, *Illustrated History of Physical Culture*, pp. 29–46.

that fitted into the imperial ideal. The body of the new imperialist was, as stated earlier, one that denoted action and also physicality – exercise could provide that. This rush for fitness was also seen via the rise of Swedish exercises that also became commonplace in schools by 1909.[45] Schools were also keen to support team sporting activities. This was especially driven by the headmaster of Rugby School, Thomas Arnold, whose ethos of promoting team sport as a way to foster ideas of honour, sacrifice, physical strength, discipline, teamwork and endurance for the elite. These ideas were embraced nationwide.[46] This notion placed athleticism over intellectual endeavour and elevated the notion of the will. It was also promulgated to less fortunate boys through clubs such as the scouting movement and through ripping yarns about young, sporty heroes.[47]Arnold was in the business of creating a stream of young men to support and run the British colonies, but the way in which the cable companies and their men embraced sport attests to its significance in their lives and their identities as well.

Depicted as a healthy body creating a sound mind, the connection between the physical (strong and athletic) and the will (courageous, disciplined etc.), was an ideal that certainly strengthened the companies' position and appealed to these middle-class men. Sport was also another medium through which imperial ideology was disseminated. Sport with its rules and regulations were seen as a "civilizing" process.[48] However, most sport documented by the telegraphers was not about imperial domination, but about winning against, and passing time with, other westerners or telegraphers. Whilst there must have been some sport played with "native" teams, *The Zodiac* and the various private documents cited in this study did not discuss this. This may have been due to a belief by telegraphers that such matches were of little importance and certainly not worthy of comment. Sport may have seemed a way to keep "Englishness" alive for the telegraphers in far flung stations and stave off boredom and isolation, rather than part of any clear-cut imperial attempt to extend the empire or "civilize" the "natives." In many ways, it could be seen as a way to keep the telegraphers "civilized." Sport was also commonly understood as a performance of the hypermasculine man where danger, controlled violence and lack of women were important markers of

45 McIntosh, *Physical Education in England since 1800,* p. 158.
46 DeGroot, *British Society in the Era if the Great War*, pp. 32–33.
47 Kantikar, Real True Boys, p. 186.
48 McDevitt, Muscular Catholicism, p. 268.

identity.[49] *The Zodiac* was full of sporting fixtures, trophies, and stories to attest to the centrality of sport to company culture and the telegrapher identity. It was also something supported by both the men and the companies. Another favourite and very English pursuit of the telegrapher was hunting. Men wrote in great detail about their hunting experiences in *The Zodiac* and depicted themselves in photographic collections. These trips could be informal affairs with just a few men taking off into the field or more traditional events with cooks and servants. The stories they told about their trips had an adventurous feel. "Exile" wrote of a hunting trip in India:

> Into the mangroves once more, we struggled through an hours agonising effort in the Turkish-bath heat of the swamped low-lands, and so back to camp.[50]

These stories of mishap and adventure were often written as a first-person narrative that helped give the reader a sense of immediacy and as many had had similar experiences, created a sense of connection too. Spoofs were also written for the men to laugh and identify with. One was written by M.G. Skipper called "The Mighty Hunter," in the form of a humorous interview from Banjoewanjie.

> We found him [a fellow employee, "The Mighty Hunter"] standing amidst a very orgy of slaughter, his face begrimed with powder and his waist garnished with spreading antlers of Mejangang, the heads of lesser deer, and the tusks of the wild babi. The bodies of two large tigers lay couchant at his feet.[51]

Whilst humour was redolent in *The Zodiac* and showed shared knowledge and identity with other telegraphers, hunting seemed to be quite a staple narrative. Such accounts could also be found in *She* by Rider Haggard:

> I had a single barrel sporting Martini with me in the litter...and espying a beautiful fat eland [...] I jumped out of the litter, and proceeded to creep as near him as I could.[52]

Whether the men sought to emulate a writing style or imitate a literary adventure, it is clear that the literary narrative is little different than that of the telegraphers'. Both epitomised the image of the imperial hunter, who was physically agile and eager to make his "civilized" kill. Both also provided moments of tension and danger.

49 Mosher and Serkin, Measuring a Macho Personality, pp. 150–152.

50 Exile, The Four Guns, p. 22.

51 Evans, Up the Mauta River, p. 206.

52 Haggard, *She*, p. 126.

Narratives of danger and incident were quite commonplace in *The Zodiac*. Some were told in humour, but most recounted as tales of derring do situate the narrator right within the heroic mould.

After fighting against the Red Indians and on board the Kearsage when it destroyed the Alabrama, working in a powder factory, sailing round the world, and surviving the San Francisco earthquake, James Winslow received his first injury in Omaha the other day by slipping on a banana peel.[53]

In sending up Mr Winslow, the writer also highlights the type of heroic, dangerous and adventurous life he had lived. Such experiences were very much part of the cableman identity and *The Zodiac* was one place in which they were able to share their stories of heroism and derring do. One such was the Philippine Revolution of 1898. The cable station at Bolinao came under siege as the local inhabitants rose up against the local Spanish colonists. On awaking to a noise at 2am in the morning, T.H. Tolandano wrote:

[...] we were all cool, and no excitement showed in my comrades, who were as collected and as calm as if nothing unusual were happing. We locked and barricaded all windows and doors. Then we overhauled our sporting ammunition, of which we had a fair amount [...] We then quietly waited. I could hear the ticking of the hall clock as distinctly as if it had been a sledge hammer. We could still hear the sounds of the mob, that characteristic clamour, and distant cries of a large body of men, accompanies by occasional shots and yells. Thus we waited, three good friends, ready for any eventuality.[54]

The situation worsened as the Spanish took refuge at the cable station. They were eventually surrounded by the rebels but continued try to maintain and protect the cables. The author did recount the subsequent "native" massacre, but it is the heroism of the men's actions whilst in extreme danger that stands out. Whilst we can argue the accuracy of the experience it is the way the narrative is conveyed that is important. The imperial adventurer had come alive and H. Rider Haggard would have been proud.

There are many such tales in *The Zodiac*, but they can also be found in private documents. James Golding recorded the chaos that followed an earthquake in Peru in 1881:

Peruvians were just outside[...] [and] bang went the rifles and the firing commenced in earnest. The people rushed out of their houses and women crying, and a wild

53 "His First Incident," p. 378.

54 Tolandano, An Episode in Rebellion in the Philippines, pp. 347–353.

scene ensued...The house was soon full of soldiers and we determined to stay in our house occasionally popping our heads outside the door, not daring to expose ourselves for bullets were flying about thick, and with a horrid noise as they rushed through the air.[55]

Within the diaries, photographs, stories, rules and lifestyles of the telegraphers stationed abroad there was a clear construction of a masculine identity that was built through both work and leisure. Both often created through isolated life experiences, where work and leisure were experienced in an often very tight-knit circle of men, and at the heart of their relationship was the job they did. Even when on leave or furlough they spent time together playing sports and in club dinners. Men connected with each other via personal communication but also through *The Zodiac* stories and narratives. They could keep tabs on old friends through transfer lists and retirement notices. "Old boys" kept in the picture through sponsoring trophies and attending dinners. This was not just a bottom up construction that created a sense of belonging, but also a top down ideal that provided financial success.

These accounts might seem to suggest that such men actually lived the life of the imperial adventure, hero and pioneer, but that would confuse reality with fantasy. For one thing, they were not there to blindly extend the British Empire, as their priority was to ensure the success of their company. Secondly, nor did the desires of the company necessarily run parallel with that of the British Government. It is clear that the E&ATC had some influence in the British government and meetings were certainly held between company directors and governmental officials. The companies were called on to help shape issues over cable security and telegraphic censorship from the early twentieth century onwards.[56] The work of Winseck and Pike shows the complex interplay between telegraphy and the British Empire and the E&ATC were shaped by imperial forces and ideas.[57] In many ways, the desires of the British government ran alongside the wants of the company. This relationship could be cordial and certainly be seen to support the economic and defensive interests of Britain and its empire, but the relationship could be more complex and tense than suggested above. One particular issue arose in regards to the employment of men abroad. The government had felt that only British subjects should be employed in skilled positions in the

55 Personal Diary of James Golding, 1880–1881. DOC/WCATC/5/3.

56 Cable Defence Minute Extracts, 1901–1915. DOC/EEACTC/1/52. Private and Confidential Papers Re Censorship etc., 1898, 1912–1926.DOC/ETC/1/4/1 .

57 Winseck and Pike, *Communication and Empire.*

British dominions to protect British interests. This was not fully embraced by the company as employing locals was often a way to appease overseas governments and could be a cheaper option.[58] In reality what suited the government did not always suit the companies and if there was one thing the directors did not forget, it was that they were running a business. Cain and Hopkins' work on British imperialism clearly show that in terms of economics the integration of gentlemanly capitalism and city financiers helped to provide a system of economic expansion. This was partly based on manly ideals of paternalism, hierarchy and honour, and a system that could be exported around the world. Such men found their structure to be highly profitable and services such as finance, and in this case, telegraphy, worked precisely because they followed a set of patterns which could, at times, either run counter to particular governmental dictates or shape future decisions.[59] What is interesting here is that Cain and Hopkins looked towards a particular gendered formation of economic prowess in their research; likewise, when looking at the E&ATC, their business achievement can be linked to the creation and support of the culture of the "manly telegrapher" within their companies. Such an ideal did not necessarily have to be fully actualised, but it had to be something that could be recreated around the world.

Thirdly, they did not always work on British territory. In many ways these realities were of little importance to this identity construction as what this fantasy ideal did was to create an idea of a shared experience that was hypermasculine in nature and built from generalised constructions of a manliness that was born out of popular culture. Thus, the telegraphic imperial adventurer was not necessarily a reality but more of a constructed and favourite shared fantasy between telegraphers and the companies, redolent of contemporary imperial clichés; adventurous, independent, brave, a desire for danger, a twist of militarism and geographically based on the edges of the world. As a shared fantasy, constructed upon notions of hegemonic masculinity and created in imperialistic popular culture, it was certainly prevalent but also intrinsically unrealistic.

58 Foreign Cables Landed on British Territory: British Operating Staff to Consist Exclusively of British Subjects, 1911– 1916. POST/30/1957B.

59 Cain and Hopkins, *British Imperialism*, pp. 17, 46 and 467.

An Imperial Adventurer Too Far

Buried in the very notion of the fantasy imperial adventurer is the idea of action and drama. In reality, such experiences can never be consistent and there were many moments of nothing much happening. John Norman wrote in 1919:

> After nearly five years out here [St Vincent] of continuous service I feel dulled with monotony and ceaseless strain [...] To stay here much longer is beyond human endurance, [...][60]

This private feeling was also echoed by A. Beach Comber who expressed the limitation of relying long-term on leisure as a means of mental and physical distraction.[61] We have seen that what made a successful identity for telegraphers was action and a spirit of adventure and seeking new ways to fulfil that need may have become a problem for some men. John Norman wrote:

> Lately have been realising how well paid we are for the work we do. Nevertheless I am not satisfied of course wanting more, and think I am wasting my time here, might be much more use to my country as a pioneer in one of the Dominions say [...] how one longs to fight against nature which develops ones manhood.[62]

Norman's discussion of his dissatisfaction in light of his non-combative role during wartime highlights the very nature of what it meant to be a "manly telegrapher" – to battle for more both physically in a fight against nature, and mentally through striving to attain manhood. He was not to give up his lucrative and secure job, but did continue to express dissatisfaction in his diary. One individual did succumb to the romanticism of pioneering and joined the gold rush in the Klondike Hills in the nineteenth century. He was fondly discussed in *The Zodiac*.[63] Others succumbed to the boredom and over-indulged in alcohol and the like, whilst some became involved in local crises.

For a few who had served in a particular locale for years, the desire to become involved became too much. One such incident was during the Anglo-Egyptian war of 1882 when a Mr Gibbs became drawn in to the armed conflict alongside the local Egyptians.[64] Often there was a fine line between

60 Handwritten Diary of John Norman, 1919–1930, 9th October 1919. DOC/ETC/5/24/1.

61 Beach Comber, Robinson Crusoes, p. 410.

62 Handwritten Diary of John Norman, 1919–1930, 8th October 1919. DOC/ETC/5/24/1.

63 Editorial, *The Zodiac*, vol. 4. 1910, p.136.

64 Minute Book Index for Eastern Telegraph Company, 1880–1882, vol. 4, under Crisis in Egypt. DOC/ETC/1/33.

becoming caught up in the drama as a company employee and identifying with the country in which a telegrapher was stationed. This type of behaviour was very much part of an imperial fantasy where an individual became involved and took action. Whilst the employees described above could be seen to be the epitome of the imperial adventurer at its zenith, other employees were unable to make the grade.

Roger Horrocks suggests that it is a mistake to see hegemonic masculine ideals as being just about male domination over women and in this case, colonial subjects. He also sees these ideals through the way men rule over each other.[65] In this case, those without heroic blood were not to be part of the collective. These men and boys were soon routed out as unable to do the job. Whether they were unable to perform the work, fit into the system or were too bothersome, they did not remain. The "manly telegrapher" identity created by the companies and the men themselves was a very particular fantasy figure who dominated their discourse.

Conclusion

This chapter has scrutinized the lives and experiences of telegraphers stationed around the world in the late nineteenth and early twentieth centuries to consider how company culture and a telegrapher identity was shaped and promulgated. This identity helped to create successful companies and was a top down and bottom up construction. It also formed the basis of a telegrapher identity that was embedded in the heroic imperial adventurer ideal. It also sought to show that although bound up in imperial masculinity, the "manly telegrapher" was not necessarily always part of a colonial project, but rather should be seen as a way to create company cohesion and to ensure the financial success of the E&ATC. As an imperial fantasy, it created a particularly clear image of hypermasculinity based within popular imperialist notions and dominant ideas about manhood and masculinity. Whilst the question of power relations between "natives" and the English imperialist is not central here, although such notions did shape the telegrapher identity, the affect of the dominant male stereotype is central to this argument.

65 Horrocks, *Male Myths and Icons*, p. 8.

This stereotype, explored in private papers and *The Zodiac* and shaped amongst the workers of the cable companies, created a worldwide bond between these men. It was also supported by the companies to encourage men in sporting and leisure activities to create discipline, teamwork and alleviate boredom. They also offered moments of "English civilization" in the "dark corners" of the world. The companies provided an oasis of "Englishness" in the form of a home and via home comforts. In tandem, rules, regulations, examinations and hierarchies served to consolidate discipline and belonging in outposts far from the company headquarters in London. Indeed, the companies presented their white, middle-class men, and fellow telegraphers offered each other, a chance for something bigger than life in Britain. They provided adventure, danger and heroism. Even if it was shaped in the imagination, the telegraphers had a chance to try to become the "manly telegrapher" and add to the ongoing narrative of that particular imperial hero and make their board members very wealthy.[66]

Bibliography

Beach Comber, A., "Robinson Crusoes of the Cable," *The Zodiac*, vol. 3, 1909, pp. 408–415.

Board Minute Books for The Eastern Telegraph Company, Vols. 1–16, 1870–1916. DOC/ETC/1/11–26, Cable and Wireless Archive.

Cable Defence Minute Extracts, 1901–1915. DOC/EEACTC/1/52. Cable and Wireless Archive.

Cain, P.J., and Hopkins, A.G., *British Imperialism: Innovation and Expansion, 1688–1914*, London 1993.

Caine, Lionel E., "An Elephant Shoot in Central Africa," *The Zodiac*, vol. 3 (1909), p. 6727.

Canning, Kathleen, "Feminist History after the Linguistic Turn: Historicizing Discourse and Experience," *Signs*, vol. 19, no. 2 (1994), pp. 368–404.

Connell, R.W. *Masculinities*, Cambridge, 1995.

Connell, R.W., and Messerschmidt, James W., "Hegemonic Masculinity," *Order and Society*, vol. 19, no. 6, December 2005, pp. 829–859.

66 This chapter is based on some of the work funded by the Arts and Humanities Research Council who funded 'Connecting Cornwall', a project between the University of Exeter and Porthcurno Telegraph Museum, which holds the Cable and Wireless Corporate Archive. http://www.porthcurno.org.uk/page.php?id=95

Dawson, G., "The Blond Bedouin: Lawrence of Arabia, Imperial Adventure and the Imagining of English-British Masculinity," in John Tosh and Michael Roper (eds.), *Manful Assertions: Masculinities Since 1800,* London 1991.

DeGroot, Gerard J., *Blighty: British Society in the Era of the Great War*, Harlow 1996.

Editorial, *The Zodiac*, vol. 4 (1910), p. 136.

Editorial, "Legion of Frontiersmen," *The Zodiac*, vol. 3 (1909), p. 290.

Evans, E.A., "Up the Mauta River from Lorenco Marques," *The Zodiac*, vol. 3, 1909, p. 206.

Exile, "The Four Guns," *The Zodiac*, vol. 6, 1915, p. 22.

Foreign Cables Landed on British Territory: British Operating Staff to Consist Exclusively of British Subjects, 1911- 1916. POST/30/1957B. BT Archives.

Haggard, H. Rider, *She*, St Ives 1887 [1994].

Handwritten Diary of John Norman, 1919–1930, 25th April 1919. DOC/ETC/5/24/1. Cable and Wireless Archive.

"His First Incident," *The Zodiac*, vol. 4 (1910), p. 378.

"History of the Atlantic Cable and Undersea Communications," accessed December 12 2010, http://www.atlantic-cable.com.

"History of the Indo-European Telegraph Department," *The Zodiac*, vol. 20 (1928), pp. 243–244.

Horrocks, Roger, *Male Myths and Icons: Masculinity and Popular Culture*, Basingstoke 1995.

Kantikar, Helen, "Real True Boys: Moulding the Cadets of Imperialisms," in A. Cornwall and N. Lindisfarne (eds.), *Dislocating Masculinity: Comparative Ethnographies,* London 1994, pp. 183–195.

Kingsley-Kent, Susan, *Gender and Power in Britain*, 1640–1990, London 1999.

List of Furniture and Outfits for Quarters of Superintendent and Assistant Superintendent and Electricians, 1910. DOC/ETC/9/1. Cable and Wireless Archive.

MacKenzie, John M., "Introduction," in John M. MacKenzie (ed.), *Imperialism and Popular Culture,* Manchester 1986, pp. 1–3.

McClintock, Anne, *Imperial Leather: Race, Gender and Sexuality in the Colonial Contest*, London 1995.

McDevitt, Patrick F., "Muscular Catholicism: Nationalism, Masculinity and Gaelic Team Sports, 1884–1916," *Gender and History*, vol. 9, no. 2 (1997), pp. 260–285.

McIntosh, Peter, *Physical Education in England since 1800*, London 1968.

Minute Book Index for Eastern Telegraph Company, 1880–1882, vol. 4 under Crisis in Egypt. DOC/ETC/1/33. Cable and Wireless Archive.

Mosher, Donald L., and Serkin, Mark, "Measuring a Macho Personality Constellation," *Journal of Research in Personality*, vol. 18, no. 2 (1984), pp. 150–163.

Mosse, George, *The Image of Man: The Creation of Modern Masculinity*, Oxford 1996.

"Notes from San Thiango," *The Monthly Correspondent of the Brazilian Submarine Telegraph Company*, 1884–1887. PUB//TCOM/48-1-4, vol. 36, 1886, Cable and Wireless Archive.

Personal Diary of James Golding, 1880–1881. DOC/WCATC/5/3. Cable and Wireless Archive.

Personal Papers of James Anderson, 1866 onwards. DOC/ETC/5/58. Cable and Wireless Archive.

Private and Confidential Papers Re Censorship etc., 1898, 1912–1926.DOC/ETC/1/4/1. Cable and Wireless Archive.

Radley, Alan, *The Illustrated History of Physical Culture: Volume One: The Muscular Ideal*, Preston 2001.

Report of work on the Telegraph System in Karachi by Charles Tooley, 1916–1918. DOC/IETC/12/3. Cable and Wireless Archive.

Rules and Regulations to be Observed by the Employee of The Eastern Extension Australia and China Telegraph Company, 1906. DOC/CW/5/186. Cable and Wireless Archive.

Said, Edward W., *Orientalism*, London 1978 [2003].

Spratt, George, "Diary of George Spratt," 9th April 1885, vol. 3, p. 4, accessed October 28 2012, http://www.porthcurno.org.uk/nerve-centre/digital-archive/george-spratt-diaries.

"The Eastern Telegraph School's Athletic Club Dinner," *The Zodiac*, vol. 3 (1909–1910), p. 294.

The Halifax and Bermuda Cable Company and Direct West India Company: Rules and Regulations no. 89, 1900. DOC//6/39-350, Cable and Wireless Archive.

The Monthly Correspondent of the Brazilian Submarine Telegraph Company, 1884–1887. PUB//TCOM/48-1-4, vol. 3, September 1886, p. 5; and vol. 47, p. 8. Cable and Wireless Archive.

Tolandano, T.H., "An Episode in Rebellion in the Philippines," *The Zodiac*, vol. 4 (1910), pp. 347–353.

Tosh, John, and Roper, Michael, *Manful Assertions: Masculinity in Britain since* 1800, London 1991.

van Hoven, Bettina, and Horschelmann, Kathrin, "Introduction: From Geographies of Men to Geographies of Women and Back Again?" in Bettina van Hoven and Kathrin Horschelmann (eds.), *Spaces of Masculinities*, London 2005, pp. 2–7.

Winseck, Dwayne R., and Pike, Robert M., *Communication and Empire: Media, Markets and Globalization, 1860–1930*, London 2007.

Use|News

Telegraphy and the Emergence of an All-India Public Sphere

Michael Mann

Newspapers deeply transformed and shaped the emerging all-India public sphere during the closing decades of the nineteenth century. At the beginning of the twentieth century this all-Indian public sphere, characterised by the emergence of critical newspapers and journals among other things, heavily attacked the British colonial regime as well as the collaborative politics of the Indian National Congress (INC). Founded in 1885, the INC initially lacked any plebicitarian or electoral vote. After World War I, it was turned into a mass organization by Mohandas K. Gandhi who, as a "media-man," drew heavily upon the telegraph as well as newspapers to create a national 'Indian' public sphere. The telegraph played a decisive role in forging public consciousness, patriotism and finally an all-India public sphere as it did in many other emerging nation-states of the world. This article aims to go beyond the established narrative of the telegraph as a means of technological invention connecting people and stresses instead its social and communicative role. Also, the article stresses the growing international and global interconnectivity facilitated-literally-by a growing world-wide cable-web.

To systematically investigate this process of transformation which lasted from the middle of the nineteenth to the beginning of the twentieth century, it is important to mark three main indicators which may provide some empirical evidence for the emergence of that new subcontinental public sphere (hereafter called the "national Indian" public sphere). Firstly, a growing number of newspapers, magazines and journals began to appear in vernacular languages as well as in English from around 1870. In Bengal, between 1818 and 1868, more than 200 periodicals were founded. Although more than 120 of them, that is over fifty percent, shut down within one to four years during the closing decades of the nineteenth century, the number of Bengal newspapers and magazines started to grow again reaching more than 250 by the turn of the twentieth century.[1] In Madras many weekly newspapers turned into dailies during the 1880s.[2] In Bombay no less than 51 newspapers in South Asian languages were published at the beginning of the 1880s.[3] In Lakhnau, north India, the number of printed periodicals rose sharply by the end of the 1860s.[4] In 1875 there were 478 newspapers in British India, and by 1911 their number had increased to 658, marking an increase of round about 30 percent.[5] The growing number of newspapers indicates the growing interest in public and civic matters. It also expressed a growing patriotism and nationalism in historical regions of South Asia like Bengal, Hindustan, Maharashtra or Tamiland.

A second indicator for the emergence of an "all-India" public sphere is the change of the focal subjects within newspapers and periodicals. The *Kayastha Samachar* of Allahabad, for instance, was first published as an Urdu monthly in 1873. It was converted into an English monthly in 1899 and was named the *Hindustan Review and Kayasta Samachar* in 1903 and, finally, the *Hindustan Review* in 1905. By 1904, it had become India's largest monthly periodical. The change of names indicates the transformation of contents, which included the alteration of the position of rubrics and articles. Originally, the *Kayastha Samachar* was the paper of the kāyastha "caste" of scribes dealing almost exclusively with "caste" matters.[6] In 1905, however, specifically "caste" matters were relegated to the rear section of the newspaper and in 1915, this rubric was covered only in some of the pages. During this time, the front section of the journal increasingly began publishing material such as

1 Numbers computed according to the list of Bengali newspapers, journals and magazines given in Chakraborti, *The Bengali Press, 1818–1868*, Appendix: "A Complete List of Bengali Newspapers and Periodicals (1818–1868)", 189–228. For a similar but extensively annotated list for the English periodicals, see Chanda, *History of the English Press in Bengal, 1780 to 1857*.

2 Sadasivan, *The Growth of Public Opinion in the Madras Presidency*, p. 61.

3 Kidambi, *The Making of an Indian Metropolis*, p. 165.

4 Stark, *An Empire of Books*, pp. 64–72.

5 Paul, *Public Opinion and British Rule*, p. 5.

6 Kāyastha or Kayasth is a Hindu caste in South Asia, whose place inside the four varnas or main castes is disputed. According to mythology, Chitragupta, Lord Bramha's 17th creation, is believed to be created from Lord Bramha's mind and soul. Chitragupta decided that his lineage would wield the pen (quill) as its weapon rather than the sword. Legend says that he is employed by Yama, the God of Death, to maintain records about every human in Mrityulok or the mortal world. Hence Kāyastha became the caste of scribes. Modern day Kayasthas probably number around 5–7 million subdivided into 12 main sub-castes.

the INC's politics, legal, educational and scientific articles as well as literary reviews. In short, the journal turned into a bourgeois periodical providing, as the contemporary Lahore *Tribune* characterised the *Kayastha Samachar* in 1900, "a general review of all matters of interest to the nation and humanity at large."[7] The journal thus evolved from a mouthpiece of kāyastha – "caste consciousness" into an interregional organ of the emerging all-Indian public sphere within a quarter of a century.

A third indicator would be the changing position of telegraphic news within such periodical press products and the network of correspondents providing newspapers, journals and magazines with telegraphic news. The widely-read *Awadh Akbar* of the Naval Kishore Press at Lakhnau, for example, a subcontinental correspondence system maintaining offices in all major cities to receive as much news as possible from British India. However, international telegraphic news remained too expensive for Indian newspapers for some time to come. It was only in 1900, decades after Reuters had opened its correspondence office in Bombay in 1866 that the first Indian-owned newspaper, *The Bengalee,* subscribed to Reuters.[8] Since copyright on telegraphic news did not exist, many newspapers which could not afford to subscribe to Reuters' services often copied the telegraphic news previously published in the British owned dailies of British India. This situation was by no means restricted to India as, globally seen, there still existed no copy-right on articles which is why many texts could be copied or taken as "quarries" for a journalist's own news-article.[9]

These possible indicators for an "all-India" public sphere may serve as empirical evidence for the strong influence of the telegraph on the emerging "national" public sphere on the subcontinent. In fact, it seems very likely that the telegraph facilitated this emergence. The story of the telegraph has, however, so far been told as an exclusively technological success story or a unique technological revolution only comparable to the revolutionary consequences of the invention of the printing press. Nevertheless, rather than speaking of revolutions taking place within a few years, it seems more appropriate to speak of "media evolutions" lasting several decades.[10] Many

7 Carroll, Kayastha Samachar, pp. 280–92, quote 285.

8 Banerjea, *A Nation in Making*, p. 157. See also the chapter of Amelia Bonea: "All the news that's fit to print?" Reuter's telegraphic news service in colonial India, in this book.

9 Cf. Amelia Bonea's article in this volume.

10 Edgerton, *The Shock of the Old. Technology and Global History since 1900*, chs. 1 and 2 passim.

histories of the telegraph also portray it as an integral part of the western understanding of modernity, efficiency, rationality, progress and so forth. This is probably the most striking example of a claimed success story according to which the "West" gained power over the "rest" of the globe.[11] However, this success story overlooks the impact of the telegraph on the so-called peripheries of global communication.

Taking a closer look at this story, three points become evident. First, the success story focuses on submarine cables connecting and thereby strengthening the trans-Atlantic relationship between Great Britain and the USA. Girdling the globe with wires within a few decades signifies the importance as well as dimension of the speedily expanding insulation and underwater cable industries.[12] These industries, along with the telegraph companies, were not controlled by British capital and politics, as it was maintained until recently, but regulated by transnationally operating cartels.[13] This, as several other incidents, among them trade, migration and communication, hints at an early phase of globalization starting in the middle of the nineteenth century. Contrary to the established master-narrative of capitalism expanding from Europe conquering the world recent research has emphasized that globalization from the middle of the nineteenth century is about qualitative changes with respect to interconnectivities and entanglements.[14]

The second component of the master narrative on modern communication technology was and still is the measuring and counting of cable mileage, followed by geopolitical concerns and questions of national-cum-imperial control.[15] Thirdly, the terrestrial telegraph network is hardly ever considered. The telegraph has perhaps not appeared as particularly spectacular, especially in contrast to the visible and audible locomotives, probably because the telegraph could be easily constructed and worked (almost) silently. There are a few studies on the USA, however, that point out the role of the telegraph as

11 A contemporary perception of this "success story" is provided by George Sauer, *The Telegraph in Europe*. *Vide* also Thompson, *Wiring a Continent*. For a recent example of the success story *vide* Standage, *The Victorian Internet*.

12 Bright, *Submarine Telegraphs*.

13 Winseck and Pike, *Communication and Empire*.

14 Matthias Middell and Katja Naumann, Global history and the spatial turn: from the impact of area studies to the study of critical junctures of globalization.

15 Headrick, *The Invisible Weapon*; Hugill, *Global Communications Since 1844*; Hills, *The Struggle for Control of Global Communication*; McMahon, *Global Control: Information Technology and Globalization since 1845*; Huurdeman, *The Worldwide History of Telecommunications*.

a means of conquering and connecting the North American continent; they highlight not only the expansionist but integrative aspects of the telegraph.[16]

Outside the trans-Atlantic sphere, "modern" means of telecommunication seem to have been non-existent. However, a couple of rather descriptive studies exist on the Brazilian and the British Indian telegraphic network that document the construction of the telegraph network and its operation and, at the same time, point to large telegraphic networks outside Europe and the USA.[17] Yet in these studies, as elsewhere, numbers and statistical data are omnipresent following the trodden path of measurable progress. So far some attention has been paid to the social consequences of the telegraph, in particular to the impact of the telegraph on newspapers.[18] With respect to British India, a couple of recent articles demonstrate the social implications of the telegraph. The strike of the Indian telegraph personnel in 1908 is a particularly lucid example of the social history of the telegraph.[19] As a contribution to the telegraph's social and cultural history, the present article will shed some light on the emergence of South Asian patriotism and the emergence of an all-India public sphere at the turn of the twentieth century. The main argument is that the telegraph, as in other parts of the world, deeply influenced existing public spheres and facilitated their transformation into or the creation of a national public sphere.

In many parts of the world there existed, or came into existence, various forms of public spheres in the eighteenth and nineteenth centuries. They were by no means restricted to the kind of *Öffentlichkeit* (lit.: publicness) which Jürgen Habermas has claimed to be elemental for the Enlightenment of western Europe and subsequently for the emergence of the bourgeois-liberal nation state. *Öffentlichkeit* was one of many forms of public spheres. As has been demonstrated in the responses to Habermas's seminal essay on

16 Field, The magic telegraph, price and quantity data, and the new management of capital, esp. pp. 406–9.

17 Berthold, *History of the Telephone and Telegraph in Brazil, 1851–1921*); Shridharani, *Story of the Indian Telegraphs: A Century of Progress*; Gorman, Sir William O'Shaughnessy, Lord Dalhousie and the establishment of the telegraphic system in India.

18 Lubrano, *How Technology Innovation Caused Social Change* and on the telephone the hitherto unique study of Fischer, *America Calling. A Social History of the Telephone to 1940*. For Germany and Europe *vide* Wilke, *Nachrichtenauswahl und Medienrealität in vier Jahrhunderten*.

19 Choudhury, India's first virtual community and the Telegraph General Strike of 1908, id., Treason of the clerks.

Öffentlichkeit/public sphere,[20] similar forms of *Öffentlichkeit* also existed in South Asia.[21] Meanwhile the concept has been re-considered and has taken on a more elaborate character.[22] The following subchapter will give a short overview of these multiple public spheres, particularly in northern India (Hindustan) and Bengal. The point here is that due to the influence of the telegraph, these domains of *Öffentlichkeit* were extended and intensified and, at the same time, partially transformed into an all-India *Öffentlichkeit*.

The Many Worlds of Public Participation and *Öffentlichkeiten* from the Late Eighteenth to the Second Half of the Nineteenth Century

Before 1880, some kind of rudimentary public spheres existed within British India as well as within its preceding regime, the Mughal Empire. Whereas performative public spheres had probably existed all over South Asia since the beginning of humankind, as was the case in most other parts of the world inhabited by humans and organised in societies, *Öffentlichkeit*, a specific elitist form of the public sphere, first came into being during the eighteenth century. At that time, the Mughal Empire experienced a process of substantial transformation from a centrally organized early-modern state to a decentralized union of semi-autonomous territorial states under the legitimizing umbrella of a growing Mughal myth.[23] During this time *Öffentlichkeit* came into existence as an elitist discourse on the role of the state and began playing its role of criticizing the organs and organisations of the emerging modern (territorializing) state. This transformation is mirrored by various developing urban-based public spheres.[24]

Events and results of this transformative process were vividly debated within and among political circles of the Mughal Empire. In particular, the Mughal elite of Delhi discussed the consequences of being placed under the

20 Habermas, The public sphere; id., *The Structural Transformation of the Public Sphere.*

21 Freitag, Enactments of Ram's story.

22 Crossley (ed.), *After Habermas.*

23 Mann, *Geschichte Südasiens 1500 bis heute*, pp. 24, 39, 66, 83–4.

24 For reasons of space, this article concentrates on northern India (Hindustan). Public spheres also existed in Calcutta and Haiderabad. It needs much more academic scrutiny to establish the public sphere particularly for south Indian regions/states.

"protection" of a non-Muslim power after Mughal Shah Alam II (r. 1759–1806) vested the (Hindu) Maratha-Raja of Sindhia with the Empire-Provinces of Agra and Delhi in 1784. The debate gained particular momentum after British troops occupied Delhi in 1803 establishing British (Christian) power in north India. Yet it was not only the political composition of the Mughal Empire that was in transition; its most influential elite, the "learned" *ulema* were also experiencing changes.[25] The debates taking place during this time of transformation must be analysed in the wider context of *ijtihad*, meaning "independent reasoning." *Ijtihad* is the basic principle of discourse within the *ulema*. Originally *ijtihad* referred to legal questions, yet according to changing circumstances its connotation expanded to include political-cum-societal problems.[26]

Emerging new urban commercial classes of merchants and traders demanded social reforms and soon looked for political participation. The growth of these new classes undermined the *ulema's* monopoly of education.[27] Further important players in this new discourse were the members of the Mughlai administrative elite. Within the emerging territorial states of Awadh and Bengal, for example, this elite formed a new class named *khanazadgi* that acted as a self-legitimizing custodian of the (Mughal) realm and, at the same time, as a keeper of its moral values. As learned men, the *khanazadg* class also acted as historiographers critically commenting on the various revolutions (*inqilab*) happening in the Mughal Empire. Public comments on political matters within a contemporary historical context is another indication of a change in South Asia's historiography which must be regarded as an important part of the emerging public spheres of Awadh (Lakhnau) and Bengal (Murshidabad-Dhaka).[28]

Of course, this gradually growing public sphere consisted of more than the self-transforming *ulema* and the awareness of the *khanazadgi* class. Public debates occurred at different places such as official court-darbars, the premises of mosques and madrasas, as well as in Sufi-order houses and private houses.[29] In Delhi and other urban centres of Hindustan like Awadhi Lakhnau, trading activities and merchant wealth also changed the pattern

25 Metcalf, *Islamic Revival in British India*.

26 Reetz, *Islam in the Public Sphere*.

27 Malik, Reformen in Südasien – das 18. Jahrhundert, esp. pp. 74–6. Cf. Reetz, *Islam in the Public Sphere*, pp. 26–9.

28 Chatterjee, History as self-representation.

29 Bayly, *Empire and Information*.

of patronage initiating a shift of patronage from a hitherto exclusively Indo-Persian courtly elite to merchants, bankers and traders who sponsored the construction of temples and mosques as well as the proliferation of Urdu as a language of literature. Famous and wealthy courtesans often maintained literary salons as places of convention and performance.[30] In Delhi and Lakhnau, Urdu, by no means a standardized language at that point, was turned into the medium of literati, intellectual and political debate.[31] Discussions on religious as well as social reforms took place at Delhi and Lakhnau, often opposing Mughal court politics.[32] Now, meeting places also included libraries, hamams and coffee-houses. Coffee-houses in particular reminded some eighteenth-century Indian travellers to Europe of their own places of public debating.[33]

In the 1870s, a double movement transformed Indian public spheres. The first movement spilled over the boundaries between languages and communities thereby creating what may be called an all-Indian public sphere and *Öffentlichkeit*. The second movement created, in the long run, strictly separated and supposedly fundamentally different communities of Hindus and Muslims. In the final days of the colonial regime, the latter limited the potential of an all-India public sphere/*Öffentlichkeit* and, ultimately, facilitated the partition of British India in 1947. At the beginning of this progress, however, stood the somewhat late start of Indian-owned printing press products, be that newspapers, magazines, pamphlets or books. Yet this late coming also indicates that, despite the absence of widespread use of printing presses, news had still been distributed with appropriate efficiency to a large number of people within a time regarded as sufficient up to the middle of the nineteenth century. Personal and institutional letter writing as well as postering and public meetings were still the substantial means of South Asia's well-functioning public spheres.[34] Soon, however, printed newspapers

30 Freitag, Enactments of Ram's story, pp. 81–3. Bayly, Town building in North India, 1780–1830.

31 Naim and Petievich, Urdu in Lucknow/Lucknow in Urdu.

32 For the Awadhi opposition at Lakhnau *vide* Cole, *The Roots of North Indian Shi'ism in Iran and Iraq*, ch. 4, pp. 92–119.

33 Malik, Reformen in Südasien.

34 Bayly, *Empire and Information*, pp. 200–3. For the literacy of the Awadhi elite *vide* Sprenger, *A Catalogue of the Arabic, Persian and Hindústány Manuscripts*. The volumes disappeared, vol. 1 was reprinted by Biblio-Verlag, Osnabrück 1979. The Awadhi readership comprised official clerks and officers, noblemen and coutesans. Yet also the middle and lower strata of the population like druggists, serving women, scribes, even dancing wom-

and journals in vernacular South Asian languages induced a process of fundamental transformation extending as well as changing the subcontinent's many public spheres.

Telegraphy and its Influence on Indian Public Spheres

As indicated in the introduction to this chapter, telegraphy had an immense impact on the layout and organization of newspapers. Since telegraphically transmitted information attracted a great readership, the *Awadh Akbar*, for example, soon placed telegraphic news under a separate heading. By 1878, the rubric "Telegraphic News" had moved to the front section of the paper, whereas local (Lakhnau) news and advertisements were relegated to the rear part of the paper.[35] A similar yet also differing process can be observed in other Indian-owned newspapers when "international" telegraphically transmitted news, which means that news from within and sometimes also without the British Empire, were relegated to the back pages of the *Amrita Bazar Patrika* of Calcutta and, at the same time, local, that is "national" telegraphically transmitted news moved to the front pages.[36] Apparently the telegraph deeply influenced the organisation of rubrics and presentation of articles. Nevertheless, political developments within British India and the political bias of a newspaper's owner also played important roles in the positioning of rubrics and articles.

To facilitate speedy transmission of news and information and to present them as "current news," journalists were not any more simply personal reporters and private correspondents but were becoming professionals. They turned into critical and investigative, publicly acting journalists occupying a rather independent role in the public sphere. Yet, only in a few countries did journalists and publishers become independent makers of opinion. In most countries, the press and as a consequence public opinion was

en, musicians and fakirs from a rather poor social background belonged to that readership: Mantl, *Aloys Sprenger.* Chaghatai, Dr Aloys Sprenger and the Delhi College.

35 Stark, *An Empire of Books*, pp. 77–9.

36 The fact is illustrated by the development of *The Times* (of London). Telegraphically transmitted international news caused such a dramatic re-orientation of the press that, for instance, the paper allocated more of its front page to foreign news in the 1890s than it does today, Winseck and Pike, *Communication and Empire*, p. 3.

severely hampered by press acts.[37] This is in particular true for British India where, after the Vernacular Press Act of 1878, an independent and investigative press emerged only slowly. Only in the USA and partially also in Britain, journalists, reporters and publicists became the *de facto* creators, bearers and, in countries like the USA guardians of public opinion after newspapers changed from monthlies into weeklies and soon after into dailies due to telegraphic transmission.[38] As Stuart Hall has pointed out "[t]he media [did] not simply report events which [were] 'naturally' newsworthy *in themselves*. 'News' [was] the end-product of a complex process which [began] with a systematic sorting and selecting of events and topics according to a socially constructed set of categories."[39]

In British India, the Vernacular Press Act of 1878 tried to curb the growing influence of the indigenous (and local language) press, because the colonial government understood itself as the sole guardian of what it regarded as public opinion. Despite this claim of the colonial state, various local and regional public opinion and public spheres still continued to thrive and to expand; most importantly, an all-India public sphere gradually emerged during the last decade of the nineteenth century. This is not only reflected by the growing number of newspapers and journals, their regular appearance and increased frequency, but also by a change in organization, layout and presentation of newspapers. Newspapers began to reconceive their task as guardians and megaphones of the national movement. In this context it seems noteworthy that particularly Reuters had a deep influence on this transformation since Reuters' agents started to package their product by attempting to make both the form and the content of news more attractive. For example, they gave preference to sensational events, such as wars, over commercial information which indicates that such news was aimed at a larger readership than just the merchant community of Bombay and other metropolises of British India.[40]

Among the early British-owned newspapers which used the telegraph are the *Delhi Gazette* and the *Lahore Chronicle*. Due to the exorbitant prices for telegrams, Indian-owned newspapers could only afford telegraphic news from Reuters at the turn of the twentieth century. Even the basically Indian-owned Associated Press of India, founded in 1899, was fairly expen-

37 Cf. Wilke, *Nachrichtenauswahl und Medienrealität in vier Jahrhunderten*, pp. 45–54.
38 Hartley, *Understanding News*. Galtung and Holmboe Ruge, The structure of foreign news.
39 Hall, quoted in Fowler, *Language in the News*, p. 12.
40 Bonea, The medium and its message, p. 180.

sive.[41] But telegraphic news copied from British owned newspapers influenced and determined the outer appearance as well as the inner organization of many Indian-owned newspapers. Two newspapers may serve as examples to demonstrate the changes in reporting as well as the layout of newspapers after the emergence of in the wake of telegraphically obtained information. The *Delhi Gazette*, on the one hand, stands for the English-owned press in British India with an almost exclusively British readership. The *Amrita Bazar Patrika* (Calcutta), on the other hand, was a Bengali owned newspaper addressing English-reading customers.

However, this also indicates that two public spheres existed in British India. First, the British public sphere which, I argue, was transformed into an "imperial public sphere," as the *Delhi Gazette* shows. Second, the *Amrita Bazar Patrika* represents a growing national Bengal and all-India *Öffentlichkeit* concerned with matters of "national" importance. The analysis of the two papers will show how telegraphy, like in Europe and the US, substantially changed the form and content of Indian newspapers and, at the same time, induced developments apparently independent of any standards of the "western press." My core argument is that it was the telegraph which transformed Indian newspapers – as it did in other parts of the world – into a medium decisively helping to forge the Indian nation.[42]

The Delhi Gazette (Delhi)

The *Delhi Gazette* appeared as a weekly for the first time in 1837. Between 1841 and 1857 it came out twice a week, mostly comprising eight pages. As already mentioned, the paper primarily addressed an English speaking/reading readership, basically the British living in Delhi. Similar to present-day advertisement-papers distributed freely to individual households, the front page consisted of announcements, advertisements for books, furniture and cosmetics like soaps and oils, commercial (bank and company) advertisements, rooms and houses to let and news on the latest fashion products from

41 In 1913 API was forced to fuse with the Indian News Agency (INA) now named Eastern News Agency (ENA), cf. Israel, Communications and Power, pp. 113–5.

42 The cultural aspect of telegraphy and the print media in the forging and forming of nations and nation states has been almost completely neglected by authors of "classical" works like Anderson, *Imagined Communities.* and Hobsbawm, *Nations and Nationalism since 1780.*

Paris. These advertisements also indicate the readerships' social class, i.e. the colonial ruling elite including the wives of civilians and military officers. The second page contained news from the "Military Establishments" indicating the special status and particular importance of the military and, more generally, the fundamental character of colonial rule. News from and about South Asia were printed on page three, followed by "Shipping News," that is news on ships' passages, arrivals, disembarking of wares and goods as well as their sales. "Domestic Occurrences" listed births and deaths, followed by "Local News" of Delhi and its environs. The last two pages, in many cases just the last page, titled "Literary Gazette," published extracts from books and other literary products. The first change in this arrangement of rubrics and layout was made in 1850 when advertisements were gradually shifted to page two.[43]

From 1877 onwards the *Delhi Gazette* was turned into a daily. Various advertisements of banks and insurance companies were now printed on page one, indicating the prominent position of the financial interests as an important part of the readership or what the editors regarded of particular interest of their readers. News of sporting events were published on page five, followed by telegraphic news on page six. They covered only a third of the last of a total of four columns. Telegraphic news was restricted to short notes transmitted by Reuters referring to events in European and American metropolises. Such news comprised political and social information, sometimes also rather banal affairs reminding one of the simultaneously emerging British and American "penny-press." In some instances, the latest telegraphic data from banks and stock exchanges were also printed again pointing towards the importance of Delhi's financial and commercial class. It is obvious that news from Reuters originated from those cities where the news agency ran offices.[44]

A major transformation of the inner and outer form of the paper occurred in 1878 when the telegraphic news was shifted from page six to the front page (which only then became the "front page") labeled as a new rubric: "Telegraphic News." In this way, international news was turned into news of everybody's interest and of general importance even if such news took the form of fairly short articles consisting of only a few lines. Such news was, again, supplemented by international exchange courses and additional commercial data. It is not possible to tell whether the brief political items are printed versions of telegrams transmitted by Reuters or whether they are

43 *Delhi Gazette*, 1837–41, 1847–57 [BL: IOR]

44 *Delhi Gazette*, Thursday, February 1, 1877, Friday, 2. February 1877 and passim 1877–1879. [BL: IOR]

short articles written in a telegraphic style to make them look like telegrams to attract special attention as the latest or current news. In any case, international news did receive special attention and henceforth obtained a prominent place on the newspaper's first page. This included the presentation of the news either authentically as telegrams or written in telegraphic style.[45]

Beginning with the year 1880, the paper's table of contents included "Telegrams" with the special note on "Latest News from the Press Commissioner." Telegraphic news was printed on page one and contained general news from British India as well as news set against the contemporary political background, such as from the Afghan theatre of war. Additionally, news from other parts of the British Empire, for instance, from South Africa – events such as Krüger's election as president of the Boerish Volksraad or the British governor's announcement that Transvaal would be given the status of a crown colony – belonged to what was now regarded by journalists and publishers as current news.[46] Apart from "empire news," information covered diplomatic affairs from European metropoles (and capitals) like Istanbul and St. Petersburg. News gradually began covering global affairs articles ranging from political and economic to cultural matters.[47] At the same time, telegraphic news was not restricted to short notes, but also included short comments of up to one column in length.[48] In spite of these minor innovations, the overall arrangement of the *Delhi Gazette* as well as its layout was not changed until the end of the century.

Taking a look at the news, reports and commentaries printed in the *Delhi Gazette* until the beginning of the twentieth century, it becomes clear that the paper catered to a British readership in South Asia that was a mirror-image of Great Britain's society. Apart from the addresses of the advertisements, the newspaper's style represented the values of the liberal-bourgeois middle class as it emerged in Great Britain during the second half of the nineteenth century. This class was the seedbed and recruitment section for the Indian Civil Service's as well as the military's personnel. Constitutional questions and political movements were the main interests of this new class, interests that were dominated, however, by a royalist rather than republican tenor. This also explains the amount of news from European princely families and courts in newspapers like the *Delhi Gazette* which was in stark contrast to most of

45 *Delhi Gazette*, Monday, November 4, 1878. [BL: IOR]
46 *Delhi Gazette*, vol. X, Friday, January 2, 1880. [BL: IOR]
47 *Delhi Gazette*, vol. XIX, December 30, 1889. [BL: IOR]
48 *Delhi Gazette*, vol. X, Saturday, January 3, 1880. [BL: IOR]

the "Indian" newspapers that, for good reason, showed rather limited interest in such news.[49] Apart from the royalist mindset, the "imperial idea" was of special interest to the British readership which is why an increasing number of articles were printed covering other parts of the empire.[50] The British "gentleman" was also a good "sportsman," and therefore telegraphic news from major sports events in Great Britain were also printed from time to time.[51]

Strikingly (as well as significantly), the *Delhi Gazette* hardly ever reported on "Indian" affairs. In this respect the *Delhi Gazette* may be seen as *pars pro toto* for all British-owned newspapers in British India. The colony's British public sphere seldom considered, if at all, political movements within British India, despite their importance. At the beginning of the twentieth century, the British seemed again to lose contact to their colonial subjects, just as they had in the decade preceding the Great Rebellion (1857–1859).[52] This disconnect is reflected in the British notion of a "public sphere" in British India in which, by definition, no "Indian public sphere" could exist. The British regarded themselves as the sole representatives of the political and, according to their understanding, democratic domain. In contrast, "India's" societies and political systems, due to their lamentable history, were still regarded as remnants of basically despotic realms where, again by definition, no "public sphere" could have come into existence.[53] Despite all efforts to civilize India's populace, her inhabitants were seen, even at the turn of the twentieth century, as unable (and therefore not allowed) to critically comment on the colonial government's political matters.

This attitude overtly contradicted the latter's press legislation, particularly the Vernacular Press Act of 1878 for controlling and suppressing any sort of critical opinion, which was evidence of both the existence of an "Indian public opinion" as well as of the colonial government's recognition of that press as a social force. Yet this contradiction seems to have been an inherent part of the colonial regime, one of its constituent inconsistencies – which may be true for all modern colonial regimes.[54] To control the emergent

49 The *Amrita Bazar Patrika* is a remarkable example for this position.

50 Read, *The Power of News*, p. 30.

51 Since 1922 the "Derby" results were wirelessly broadcasted from Oxford to Karachi, from where news were transmitted by telegraph, cf. Annual Report on the Posts and Telegraphs of India for the Year 1922–23. Delhi: Government Central Press 1923, p. 50. [BL: IOR]

52 This is also indicated by Bayly, Empire and information in India.

53 This vivid public sphere *vide* Paul, *Public Opinion and British Rule.*

54 The press legislation is dealt with by Barns, *The Indian Press* and Paul, *Public Opinion and British Rule.*

Bengal, Hindustani or even "Indian" "public sphere" or, at least, to check the activities of the vernacular press, the British government of the Panjab surveyed the native newspapers of the Panjab Province through regular reports from its press. Beginning with 1896, the collected information was published in the *Punjab Native Newspaper Reports, 1896–1924.*[55] The main purpose of the information gathering was to detect and pursue seditious articles or articles which could cause sedition. As can be easily imagined, the instrument was far from being efficient, let alone successful.

The Amrita Bazar Patrika (Calcutta)

With regard to content, layout and overall arrangement, the *Amrita Bazar Patrika* is a fine example for a slightly, but decisively, different development of an "Indian" newspaper in British India. In 1868, the *Amrita Bazar Patrika* appeared for the first time as a Bengali weekly. After one year, some articles were also published in English. From 1875 onwards, an English version of the paper was distributed beyond Bengal, obviously addressing an English speaking "Indian" readership. To circumvent the stipulations of the Vernacular Press Act, the hitherto bilingual paper was, overnight, turned into an exclusively English newspaper.[56] In 1891, the *Amrita Bazar Patrika* was converted into a regular daily newspaper.[57] Apart from conventional commercials and advertisements as well as announcements of events, a separate rubric, "Scraps and Comments," printed news based on telegraphically transmitted information and short telegrams sent within British India. From time to time, telegraphic news copied from other newspapers was also reprinted. During 1890, this rubric grew from half a column to one and a half columns. A second rubric also based on telegrams, called "Telegraphic Summary" and placed on page three, printed exclusively international news.[58]

This layout-order changed at the beginning of the twentieth century. During the Russo-Japanese war of 1904–5, the *Amrita Bazar Patrika* re-

55 *Punjab Native Newspaper Reports, 1896–1924. Selections from the Native Newspapers Published in the Punjab* and *Punjab Native Newspaper Reports, 1896–1924. Punjab Press Abstract* [BL: IOR]

56 Iyer, *The Indian Press*, p. 4.

57 A short overview covering the first three decades of the *Amrita Bazar Patrika* is provided by http://www.economy-point.org/a/amrita-bazar-patrika.html.

58 *Amrita Bazar Patrika*, vol. XXIII, January 2, 1890, pp. 1 and 3; Ibid., March 27, 1890, p. 1 and 3. [BL: IOR]

ported extensively and almost every day on the events of the war, at times covering an entire page. Of particular interest was the fact that an "Asiatic" power fought, in the end, a successful war against a "European" imperial power.[59] Yet, Reuters' telegrams were not printed on the front page because that still did not exist in the layout of the *Amrita Bazar Patrika*, but all telegraphic news was placed on page four. There, however, it sometimes extended to more than four columns. Comments on the Russo-Japanese war as well as the revolution in Russia in 1905 became the object of separate comments allocated on page seven, from time to time covering a page.[60] "Indian Telegrams" reporting on occurrences within British India were now printed on page four. Special attention was given to the annual meetings of the INC taking place at every turn of the year. From the very beginning, reporting of these meetings was extensive and detailed, written by correspondents of the paper and forwarded telegraphically.[61]

The INC's annual meeting in Bombay in 1904 is of particular interest, as it demonstrates the interconnectivity of international reporting, press politics and the role of news agencies which all helped to shape an all-Indian political consciousness.

> The Bombay Congress of 1904 has not passed unobserved by the British Press. It may be that the Presidency of so able and so distinguished an Anglo-Indian as Sir Henry Cotton has had something to do with the Publicity the meetings have received both in London and provincial Journals. It is to be noted that each day Reuter has sent long telegrams, which contrast strikingly with the meagre reports that have appeared on previous occasions. Not only have the chief points [...] been cabled, but the reception speech of Sir P.M. Mehta was accorded a fairly long report, and a summary has also been sent of the speeches delivered by Mr. Surender Nath Banerjee, tha Hon. Mr. Gokhale, and Mr. Tilak; [...]. The Resolutions, too, have not been unduly condensed.[62]

59 Japan's victory lent wings to the Indian national movement, cf. Dharampal-Frick, Der Russisch-Japanische Krieg und die indische Nationalbewegung. See also Kowner (ed.), *The Impact of the Russo-Japanese War* And Wolff (ed.), *The Russo-Japanese War in Global Perspective.*

60 *Amrita Bazar Patrika*, vol. XXXVI, Thursday, January 26, 1905, p. 4. [BL: IOR]

61 Ibid., Tuesday, January 7, 1905, pp. 1–7; Between 29 December 1905 and 1 January 1906 the paper reported daily from the annual meeting of the INC, ibid., Friday, 29 Dec. – 1905 to Monday, 1 January – 1906, each p. 3. Reporting from the annual meeting of the INC at Allahabad 1910 covered significantly more of the newspaper's space, cf. ibid., Wednesday, Dec. 28, 1910 until Saturday 31, 1910, p. 6.

62 *Amrita Bazar Patrika*, vol. XXXVI, Tuesday, January, 7, 1905, p. 5.

At the same time, the partition of Bengal, publicly discussed from late 1903 onwards, announced in mid-July 1905 and implemented October 17, was debated lengthily and heftily in plenty of articles which were sent by telegram and commented on in separate articles.[63] Additionally, various mostly short articles reported on the occurrences and events of the Swadeshi-Movement[64] which soon turned into a popular resistance agitation campaign reaching far beyond Bengal.[65] To commemorate the first anniversary of the Swadeshi Movement, the *Amrita Bazar Patrika* reported on a mass meeting at Calcutta Town Hall. The meeting celebrated the adoption of Swadeshi resolutions (including the boycott of British goods until the demand for the reversal of partition was met) and ended in a procession "through public streets singing national songs."[66] Following the article, the newspaper printed 27 telegrams about anniversary celebrations all over British India; the telegrams covered almost an entire page of the newspaper.[67]

The partition of Bengal was the major event which triggered the Swadeshi Movement expanding far beyond Bengal within a short time. It also affected the INC which, according to the opinion of an organizer of the movement, "was looked upon as harmless by the Government." In September 1906 in Dhaka, the provincial capital of recently created East-Bengal, an assembly of about 15,000 people was addressed by Bipin Chandra Pal, an eminent Indian nationalist, "to make the Congress a power by securing for it a proper constitution based upon popular election." Additionally, the INC was to be urged to "[make] the Swadeshi and the boycott, which were like the two poles of the same battery, the central plank of the Congress."[68] The Swadeshi Movement, in accordance with the "nationalistic press," brought to the public light that the INC, firstly, lacked the respect of the colonial

63 Ibid., Wednesday, July 19, 1905, pp. 3–4.

64 *Amrita Bazar Patrika*, vol. XXXVI, Wednesday, July 19, 1905, pp. 3–4; Ibid., Monday, August 21, 1905, p. 3; Ibid., Tuesday, August 22, 1905, p. 3; Ibid., vol. XXXVII, Thursday, August 2, 1906, p. 3. Swadeshi actions and agitation was reported upon almost every day, ibid., passim. [BL: IOR]

65 The Swadeshi- (*swa*=self, *desh*=country, demanding the country's own and selfsustaining rule) movement organised protests after the Province of Bengal had been divided due to alledgedly administrative reasons. The protests included the burning of foreign particularly British textiles. At the same time imported goods were to be boykotted, cf. Sarkar, *The Swadeshi Movement in Bengal, 1903–1908*, and Prasad Basu, *Swadeshi Movement in Bengal and Freedom Struggle of India*.

66 *Amrita Bazar Patrika*, vol. XXXVII, Thursday, August 2, 1906, p. 3

67 Ibid., p. 3–4.

68 Ibid.,Tuesday, September 10, 1906, p. 3.

government. Governor General Curzon ridiculed the INC's influence. Secondly, the Swadeshi Movement showed that the INC lacked legitimacy as an organization supposedly representing the Indian nation. To give Congress some kind of preliminary or temporary all-India legitimacy, Swadeshi and boycott were to be put on top of the political agenda.[69]

Governor General Curzon's paternalistic rule disrespecting "Indians" and "Indian" institutions was often cause for open protest in the press as elsewhere. Shortly before Curzon came back from a visit to England in early 1905, the *Amrita Bazar Patrika* published a short yet telling note stating that "[i]n spite of what Reuter's correspondent may telegraph, and in spite of what inspired oracles in Printing House square may proclaim, the fact remains [...] that there is no enthusiasm in India over the return of Lord Curzon. The Indian community has frankness itself in expressing its dissatisfaction."[70] Against this background, it is no wonder that Curzon's downfall from office was commented on at length on the day of his resignation, August 21.[71] Apparently Curzon had overestimated his role and, to add to that, the weapon of threatening resignation had turned blunt after Curzon had used it with success on two previous occasions. Yet it was not the partition of Bengal which brought about Curzon's resignation but his insistence on his personal selection of officers in the course of the intended military reform in British India. Assuming that his stepping down from office would not be accepted by His Majesty's Government, Curzon sent a telegram in which he declared his resignation. It was a short telegram by the King stating "[w]ith deep regret I have no other alternative but to accept your resignation at your urgent requests."[72]

---

69 Yet, in vain. The leaders of the INC refused such far reaching demands fearing Congress would turn into a radical, non-cooperating organisation. Bal Gangadhar Tilak opposed the moderate views of Gopal Krishna Gokhale and was supported by fellow Indian nationalists Bipin Chandra Pal in Bengal and Lala Lajpat Rai in the Panjab. They were referred to as the Lal-Bal-Pal triumvirate. In 1907, the annual session of the Congress Party was held at Surat. Trouble broke out between the moderate and the extremist factions of the party over the selection of the new president of the Congress. The party split into the "Jahal matavadi" or "Garam Dal" ("Hot Faction," or extremists), led by Tilak, Pal and Lajpat Rai, and the "Maval matavadi" or "Naram Dal" ("Soft Faction," or moderates). It was only after WWI and the radicalisation of Indian politics that the INC included Swadeshi and boykott as weapons of resistance into its political programme. For a brief overview *vide* Pande et al., *Concise History of the Indian National Congress, 1885*–1947.

70 *Amrita Bazar Patrika*, vol. XXXVI, Monday, January 23, 1905, p. 6.

71 Ibid., August 21, 1905, p. 4.

72 Goradia, *Lord Curzon,* esp. pp. 231–6, quote 236.

Curzon's fall seems to be one of the very few instances when telegraphic correspondence actually sped up decision-making processes. Contrary to the general supposition, the telegraph did not have an immediate impact of this nature on administrative and diplomatic matters.[73] The case-study of colonial Ceylon demonstrates that telegrams were only a small part of the overall correspondence. In many cases its contents were repeated or quoted in the official correspondence which was organized along strict rules of writing and editing. This also indicates that telegrams did not possess authoritative status, meaning that they were just means of information and that telegrams still deserved written and signed documents to become legally binding.[74] Against this background, it is interesting to note that the role of the telegraph has been exaggerated with respect to official correspondence yet underestimated regarding its importance of extending public opinion as well as creating an all-India public sphere. The events of 1905 point out the importance of a revision and, consequently, reassessment, and most likely, a re-writing of the history of telecommunications in British India and the British Empire.

Finally, important changes within the layout of the *Amrita Bazar Patrika* took place in 1907 when the new rubric, "Latest Telegrams," was introduced and placed on page four. Another innovation took place in October of that year when the rubric, "Telegraphic Summary," on pages three and four was swapped with "Indian Telegrams" on pages five and six. Apparently, reports of "national" importance were given priority over "international" news.[75] Reaching far beyond Bengal, the *Amrita Bazar Patrika* addressed an increasingly "Indian" readership thus trying to create enthusiasm for matters of national importance. This was in stark contrast to *The Times* (of London) where more articles on international affairs were allocated on the front page in the 1890s than is the case nowadays.[76]

The increasingly patriotic-sounding *Amrita Bazar Patrika* emphasised the national Indian case simply by placement of articles and, at the same time, relegating international news as well as news from the British Empire to the back pages of the newspaper. The extensive reporting on the INC's annual meetings indicates just one of many important aspects of how the "Indian nation" was forged and mobilised during the first two decades of the twentieth century. This was especially true for the years following the end of World

73 Nickles, *Under the Wire.*

74 Fletcher, The uses and limitations of telegrams in official correspondence.

75 *Amrita Bazar Patrika,* vol. XXXVIII, March 1907 and ibid., October 1907.

76 Winseck and Pike, *Communication and Empire,* p. 3.

War I when failed constitutional reforms caused much discontent among several parts of British India's population. The annual meeting of the INC in Delhi in 1918 provides ample evidence of the growing discontent on the one hand, and, on the other hand, growing self-understanding. Since the American President, Woodrow Wilson, in his 14 Point Program of January 8, 1918, had promulgated the right to national self-determination, the subject was on the political agenda of the INC. The *Tribune* from Lahore reported extensively on the INC's meeting particularly on the statement of India's self-determination as a nation.

> A nation, as we understand it to-day, is made by two things and only two. One is the sense of oneness of life, of civil ideals and aspirations, in a people, a degree of cohesion among its component parts. [...] The other is the sense of its being distinct from all similar collections of men with a definite place in the life of humanity. In both respects we have been going rapidly forward for some years.[77]

Progress manifested itself in the fact that more than one hundred representatives of the agriculturist class, meaning landowning peasants, were invited, for the first time ever in the history of the INC, which "marked the beginning of the last stage in the progress of nation-formation, the putting of the cope-stone on that noble edifice."[78] For the first time, leading members of the INC became aware of the importance to define a nation which, then, could demand self-determination on its own right. Yet, according to the *Bulletin Weekly*, the INC's Note on Self-Determination "only echoed the sentiment of a nation" and confirmed that the principle of self-determination as formulated "by Mr. Lloyd George and President Wilson applies to India as well."[79]

And last but not least, the *Amrita Bazar Patrika* had its own extensive articles on this year's INC annual meeting and the question of self-determination. Reporting telegraphically from Lakhnau, a meeting of citizens had "demand[ed] the right of Self-Determination for India along with other countries in Europe at the time of the Peace Conference." The meeting resolved "that the great Communities of India are the inheritors of ancient civilization and have shown great capacity for government and administration and to progress in education and public spirit made by them during a

77 The *Tribune* (Lahore), 29 the December 1918, *Punjab Native Newspaper Reports, 1896–1924. Selections from the Native Newspapers Published in the Punjab* Received up to 4th January 1919, vol. XXXII, No 1, p. 151, para. 5a, p. 4. [BL: IOR].

78 Ibid.

79 The *Bulletin Weekly* (Lahore), 29th December 1918, ibid., para. 5c.

century of British Rule."[80] Again, by self-definition the civilizing mission of the British had come to an end also because the British "had done a good job" and accordingly Indian "communities" possessed the right of self-determination as a nation. Two months later, a couple of articles and comments still dealt with the question of India's self-determination.[81] Generally seen, the press was highly involved in an ongoing debate on the right of self-determination questioning whether that right applied *eo ipso* to India as a British colony or whether India, her politicians and people still needed to forge an Indian nation. In any case, there was no doubt that something like an Indian nation existed, in whatever form and stage of development.

Conclusion

From the sketchy evidence at hand, two points are evident. Firstly, telegraphic news and information transformed the media landscape of British India, be it British or Indian owned newspapers, journals and magazines. This change was underway parallel to the transformation in the presentation of information in press media in Europe and other parts of the world, and the emergence of the concept of "news" during the second half of the nineteenth century. News agencies like Reuters, Associated Press in the US, Havas in France and Wolff in Germany soon divided the globe into spheres of interest supplying primarily newspaper with pre-arranged sets of information, i.e. "news."[82] British India was firmly incorporated into this global media setting, mobilizing a hitherto unknown number of people forming a public sphere that has grown ever since.[83]

80 *Amrita Bazar Batrika*, Tuesday, December 24, 1918, p. 4.

81 Ibid., Monday, February 3, 1919, p. 3 and ibid., Tuesday, February 4, 1919, p.4.

82 To facilitate a global flow of news, Reuters, the 1835 founded French Agence Havas and Wolffs Telegraphenbureau, founded in Berlin in 1848, agreed on January 17, 1870 to establish a worldwide news exchange by dividing the globe in zones of collecting and distributing news. The cartel was revised and joined by Associated Press in 1893, cf. Anton A. Huurdeman, *The Worldwide History of Telecommunications* (Hoboken, NY: John Wiley & Sons, 2003), 88. and Winseck and Pike, *Communication and Empire*, 5. See also Wilke, *Nachrichtenauswahl und Medienrealität*, ch. 4, 75–180.

83 Yongming Zhou, *Historicizing Online Politics. Telegraphy, the Internet, and Political Participation in China* (Stanford, CA: Stanford University Press, 2006).

Secondly, in contrast to the British-owned press which emphasised news from the British Empire and beyond thus forming an "imperial" *Öffentlichkeit*, Indian-owned newspapers gradually emphasised national themes, thereby transforming and/or extending the existing public spheres of various South Asian historical regions into a trans-subcontinental all-Indian public sphere. This transformation was part of the larger process which helped to forge an Indian nation. Access to telegraphically transmitted information and news of "national importance," such as reports from the INC's annual meetings, speeches of its prominent members and of particular importance, and overall critical comments on its activities, were printed and widely distributed by "Indian" owned newspapers from the end of the nineteenth century onwards. Telegraphy and the emerging all-India public sphere went hand in glove, giving a decisive impulse to the nascent freedom movement.

Yet it has to be remembered that the national all-India *Öffentlichkeit* had its contemporary counterpart in the growing "communalist" public sphere that emphasized not the all-India nation, but a Muslim and a Hindu nation as imagined distinct communities. As can be seen from the *Amrita Bazar Patrika*, Hindu-biased reports and anti-Muslim sentiments gradually grew over the decades following the turn of the twentieth century and gaining momentum in the 1930s. In search for an Indian nation, the all-India *Öffentlichkeit* fell apart into at least two increasingly clearly distinguishable *Öffentlichkeiten*, its political and media members strongly mobilizing their supporters for two separate nations based on religious rather than historical grounds in the 1940s. Although their actual role has yet to be worked out, these two *Öffentlichkeiten* ultimately played a decisive part in creating Pakistan and India.

Bibliography

Amrita Bazar Patrika, vol. XXIII (1890); vol. XXXVI (1905); vol. XXXVII (1906); vol. XXXVIII; (1907); vol. XLI (1910); vol. XLIX; vol. L (1919) [BL: IOR].

Anderson, Benedict, *Imagined Communities. Reflection on the Origin and Spread of Nationalism*, London/New York 1983, revised edition 2006.

Annual Report on the Posts and Telegraphs of India for the Year 1922–23. Delhi: Government Central Press 1923 [BL: IOR].

Banerjea, Surendra Nath, *A Nation in Making. Being the Reminiscences of Fifty Years of Public Life*, London 1925, repr. Calcutta 1998.

Bayly, Christopher A., *Empire and Information. Intelligence Gathering and Social Communication in India, 1780–1870,* Cambridge UK 1997.

—, "Knowing the Country: Empire and Information in India," *Modern Asian Studies* 27 (1993), pp. 3–43.

—, "Town Building in North India, 1780–1830," *Modern Asian Studies* 9 (1975), pp. 483–504.

Barns, Margarita, *The Indian Press: A History of the Growth of Public Opinion in India,* London 1940.

Berthold, Victor Maximilian, *History of the Telephone and Telegraph in Brazil, 1851–-1921,* New York 1922, reprint Whitefish, MT 2009.

Bonea, Amelia, "All the News That's fit to Print?" Reuter's Telegraphic News Service in Colonial India, in this book.

—, "The Medium and its Message: Reporting the Austro-Prussian War in the *Times of India,*" in Roland Wenzlhuemer (ed.) *Global Communication: Telecommunication and Global Flows of Information in the Late 19th and Early 20th Century), Special Issue of Historical Social Research,* Köln 2010, pp. 167–87.

Bright, Charles, *Submarine Telegraphs. Their History, Construction and Working,* London 1898, reprint New York 1974.

Brown, Lucy, "The Treatment of News in Mid-Victorian Newspapers," *Transactions of the Royal Historical Society,* vol. 27 (1977), pp. 23–39.

Carroll, Lucy, "Kayastha Samachar: From a Caste – to a National Newspaper," *Indian Economic and Social History Review,* vol. 10 (1973), pp. 280–292.

Chaghatai, M. Ikram, "Dr Aloys Sprenger and the Delhi College," in Margrit Pernau (ed.), *The Delhi College. Traditional Elites, the Colonial State, and Education before 1857,* Delhi 2006, pp. 105–124.

Chakraborti, Smarajit, *The Bengali Press, 1818–1868. A Study in the Growth of Public Opinion,* Calcutta 1976.

Chanda, Mrinal Kanti, *History of the English Press in Bengal, 1780 to 1857,* Calcutta/New Delhi 1987.

Chatterjee, Kumkum, "History as Self-representation: The Recasting of a Political Tradition in Late Eighteenth Century Eastern India," *Modern Asian Studies,* vol. 32 (1998) 4, pp. 913–948.

Choudhury, Deep Kanta Lahiri, "India's First Virtual Community and the Telegraph General Strike of 1908," in Aal Blok and Greg Downey (eds.), *Uncovering Labour in Information Revolutions, 1750–2000. International Review of Social History,* vol. 48 (2003), Supplement 11, Cambridge UK 2003, pp. 45–71.

—, "Treason of the Clerks. Sedition and Representation in the Telegraph General Strike of 1908," in Crispin Bates (ed.), *Beyond Representation. Colonial and Postcolonial Constructions of Indian Identity,* Delhi 2006, pp. 300–321.

Cole, Juan R.I., *The Roots of North Indian Shi'ism in Iran and Iraq. Religion and State in Awadh 1722–1859,* Berkeley, CA 1988.

Crossley, Nick (ed.), *After Habermas. New Perspectives on the Public Sphere,* Oxford 2004.

Delhi Gazette, vol. VII–X (1877–1880); vol. XIX (1889) [BL: IOR].

Dharampal-Frick, Gita, "Der Russisch-Japanische Krieg und die indische Nationalbewegung," in Maik Hendrik Sprotte, Wolfgang Seifert und Heinz-Dietrich Löwe (eds.), *Der Russisch-Japanische Krieg 1904–1905. Anbruch einer neuen Zeit?*, Wiesbaden 2007, pp. 259–275.

Edgerton, David, *The Shock of the Old. Technology and Global History Since 1900*, Oxford 2007.

Field, Alexander J., "The Magic Telegraph, Price and Quantity Data, and the New Management of Capital," *Journal of Economic History*, vol. 52, no. 2 (1992), pp. 401–413.

Fischer, Claude S., *America Calling. A Social History of the Telephone to 1940*, Berkeley/Los Angeles/London, 1992.

Fletcher, Paul, "The Uses and Limitations of Telegrams in Official Correspondence Between Ceylon's Governor General and the Secretary for State for the Colonies, circa 1870–1900," in Roland Wenzlhuemer (ed.) *Global Communication: Telecommunication and Global Flows of Information in the Late 19th and Early 20th Century), Special Issue of Historical Social Research*, Köln 2010, pp. 90–107.

Fowler, Roger, *Language in the News: Discourse and Ideology in the Press*, London/New York 1991.

Freitag, Sandria B., "Enactments of Ram's story and the Changing Nature of 'The Public' in British India," *South Asia*, vol. 14, no. 1 (1991), pp. 65–90.

Galtung, Johan and Holmboe Ruge, Mari, "The Structure of Foreign News: The Presentation of the Congo, Cuba and Cyprus Crisis in Four Norwegian Newspapers," *Journal of Peace Research*, vol. 2, no. 1 (1965), pp. 64–91

Goradia, Nayana, *Lord Curzon. The Last of the British Moghuls*, Delhi 1993.

Gorman, Mel, "Sir William O'Shaughnessy, Lord Dalhousie and the Establishment of the Telegraphic System in India," *Technology and Culture*, vol. 12, no. 4 (1971), pp. 581–601.

Habermas, Jürgen, "The Public Sphere: An Encyclopedia Article" (1964), *New German Critique*, vol. 3 (1974), pp. 49–55.

—, *The Structural Transformation of the Public Sphere. An Inquiry into a Category of Bourgeois Society*, Cambridge, MA 1989.

Hartley, John, *Understanding News*, London 1990.

Headrick, Daniel R., *The Invisible Weapon* Telecommunications *and International Politics, 1851–1945*, Oxford 1992.

Hills, Jill, *The Struggle for Control of Global Communication: The Formative Century*, Urbana, ILL 2002.

Hobsbawm, Eric, *Nations and Nationalism since 1780. Programme, Myth, Reality*, Cambridge 1990.

Hugill, Peter, *Global Communications since 1844. Geopolitics and Technology*, Baltimore, MD 1999.

Huurdeman, Anton A., *The Worldwide History of Telecommunications*, Hoboken, NY 2003.

Israel, Milton, *Communications and Power. Propaganda and the Press in the Indian Nationalist Struggle, 1920–1947*, Cambridge 1994.
Iyer, Vishwanath, *The Indian Press*, Bombay 1945.
Kidambi, Prashant, *The Making of an Indian Metropolis. Colonial Governance and Public Culture in Bombay, 1890--1920*, Aldershot/Burlington 2007.
Kowner, Rotem (ed.), *The Impact of the Russo-Japanese War*, London 2007.
Lubrano, Annteresa, *The Telegraph: How Technology Innovation Caused Social Change*, New York/London 1997.
Malik, Jamal, "Reformen in Südasien – das 18. Jahrhundert," in Karin Preissendanz and Dietmar Rothermund (eds.), *Südasien in der 'Neuzeit'. Geschichte und Gesellschaft, 1500–2000*, Wien 2003, pp. 72–89.
Mann, Michael, *Geschichte Südasiens 1500 bis heute*, Darmstadt 2010.
Middell, Matthias and Katja Naumann, "Global History and the Spatial Turn: From the Impact of Area Studies to the Study of Critical Junctures of Globalization", *Journal of Global History*, vol. 5 (2010), pp. 1–22.
Mantl, Norbert, *Aloys Sprenger: Der Orientalist und Islamhistoriker aus Nassereith in Tirol. Zum 100. Todestag am 19. Dezember 1993*, Nassereith 1993.
McMahon, Peter, *Global Control: Information Technology and Globalization since 1845*, Cheltenham 2002.
Metcalf, Barbara D., *Islamic Revival in British India: Deoband, 1860–1900*, Princeton, NJ 1982.
Naim, C.M., and Petievich, Carla, "Urdu in Lucknow/Lucknow in Urdu," in Violette Graff (ed.), *Lucknow. Memories of a City*, Delhi 1997, pp. 165–180.
Nickles, David P., *Under the Wire. How the Telegraph Changed Diplomacy*, Cambridge, MA. 2003.
Pande, B. N., Ray, Nisith Ranjan, Kumar, Ravinder and Das, Manmath Nath, *Concise History of the Indian National Congress, 1885–1947*, Noida 1985.
Paul, Suchitinder Nath, *Public Opinion and British Rule: A Study of the Influence of Indian Public Opinion on British Administration and Bureaucracy, 1899–194l*, New Delhi 1979.
Prasad Basu, Sankari, *Swadeshi Movement in Bengal and Freedom Struggle of India*, Calcutta 2005.
Punjab Native Newspaper Reports, 1896–1924. Selections from the Native Newspapers. Published in the Punjab Press Abstract [BL: IOR].
Reetz, Dietrich, *Islam in the Public Sphere: Religious Groups of India, 1900–1947*, Delhi 2006.
Sadasivan, D., *The Growth of Public Opinion in the Madras Presidency (1858–1909)*, Madras 1974.
Sarkar, Sumit, *The Swadeshi Movement in Bengal, 1903–1908*, Calcutta 1973.
Sauer, George, *The Telegraph in Europe.* Paris 1869, Printed for Private Circulation.
Shridharani, Krishnalal, *Story of the Indian Telegraphs: A Century of Progress*, New Delhi 1953.

Sprenger, Aloys, *A Catalogue of the Arabic, Persian and Hindústány Manuscripts in the Libraries of the King of Oudh, Under the Orders of the Government of India*, Calcutta 1854.

Standage, Tom, *The Victorian Internet. The Remarkable Story of the Telegraph in Nineteenth Century's On-line Pioneers*, New York 1999.

Stark, Ulrike, *An Empire of Books. The Naval Kishore Press and the Diffusion of the Printed Word in Colonial India*, Ranikhet 2007.

Wilke, Jürgen, *Nachrichtenauswahl und Medienrealität in vier Jahrhunderten. Eine Modellstudie zur Verbindung von historischer und empirischer Publizistikwissenschaft*, Berlin/New York 1984.

Winseck, Dwayne R. and Pike, Robert M. *Communication and Empire. Media, Markets, and Globalization, 1860–1930*, Durham, NC and London 2007.

Wolff, David (ed.), *The Russo-Japanese War in Global Perspective*, Leiden 2007.

Zhou, Yongming, *Historicizing Online Politics. Telegraphy, the Internet, and Political Participation in China*, Stanford, CA 2006.

“All the News That’s Fit to Print”? Reuters’ Telegraphic News Service in Colonial India

Amelia Bonea

In the 1860s, a new category of news appeared in the pages of some of the English-language newspapers published in colonial India. Known as “Reuter’s Telegrams,” the items published under this heading contained political and commercial intelligence from Britain, Europe and, occasionally, from other parts of the world. Reuters’ news came to occupy an increasingly prominent place in the pages of the English-language press, so much so that by the end of the century its telegrams spread over entire columns of the newspapers. The agency’s expansion into Asia and its monopoly on news exchange between Britain and India were directly connected to the development of a global network of communication, in particular the proliferation of telegraph technology. In this paper, I trace Reuters’ ascendancy to the position of news agency of the British Empire by focusing specifically on the circulation of foreign intelligence between Britain and India during the second half of the nineteenth century. My analysis highlights the complex nature of news in the colonial environment of India as a contested field in which power relations were constantly played out, as well as the role of technology in the transmission of intelligence. I argue that the negotiations between Reuters, the colonial government of India and newspaper editors were not only disputes over access to technologies of communication, but also attempts to define the very category of news. In short, news in nineteenth-century India was a function of both available technology and socio-economic and political factors.

The benefits of this historical exercise are particularly relevant in an age when we are constantly grappling with the changes spawned by new technologies of communication such as the Internet. Understanding how technology mediated news flows during the nineteenth century and how news agencies such as Reuters established themselves as globally relevant suppliers of intelligence provides a much needed historical perspective on contemporary patterns and ideas of journalism and news reporting. Since the 1960s, exponents of the theory of media and cultural imperialism have discussed the manifold ways in which Western media organizations have come to dominate global exchanges of news. This situation has been documented at various levels of news production and distribution.[1] In the Indian case, more than two decades ago, Somnath Sapru also made a case for media imperialism when he drew attention to the one-sided nature of international news flows and their inherent bias in reporting about the non-Western world:

> The complaint of most of the former colonies, now loosely grouped together as the Third World, is that though they represent two-thirds of the world's population and area, events concerning the West get the major news coverage, and whatever they see of themselves in the First World media is mainly as part of disasters, coups, epidemics and the like.[2]

Scholars have challenged theories of media imperialism in recent years. Often inspired by the work of Arjun Appadurai, they have emphasized, among other things, the multi-directionality of global information flows.[3] Indeed, the last few decades have witnessed a considerable diversification of media outlets and products both with a regional and global reach. Much of this diversification has been spurred by developments in telecommunication technologies. In India, this situation has been reflected in the increasing regionalization and commercialization of the newspaper press, two phenomena which have become particularly conspicuous in the wake of the economic liberalization of the 1990s.[4] Yet, the multi-directionality of news flows does not necessarily translate into the absence of more or less subtle hegemonic relations between various participants in the process of news exchange. As far as India is concerned, it is interesting to note that a recent study by Kapil Arya found that English-language newspapers published in this country continue to rely heavily on global news agencies such as Reuters and the Associated Press for their access to foreign news. As Arya's analysis suggests, the international news content published by India's main English-language dailies, the *Times of India*, the *Hindustan Times* and *The Hindu*, reflects the

1 Boyd-Barrett and Galtung's works are among the most influential in this field. See, for example, Boyd-Barrett, Media Imperialism, and Galtung, A Structural Theory of Imperialism.

2 Sapru, *The News Merchants*, p. 2.

3 Appadurai, *Modernity at Large*; for two examples of recent reassessments of the media imperialism thesis see Kperogi, Guerrillas in Cyberia, and Chadha/Kavoori, Media Imperialism Revisited.

4 Jeffrey, *India's Newspaper Revolution*.

agendas and trends set by global news agencies.[5] By drawing on the nineteenth-century experience of telegraphy and news and, more specifically, on Reuters' activity in colonial India, the present paper aims to show not only that certain categories of news have always been more "fit to print" than others, but also why. As the paper suggests, the use of telegraphy facilitated the integration of India into an increasingly global network of news and journalistic standards, while also creating new distinctions and differences between those who could access the news and those who could not.

Reuters' Indian Story, 1858 – 1900

During the nineteenth century, Reuters emerged as the main purveyor of intelligence in the British Empire and beyond. The early beginnings of Julius Reuter's agency were inextricably linked to developments in telecommunication technology, in particular the increasing use of electric telegraphy for the exchange of financial information between various European cities in the 1840s. In Europe, Reuter's emerging news trade drew on a variety of means of communication which complemented the expanding, albeit imperfect telegraph network. His Indian venture also used the available means of transport of the day, such as steamers, well before functional telegraph lines brought the subcontinent in direct communication with Europe. Thus, although the first overland telegraph routes between Britain and India were opened in the mid-1860s and the direct submarine route via the Red Sea was inaugurated in 1870, in 1858 Reuters was already providing Indian news to the British public.[6]

The information transmitted reflected contemporary political and mercantile preoccupations, being predominantly of a military and commercial nature and originating primarily with the English-language newspapers of India, in particular the vibrant port-city of Bombay.[7] For example, the earliest surviving Reuters telegram from India, datelined Marseilles, 18 October

5 Arya, *The Over-Dependence of Indian English Newspapers*, especially pp. 45–46.

6 For an excellent account of the British press' coverage of colonial India, including Reuters' role in supplying Indian intelligence to the British public, see Kaul, *Reporting the Raj*. Unlike Kaul's study, the present article is concerned predominantly with the flow of information from Britain to India, as mediated by telegraphy and Julius Reuter's news agency.

7 Reuters Archives (hereafter RA), Read, Reuters and India, p. 2.

1858, contained a comprehensive report (approximately 150 words) about the unrests in the Punjab, Gwalior and Awadh, as well as news about the establishment of a submarine telegraph connection between India and Ceylon.[8] The source of this information was the *Bombay Times* and the message had been sent by ship to Marseilles and thence by telegraph to Reuter's Office in London. The first extant telegram datelined Calcutta arrived in London in October 1858, shortly after the first "Electric News" from Bombay, and reported on the state of the market in Bengal: "During the last fortnight there have been only small arrivals by the rivers. The insufficiency of provisions has caused the prices of all articles to rise excessively. It is hoped that the victories will ameliorate the state of affairs."[9]

Although the number of telegrams transmitted from India to Britain was still limited at this early stage in Reuters' history, the supply was fairly regular and followed the fortnightly arrival of the Bombay and Calcutta mails to Alexandria. The latter was an important node in the communication with the "East:" newspapers, letters and other printed matter dispatched from various Indian ports were forwarded from Alexandria to Europe either via Marseilles by the Peninsular and Oriental Steam Navigation Company (hereafter P&O) or via Trieste by the Austrian Lloyd's Steam Navigation Company.

These early attempts to dispatch Indian intelligence to Britain were not yet part of a formal system of news distribution. However, the pattern of transmission already testifies to Reuters' commitment to use all extant routes of communication with India in order to expand its service. In October 1859, two telegrams from Calcutta and Ceylon were advertised as "the first messages forwarded by [Mr. Reuter's] newly-established special service between England and the East."[10] From this moment on, the agency's determination to establish itself formally as the main supplier of news between Britain and India became increasingly manifest. Telegrams printed at "Mr. Reuter's Office" at the Royal Exchange Buildings in London sported conspicuous headlines which drew attention to "Reuter's Special Indian Service" and began to emphasize the timeliness of the news transmitted. For example, on October 26, 1859, Reuter's "Latest Intelligence" section contained market quotations from Calcutta and Colombo received only "six days later than the departure of the mail from Calcutta."[11] The advantages of transmitting

8 RA, LN976, Reuters Telegram Books to 1881, 18 October 1858.
9 Ibid., 29 October 1858.
10 Ibid., 11 October 1858.
11 Ibid., 26 October 1858.

intelligence by telegraph, as opposed to mail steamers, were becoming increasingly evident.

Although Reuters' Indian telegrams were an amalgam of intelligence conveyed by a variety of agents and means of communication, the layout of these dispatches already betrays the emergence of a certain hierarchy of news based on the mode of communication employed to transmit it. As a rule, telegrams summarizing the main items of intelligence delivered by the P&O's regular mail received less emphasis than the "Special Service" news, whose transmission routes clearly followed the latest developments in telegraphic communication with India. Thus, if in 1861 Reuters was able to announce the receipt of intelligence "by telegraph to Bombay" and two years later "by telegraph from Suez," by 1866 its bylines proclaimed proudly that the news had been received "by direct overland telegraph." The advantages of telegraphy, as opposed to steamers and other means of communication, had led to the identification of telegraphic news with timeliness and speed.[12]

Reuters' aim in establishing its "Special Service with the East" was not, however, limited to providing Indian news to subscribers in Britain: the agency was equally desirous of selling British and European intelligence to potential clients in India. The Government of India, the mercantile community and the growing newspaper press were among the agency's first and most important subscribers. The service was based on an agreement with the *Bombay Times*, the first newspaper in India to publish Reuters telegrams.[13] Like in Britain, the agency's early forays into the Indian news market depended to a great extent on its ability to establish a network of potential providers and consumers of news. Its focus on the port-city of Bombay seems natural in light of that town's emerging position as the main port of entry of information into India. Despite the best efforts of its mercantile community, by the 1860s it had become increasingly evident that Calcutta's importance as a mail port was rapidly declining. The completion of railway communication between Bombay and Calcutta, as well as between Bombay and Madras, eventually sealed the long-lasting rivalry between the two port-cities over the delivery of mails. A select committee of investigation set up in 1866 by the British Government discovered that a weekly mail service to Bombay had become more efficient than the previous system of dispatching mails fortnightly to Calcutta and Bombay. The recommendations of the committee

12 On the increasing importance of telegraphy, see also Sprenger's contribution in the present volume.

13 Sapru, *The News Merchants*, p. 9.

were soon put into practice and Bombay became the main port of entry of information into India.[14]

Reuters' early activity in the subcontinent was thus intimately connected to Bombay, in particular to the *Bombay Times* and its direct successor, the *Times of India*. Conspicuous among the agents who mediated this early exchange of news between India and Britain was Robert Knight, an Anglo-Indian editor associated with both newspapers. Now hardly remembered – despite the fact that he founded two of India's most enduring dailies, the *Times of India* and the Calcutta *Statesman* – Knight went to Bombay in 1847 as a merchant, not as a journalist. He dabbled in various occupations before turning to journalism in 1857. His four-year employment with Waghorn and Company (1852–1856), the pioneer of the overland route to India, testifies to his early interest in the news trade.[15] As Knight's correspondence suggests, in 1860 Reuters offered him "his sole agency, not in the Western presidency only, but in all of India."[16] A year later, upon the reorganization of his newspaper venture into the *Times of India*, Knight sought to establish himself as a supplier of telegraphic news to the Anglo-Indian press. To this end, he organized his own news agency, known as the "*Times of India* Telegraphic Agency" and began to sell intelligence to the major Anglo-Indian newspapers for a monthly subscription fee of Rs. 500. Eventually, Reuters took over Knight's telegraphic agency: there is confusion regarding the date, but Knight's biographer infers from archival records that this happened either in 1862 or in 1864.[17]

My own examination of *The Englishman*, an Anglo-Indian daily published in Calcutta during the nineteenth century, suggests that Robert Knight might have relinquished his news agency to Reuters in 1866. According to John O'Brien Saunders, editor and proprietor of *The Englishman*, at the beginning of that year, the *Times of India* "got the majority of the Press of the three Presidencies to enter into a confederacy, whereby they would all receive the same telegrams at a rate considerably less than that which each journal was paying before the amalgamation."[18] The news was transmitted

14 Parliamentary Papers (hereafter PP), *East India Communications*, pp. vi–vii; Harcourt, *Flagships of Imperialism*, p. 105; *Gazetteer of Bombay City and Island*, vol. 1, pp. 377–378.

15 Hirschmann, *Robert Knight*, pp. 18–21.

16 Ibid., p. 74.

17 Ibid.

18 *The Englishman*, 13 July 1866. Saunders writes *Bombay Times* instead of the *Times of India*, but the newspaper had ceased to exist under that name in 1861, when it incorporated the

by the London agents of the *Times of India*, most likely by Robert Knight himself, who was on an extended stay in Britain at the time. Saunders' claim that the Viceroy of India himself had been part of this scheme is confirmed by official correspondence which indicates that the "manager" of the *Times of India* supplied the Viceroy with telegraphic summaries of mails. However, this arrangement was discontinued in July 1866, when Henry Collins, Reuters' agent at Bombay, undertook to supply the Viceroy with a daily message in exchange for a monthly subscription fee of Rs. 400. [19]

Known as "Subscription Telegrams," the news sent via the *Times of India* was published in 1866 in a handful of Anglo-Indian newspapers along with "Reuter's Telegrams."[20] Unlike Reuter's messages which emphasized political events, "Subscription Telegrams" focused on the reporting of commercial intelligence, which was always published before other types of information. This suggests that the telegrams were aimed primarily at the mercantile community of India, and especially the merchants of Bombay, many of whom were also involved in the publication of the newspaper press. On the other hand, the content of Reuters' telegrams suggests a more inclusive idea of news, as a product which appealed to a broader audience, despite the fact that it was itself focused on the transmission of political and commercial intelligence.[21] As far as the *Times of India* Telegraphic Agency was concerned, prioritizing commercial news proved to be unsuccessful in the long run, as Anglo-Indian editors soon began to complain that "Subscription Telegrams" were "crammed" with cotton quotations and other commercial news which could only interest the merchants and speculators of Bombay. To overcome this criticism, the agency reassured its subscribers that proper attention would be given to quotations from Madras and Bengal as well. However, this failed to convince the editors, whose main concern was not the inadequate geographical distribution of commercial news, but the sparseness of political intelligence; as one editor put it, this was a "serious drawback to

Bombay Standard and became the *Times of India*. In this connection, see Watson, Growth of the Press, p. 124.

19 National Archives of India (hereafter NAI), Home Department, Public, July 1866, no. 192, Supply of daily message by Reuter's Company Bombay.

20 These "Subscription Telegrams" circulated as far as Japan. For example, the *Japan Times* of 26 January 1866 contains telegrams originally published in *The Englishman* of Calcutta and brought to Japan via Hong Kong by the Japanese steamer *Fusiyama* (sic).

21 For a detailed discussion of "Subscription Telegrams" and "Reuter's Telegrams" see Bonea, The Medium and Its Message.

the successful working of the joint system."[22] It was against this background that Robert Knight decided to sell the whole "confederate agreement" to Mr. Reuter, much to the dismay of the other Anglo-Indian editors.[23]

According to Saunders, some editors refused to accept the terms proposed by Reuters, a situation which resulted in a temporary lack of telegraphic intelligence for these newspapers. *The Englishman* itself was temporarily deprived of intelligence due to its editor's refusal to "bind" himself to Mr. Reuter until he was fully aware of the details of the agreement. But this opposition was short-lived, for even the critical Saunders, pressured by the lack of any viable alternatives, was soon forced to admit that *The Englishman* was "willing to pay any charge which the agency chose to make for such messages as they may send to us."[24] Saunders depicted Robert Knight as an unscrupulous person, an evaluation which begs scrutiny, particularly in light of the notorious rivalries between Anglo-Indian editors and Knight's reputation of being "pro-native" and even a "socialist." Although it is difficult to judge his character and motives based on the scant information available to us, we do know that the period when this transaction took place was a particularly difficult one for the young editor and his family. During his time in London from 1864 to 1868, Knight suffered significant financial losses, first in a coffee plantation in South India and later when the Bank of Bombay collapsed due to the sudden drop in cotton prices, which brought to an end the prosperity enjoyed during the American Civil War.[25]

As I have suggested above, the expansion of the telegraph network was essential to Reuters' success in South Asia. But equally important was the more or less overt support of the colonial administration and the ability to secure the services of news agents at various locations across the Indian subcontinent, in particular at the main economic and political centers. In its early days Reuters relied on local traders, Christian missionaries and employees of the steamship companies for the collection and supply of information. According to Chandrika Kaul, in 1862 Edwyn Dawes of the P&O was the first "named agent" to work for Reuters in Bombay.[26] As far as Calcutta is concerned, new evidence suggests that Robert Campbell and Co. worked

22 *The Englishman*, 13 July 1866.
23 Ibid.
24 Ibid.
25 Hirschmann, *Robert Knight*, p. 99, but also pp. 50–54, 209–24.
26 Kaul, Imperial Communications, p. 67.

as agents for Reuter's Telegraphic Office as early as 1860.[27] The activity of these agents was part of Reuters' early efforts to establish a network of correspondents in India's main port-cities and became the foundation on which the company established its first formal office in the subcontinent. This was opened in 1866 in Bombay by Henry Collins, who succeeded in expanding the agency's reach as far as Japan in only six years.[28]

Apart from the newspaper press, many of Reuters' subscribers in India were members of the mercantile communities of Bombay and Calcutta, who were particularly interested in market quotations, but also in international political events such as wars, which had the potential to disrupt trade (the American Civil War is a case in point). As we have seen, the Government of India was also amongst the earliest subscribers. As part of its special agreement with the news agency, the Government was allowed to distribute free copies of telegrams to a select number of colonial officials. For example, in the 1870s, Reuters messages were received by the Viceroy of India and seventeen officers of the Government of India, such as the Governors of Bombay and Madras, the Lieutenant-Governor of Bengal, the Chief-Commissioners of Oudh, Central Provinces and Mysore and the Residents at Hyderabad and Bushire (Bushehr, in present Iran).[29] Timely and regular access to the latest political and commercial intelligence was essential for the discharge of administrative duties. However, the list of officers to whom the Indian Telegraph Department sent, at its own expense, copies of Reuters telegrams, was restricted to the most senior officials and the Government of India frequently received petitions for access to these telegrams, especially from officers stationed in remote areas such as Arraccan (now Rakhine State, in Myanmar). Requests to be added to the official list of officers entitled to receive free copies of Reuters telegrams were usually couched in a language which emphasized the interest to the public of such news items, an argument which the Government of India often rejected.[30] This suggests that access to information was asymmetrical even among the ranks of the colonial administration, not only for reasons of financial exigency, but also because higher

27 Half-Yearly Report of the Bengal Chamber of Commerce, November 1860 – April 1861, Appendix L, Robert Campbell and Co. to H. W. I. Wood, 27 November 1860.

28 Storey, *Reuters' Century*, p. 62. For an account of Reuters' expansion into South and East Asia, see Henry Collins' autobiography, *From Pigeon Post to Wireless*.

29 British Library, Oriental and India Office Collections (hereafter OIOC), L/PWD/7/1056, File 368, Press Messages Reuters, Telegraph Department to the Secretary to the Government of India, 23 September 1876.

30 Ibid.

officials jealously guarded their privileged access to news. One indication of this is that the telegrams, dispatched each week alternatively to Bombay or Galle, were forwarded to Simla, the newly-established summer capital of British India, "under the Clear the Line Signal," a move which allowed the Government of India to keep abreast of relevant political developments, both in the British Empire and internationally.[31]

The hierarchy which characterized Reuters' news service was also reflected in the range of subscribing newspapers, most of which were the major Anglo-Indian, English-language dailies of India's main cities. This is hardly surprising, given the fact that price drastically limited access to Reuters news and only the biggest newspapers could afford the subscription fee of Rs. 400 per month. According to the contracts signed in October-December 1879 by *The Englishman*, the *Bombay Gazette*, the *Civil and Military Gazette*, *The Pioneer* and the *Madras Times*, this sum was paid in monthly installments and the news had to be acknowledged as "Reuter's Telegrams" when published in these newspapers.[32] In 1890–1891, upon the renewal of the contracts, the subscription fee remained unchanged, but a new stipulation was introduced which entitled Reuters to increase the sum to Rs. 6,000 "in times of war, civil or otherwise, in which England, or any great European power or America, may become involved, or during any prolonged political excitement affecting the international relations of any two of such Powers, and during a General election in the United Kingdom."[33] Apart from restricting access to information for newspapers which could not afford these fees, the contracts also effectively consecrated certain categories of events as news-worthy and important to the reading public in India.

The "subsidy" which the Government of India paid to Reuters for the regular delivery of telegraphic news continued to increase during the nineteenth century. Like the other subscribing newspapers, the Government had also begun by paying a monthly fee of Rs. 400. Unlike the newspapers, however, this sum had increased to Rs. 600 by 1873, while a few years later the agency asked the Indian authorities to double their contribution.[34] The

31 Read, Reuters and India, p. 5.

32 RA, LN435, 1879, Agreements between Reuters and *The Englishman*, *Bombay Gazette*, *Civil and Military Gazette*, *The Pioneer* and *Madras Times*. The contract with the *Times of India* has not survived, but it is known that the paper was among the subscribers.

33 RA, LN247, 1890 – 1891, Agreements for supply of news between Reuters and the *Times of India*, *Civil and Military Gazette*, *Madras Mail*, *Rangoon Gazette*, *The Pioneer*, *Bombay Gazette*, *Madras Times*, *Morning Post* and *The Englishman*.

34 RA, LN245, Agreement with the Indian Government, May 1873.

relationship between the Government of India and Reuters is significant in light of the fact that the flow of information between these two entities was bi-directional. Thus, it was not only Reuters who supplied news to the Government and the public of India, but the Government itself was an important supplier of news to Reuters and, by extension, the British public. The interaction between the two indicates that their interests converged on one point, namely the desire to achieve a certain monopolistic control of the flow of intelligence between Britain and India. Reuters' attempts to secure privileged access to official information aimed to brush aside potential competitors in the news business, such as newspaper correspondents and merchants, who also furnished the British press with information on Indian affairs. For example, as early as September 1867, the Government of India replied positively to Reuters' petitions by agreeing to supply Henry Collins with a précis of intelligence containing "such important movements, orders, and events as are suitable and of sufficient interest to be given to the public [in Britain]."[35] Yet, there were limits to the Government's benevolence: for example, in 1882 it declined categorically the request by Reuters' agent at Bombay that "ordinary news telegrams from the Government of India to Her Majesty's Secretary of State for India should, in the future, be transmitted through the agency of Reuter's Telegram Company."[36]

Such occasional drawbacks notwithstanding, Reuters was certainly successful in securing various privileges from the Indian Government. In August 1895, for example, it managed to obtain "a very substantial concession" which allowed it to send telegrams addressed to the Government and the press at deferred rates, although such messages were classed as urgent.[37] From this moment on, Reuters telegrams were no longer regarded as "Press telegrams," but as "News-free telegrams," in the same category as mail steamer and meteorological reports. According to C. H. Reynolds, Director General of Telegraphs, this concession was sanctioned "in the public interest [...] Reuter's press messages addressed to newspapers are carried at *deferred* rates though transmitted as *urgent*. Certain meteorological messages have

35 NAI, Foreign Department, General A, Nos. 7 – 10, October 1867, Application from Reuter's Telegram Co. for the Aid of the Government, Resolution of the Government of India, 26 September 1867.

36 RA, LN245, Letters from Government of India and Reuters Telegram Company Ltd re News Telegrams, 1882, Under-Secretary to the Government of India to the Secretary of Reuter's Company, Bombay, 26 April 1882.

37 OIOC, V/24/4287, Administration Report of the Indian Telegraph Department for 1896 – 1897, p. 21.

a somewhat similar privilege and are transmitted under the urgent signal though charged as ordinary [...]" (emphases added).[38] Essentially then, this was a struggle over the definition of news, especially important news, since the more "public" a certain category of intelligence was deemed to be, the more likely it was to be transmitted at a cheaper rate, a situation which Reuters correctly recognized.

As I have pointed out above, there was a clear rift between the various sections of the press in colonial India with regard to their ability to subscribe to Reuters telegraphic news. The system of news distribution via Reuters clearly privileged the major daily newspapers, usually published by Anglo-Indians in the English language, at the expense of smaller and less economically viable newspapers published by Indians in the vernacular. In 1900, Surendranath Banerjea's *Bengalee*, an English-language newspaper published in the urban environment of Calcutta which had recently turned into a daily, became the first Indian-owned newspaper to subscribe to Reuters services.[39] It is obvious from this account that telegraphic news was a commodity not all newspapers could afford and that timely, regular installments of telegraphic intelligence were particularly indispensable to the daily press rather than the periodical press. This, in turn, supports Menahem Blondheim's argument in a different context that the effects of electric telegraphy were hardly confined to connecting and "democratizing" the world, but that they also included the creation of new boundaries and distinctions.[40]

In India, such distinctions were particularly significant if we keep in mind that by the second half of the nineteenth century the ability to access information in a timely and regular manner came to be increasingly identified with "good journalism." In this context, the telegraph became another tool which, albeit indirectly, shaped colonial perceptions of the "native press" and helped to define it as lacking in accurate and timely intelligence: in fact, even lacking the interest in such intelligence. To begin with, colonial administrators and commentators rarely regarded Indian-owned newspapers as exponents of good journalism, although some allowance was made for journals published in English by educated Indians – "the better class" of the "native press," represented by a handful of papers such as the *Hindoo Patriot* (Cal-

38 OIOC, V/24/4287, Administration Report of the Indian Telegraph Department for 1895 – 1896, p. 18.

39 Banerjea, *A Nation in the Making*, p. 157. See also *The Bengalee*, 6 January 1900.

40 For a discussion of the dual role of telegraphy in the United States see Blondheim, "Slender Bridges" of Misunderstanding, pp. 153–170.

cutta), the *Indian Spectator* (Bombay) and *The Hindu* (Madras).[41] Observers complained about the "free-and-easy principles" which guided Indian journalists and argued that the range and content of their news clearly reflected their ideological leanings. Conveniently overlooking the hierarchies of news distribution in India, as well as the fact that the Indian press consisted predominantly of non-daily newspapers which were unable to compete financially with their Anglo-Indian counterparts,[42] G. M. Chesney argued, for example, that the Indian public was notoriously uninterested in news on foreign politics, sports, social amusements and drama. This, he claimed, was due to the fact that most Indian newspapers devoted their energy to criticizing the colonial government.[43] In this way, Indian newspapers came to be regarded predominantly as useful barometers of social and political discontent, not as trusted purveyors of accurate information.

Yet, as I have already suggested, attitudes towards news reporting were more likely shaped by the economic standing of a particular newspaper, its location and frequency of publication, as well as the socio-economic background of its managers and editors, rather than by perceptions of news which were inherently Indian or British. In her important study of the *Avadh Akhbar*, an Urdu-language weekly published in Lucknow by Munshi Naval Kishore, Ulrike Stark provides information about the alternative modes of news collection and dissemination employed by the vernacular press.[44] According to her, the *Avadh Akhbar* published a great variety of domestic and foreign news items, reprinting many of them from English or Anglo-Indian newspapers such as *The Times*, *The Pioneer* and the *Friend of India*, or from Persian and Arabic newspapers. The rest of the news was obtained by means of an extensive system of correspondents maintained by Munshi Naval Kishore or by enlisting the help of the reading public who received free copies of the newspaper in exchange for items of intelligence of par-

41 Wright, *Baboo English*, p. 16; Lethbridge, Vernacular Press in India, p. 462.

42 In this connection, see Raghavan, *PTI Story*, pp. 50–51. In fact, the hierarchies of telegraphic news were not restricted to foreign intelligence, but affected domestic intelligence as well, in particular official information. For example, the Press Commissioner scheme implemented in the late 1870s, explicitly excluded vernacular messages from the "Press Message privilege," which would have allowed them to receive news at a cheaper rate. Exceptions seem to have been made, but they depended entirely on the good will of the Director General of Telegraphs. See NAI, Home Department, Public A, June 1878, no. 14, Rules for "Press" messages in India, Burmah and Ceylon.

43 Chesney, Native Press in India, pp. 269–271.

44 Stark, *An Empire of Books*, p. 365.

ticular interest. Although the newspaper did not subscribe to Reuters, it did publish telegraphic news and, unlike the majority of English-language newspapers in the 1870s, it published telegrams on the front page of the newspaper, at the expense of advertisements and local news which was moved to the last pages.[45]

In conclusion, Reuters' expensive monopoly over the circulation of news between India and Britain left smaller newspapers such as many of the vernacular papers with little choice other than to translate and publish items of intelligence previously printed in the Anglo-Indian press. As Lucy Brown has pointed out with regard to Victorian Britain, in the 1870s this practice was widespread among British newspapers as well, a situation which often attracted resentful complaints from the major London dailies from whose pages the telegrams were usually lifted.[46] Yet, this was not a simple matter of "stealing" news: rather, it was a practice which stemmed from a different understanding of journalism, one which did not condemn the clipping of news from other journals, particularly if the source was acknowledged. In the same manner, the argument that news was a "public good" rather than a commodity was used in India by less privileged journalists, be they Anglo-Indian or Indian, who clipped Reuters telegrams without actually paying subscription fees. Occasionally, this argument was also employed by members of the colonial administration who found themselves outside the purview of a government policy that sanctioned the free distribution of Reuters telegrams only to a select group of high-ranking colonial officials.

Thus, apart from outlining Reuters' activity in South Asia during the nineteenth century and the strategies it used in order to establish a monopoly over news exchange between Britain and India, this section has also begun to show that the agency was instrumental in promoting an increasingly pervasive vision of news as commodity. Despite its own constant attempts to obtain concessions from the Government of India, Reuters clearly subscribed to the idea that only those who paid for its news were entitled to access it; as a result, the news needed to be protected from those who failed

45 Ibid. It is not clear from Stark's account whether the *Avadh Akhbar* reprinted Reuters telegrams from the English-language journals, but it does appear that the telegraphic news published on the front page contained international news. If this is the case, it would be interesting to know exactly what was the source of this news, especially in light of Donald Read's statement that "Whole pages of the Indian newspapers, English-language and vernacular, were filled with Reuter telegrams." See Read, *The Power of News*, p. 87.

46 Brown, Treatment of News in Mid-Victorian Newspapers, p. 39.

to pay for it. The problem of efficiently protecting such an elusive product as news was to challenge the agency for many years to come, not only in India, but also in Britain. The following section takes a look at some of the strategies adopted by Reuters in order to achieve this goal in colonial India.

Protecting Reuters' News

Since Reuters' entire trade depended on its ability to attract subscribers to its news service, it was only natural that the agency would become more and more protective of the intelligence it sold. By the end of the nineteenth century, it was clear that the issue of protecting news through copyright had become more and more pressing. Reuters telegrams were registered for copyright at Stationer's Hall, London, in August 1870, but this measure failed to offer the desired protection in Britain, let alone in India.[47] In 1881, in a famous decision in the case of Walter vs. Howe, Sir George Jessel declared that newspapers were to be classed as books under the British Copyright Act of 1842, but a decade later it was still not clear what categories of newspaper content were to be protected and how.[48] In the absence of a satisfactory solution, Reuters' agent in India skillfully negotiated the company's position with the colonial administration, constantly pressuring the Government to cave in to its demands.

Since 1874, Reuters had twice attempted to push for a change in legislation which would offer copyright protection to telegrams for at least 48 hours after their publication.[49] Twice their proposal met with the rejection of the Government of India and a new attempt launched in 1878 encountered a similar fate. On 17 April 1878, F. J. Griffiths, Secretary to Reuter's Company at Bombay, wrote to S. C. Bayly urging the Government to grant copyright protection to telegrams published in India. Griffiths argued that telegram piracy was a serious "abuse" to those newspapers which paid for their telegraphic news: "On the one hand, they [newspapers] are deprived in a measure of the value of the information which they receive from their own

47 Read, *The Power of News*, p. 59.

48 Brake/Demoor, *Dictionary of Nineteenth-Century Journalism*, s.v. "Copyright," p. 143; Low, Newspaper Content.

49 NAI, Home Department, Public A, June 1878, Nos. 36 – 39, Copyright of Telegrams, Minute of C. Bernard, 11 May 1878.

correspondents; on the other, as regards a subscription service like Reuter's telegrams, they are liable to pay a higher rate than might be necessary if all newspapers or individuals now publishing these telegrams had to purchase a license to do so."[50]

The practice was all the more objectionable, Griffiths argued, at a time when important political events took place in Europe and the "extras" or news bulletins issued upon the receipt of telegrams were "unscrupulously" reproduced by newspapers which did not subscribe to Reuters.[51] Griffiths' complaints suggest that the increasing commoditization of news during the last decades of the nineteenth century had led to a shift in perceptions regarding news reporting. Clipping intelligence and articles from other newspapers, once a widely accepted journalistic practice, was now increasingly described by some people as "piracy." In other words, Reuters argued that by paying the subscription fee, newspapers in India had acquired the right of property over the telegrams, a right which had to be protected against abuse. In support of this request, Griffiths cited a legislative act and two court cases from Victoria and Singapore which offered copyright protection to telegrams.[52] What Griffiths failed to mention was that all these acts had been temporary measures. For example, Act no. 414, which offered copyright protection to telegrams for 24 hours, was a year-long measure passed on November 23, 1871 by the Victorian Legislature which expired on December 31, 1872.[53]

The response of the Government of India betrayed Reuters' expectations. While admitting that the piracy of telegrams was proliferating, the Government rejected the claim that telegraphic intelligence was, in any way, private property.[54] Ironically for a Government which had always tried to curb the right of newspapers to publish and comment freely on official information, when confronted with Reuters' request, most colonial administrators argued that the information contained in telegrams was "public good." For example, one official pointed out that, unlike works of literature and art, where there was "a distinct public advantage in granting protection," the dissemination of telegraphic information benefited the public interest, while another claimed

50 NAI, Home Department, Public A, June 1878, Nos. 36 – 38, Copyright of telegrams, Griffiths to Bayly, 17 April 1878.

51 Ibid.

52 Ibid.

53 Ibid., Minute of W. Stokes, 27 May 1878.

54 Ibid., Minute of C. Bernard, 11 May 1878.

that "the benefit produced by the prompt dissemination of news outweighs the mischief (if any) done to Reuter's Company and the newspapers paying for telegrams."[55] Since no public interest was harmed, Government logic went on, the piracy of telegrams could not be declared a criminal offence. Furthermore, since the piracy of telegrams did not constitute a criminal offence in any other country in the world, such legislation could not be passed in India as well, as the colony "should follow rather than lead more advanced countries in legislating on such matters as telegrams."[56]

By the 1890s, the issue of news copyright had become an important topic of discussion in Britain, especially after the publication of Sidney Low's article *Newspaper Copyright* in the *National Review* of July 1892. Low claimed that treating newspapers as books for purposes of copyright was inappropriate and argued for a separate Newspaper Copyright Act which, 1) vested copyright in the registered proprietors or publishers of a paper, and not its "authors," 2) allowed only owners of registered newspapers to sue for infringement of copyright, 3) allowed owners to sue for infringement only if they had expressly prohibited the reproduction of the whole newspaper or certain parts of its literary articles,[57] 4) prescribed that such literary articles should be protected in the same way as those contributed to monthly magazines or similar periodicals, and 5) granted no copyright in news.[58]

Opinions were divided regarding the possibility of news copyright. As Low explained metaphorically, this was a battle between "the austere vindicator of the sanctity of news, who would have every 'pirate' promptly riddled below the water-line and sunk" and "the bold 'free-trader' who thinks that the newsmonger may cast his nets into every sea and land whatever fish he can draw into them."[59] The issue at stake was whether it was indeed possible to protect the content and form of news. In a lawsuit against the *St. James*

55 Ibid., Minute of C. Bernard, 14 May 1878, ibid., Minute of W. Stokes, 27 May 1878.

56 Ibid.

57 This was in accordance with Art. 7 of the Berne Convention for the Protection of Literary and Artistic Works (1886) which stated that, "Articles in newspapers or magazines published in any country of the Union may be reproduced, in original or in translation, in the other countries of the Union, unless the authors or publishers have expressly forbidden it. For magazines, it is sufficient if the prohibition is made in a general manner at the beginning of each number of the magazine. No prohibition can in any case apply to articles of political discussion or to the reproduction of news of the day or miscellaneous items (notes and jottings)." Bentley/Kretchmer (eds.), Primary Sources on Copyright.

58 Low, Newspaper Copyright, p. 666.

59 Ibid., p. 662.

Gazette (Walter v. Steinkopff, June 3, 1892), *The Times* claimed that the former had unlawfully reproduced three "copyrighted" news paragraphs, as well as a descriptive article contributed by Rudyard Kipling. Justice North, while recognizing that copyright was infringed with regard to Kipling's article, ruled that there was no copyright in the three news paragraphs, since news, once published, belonged to the public domain and was available to everyone. North's decision was interpreted by legal practitioners to mean that there was no copyright in the content of news, only in the form in which it was expressed.[60]

Reuters' archival holdings – copies of copyright-related Indian legislation, such as the Act XX of 1847 (The Copyright Act) and Act XXV of 1867 (Press and Registration of Books Act) – suggest that by the 1890s, the agency's determination to protect its telegrams both in Britain and India had become stronger.[61] To achieve this, Reuters resorted to an ingenious strategy which involved the publication of its telegrams as part of a daily newssheet known as *Reuter's Journal*.[62] According to Graham Storey, one of Reuters' official historians, this bulletin contained an assortment of Reuters' most important daily telegrams and was partially successful in preventing other newspapers from "illegally" reproducing them in their pages. Both the *Western Daily Mercury* and the *Belfast News-Letter*, two of the more notorious offenders, were forced to renew their subscriptions to Reuters as a result of this measure.[63]

A similar publication was issued five years later in India under the name *Reuter's Indian Journal*. On 30 August 1895, the agency's solicitors at Calcutta submitted the first copy of the journal for registration under Section 11 of Act XX of 1847. In this way, Reuters' ownership in the content of the journal was also recognized. The copy submitted shows that the journal was published at Bombay and it contained, under the heading "Parliamentary News," a digest of Reuters' London telegrams on topics such as: "Mr. Stanley on Uganda Railway," "France and the Niger," "Mr. Curzon on Egypt," "The Siam Buffer State" and "Lancashire Mill Owners and Cotton Duties."[64] In

60 Ibid., p. 658.

61 See, for example, RA, LN1036, 12 June 1896, Copyright Act (India) 1847, certified copy obtained by Reuters in 1896.

62 Read, *The Power of News*, p. 59.

63 Storey, *Reuters' Century*, p. 112.

64 NAI, Home Department, Books and Publications, September 1895, Nos. 114 – 118, Copyright registry of Reuter's Indian Journal.

addition to this measure, Reuters requested all agents and correspondents, regardless of location, to sign a form by which they "invested [the Agency] with the copyright in telegrams and news communications supplied by them to us." Edward J. Buck, Reuters' agent at Simla, relinquished his property rights in the news supplied to London on 23 February 1894.[65]

Conclusion

As this paper has demonstrated, news was a contested domain in nineteenth-century India. Previous histories of the press in colonial South Asia have focused predominantly on the role of various sections of the newspaper press – such as vernacular and Anglo-Indian – in promoting or opposing nationalist or imperialist agendas, thus placing particular emphasis on the ability of the press to shape public opinion. By contrast, this paper has focused on the intricacies of the actual process of foreign news distribution and transmission, aiming to explain the various mechanisms which underlined the flow of news between India and Britain and which provided, to a great extent, the informational basis on which administrative decisions were made and opinions were forged. The paper focused on Reuters' activity in nineteenth-century India, tracing its rise to prominence as a supplier of news from Britain to India and pointing out that its constant attempts to monopolize the flow of intelligence did not go unchallenged. Among the other competitors in the race for news were journalists such as Robert Knight of the *Times of India*, who set up his own news agency, as well as correspondents for various British newspapers and members of the mercantile communities of Calcutta and Bombay.

It is against this background that Reuters, mainly through its agents in India, worked to establish itself as the main purveyor of intelligence between Britain and its colony. As we have seen, Reuters had strong connections with the colonial government of India, whose benevolent attitude was essential to the agency's success. This position, simultaneously privileged and vulnerable, required a fine balancing act between the need to support the imperial project in order to gain monopolistic access to the Indian news market and the more general goal of surviving in the global telegraphic news market. In particular, overt support of the colonial government could seriously com-

65 RA, LN39, 26 January 1894, Copyright agreements of agents and correspondents.

promise the agency's claims of independent reporting, an ideal which had gained increased recognition as a mark of good journalism during the nineteenth century, in particular under the influence of American journalism.[66]

Reuters' activity in the subcontinent, technologically mediated by electric telegraphy, fostered global connections and interactions between English-language journalism in India and other parts of the world, in particular Britain and the United States, but also East Asian countries such as Japan.[67] Globality in news and journalism was predicated upon the increasing commercialization and commodification of news, a position to which not all journalists or consumers of news subscribed. As a technology, the telegraph was associated with speed, timeliness and concise, factual reporting. Through its form and content, the telegraphic news published by Anglo-Indian newspapers proclaimed not only Reuters' monopoly of news distribution, but also London's centrality in an emerging global network of communication, since the news was almost invariably published as having originated in London, despite the fact that it reported on various parts of the world.[68] The form and content of the telegrams thus promoted a vision of news and journalism which was essentially Anglo-American and which was also reflected in the English-language press in other parts of the world.[69] To give but two examples, the following Reuters telegram announcing that "General Skobeloff has dispatched a column to occupy Kizilarvat" was published in this form in the *Bombay Gazette* of 7 July 1880 and *Japan Gazette* of 12 July 1880, while on 7 July 1890 both *The Englishman* and *The Mercury* of Hobart announced, respectively, that "In the House of Lords this evening the Bill for the cession of Heligoland to Germany was read the first time," and "The bill legalising the cession of Heligoland to Germany has been read a first time in the House of Lords."

As the present paper has shown, the emerging vision of news as commodity was often challenged in the language of public interest, which gen-

66 A long-overdue study of the interactions between American and British journalism during the nineteenth and early twentieth centuries, including a discussion of the notion of objectivity, can be found in Wiener, *The Americanization of the British Press*.

67 A comparative analysis of news reporting in nineteenth-century Indian and Japanese newspapers was recently presented by the author as Modernity, Technology and Speed: Insights from South and East Asian Journalism, International Conference "Technology, Media and Journalism," Saarland University, 22–23 March 2013.

68 In this connection, see also Wenzlhuemer, London in the Global Telecommunication Network.

69 For interactions between British and American journalism during the nineteenth century, see also Hampton, *Visions of the Press in Britain*.

erally downplayed Reuters' claims that it had invested considerable effort and expense in the collection and distribution of intelligence. Apart from clipping news from bigger and more established newspapers, the alternative, for those papers which could not afford the costly services of Reuters and were often overwhelmed by competition even on the domestic news market, dominated by journals such as *The Pioneer*, *The Englishman* and *The Statesman*, was to organize an independent news agency. The Associated Press of India, founded in 1908 and modeled after the Associated Press of New York, was the first, albeit short-lived, effort in this direction.[70]

Thus, during the nineteenth century Reuters played an important role in facilitating the circulation of information between Britain and India and helped to integrate these regions into a network of news which extended well beyond their geographical confines. At least in the field of English-language journalism, the circulation of Reuters telegrams also facilitated the circulation of certain forms and standards of reporting which were reflected in the newspapers examined. Yet, as the present paper has shown, the globalization of news via the electric telegraph and Reuters' news agency was far from being a straightforward process: it also involved continuous negotiations of what counted as "news," who could access news and who could be regarded as a trusted purveyor of news.[71] In this sense, the telegraph created new distinctions and boundaries as much as it facilitated integration and connections. The dependence on foreign news provided by Reuters gave birth to certain asymmetries in communication which have persisted long after the demise of colonialism in India. It remains to be seen whether the telegraph of the twenty-first century, the Internet, can help to correct this situation, as some technology enthusiasts seem to believe.

Bibliography

Archives and Libraries consulted

British Library, Oriental and India Office Collections, London
National Archives of India, New Delhi
Reuters Archives, London

70 Raghavan, *PTI Story*, pp. 54–55.
71 In this connection, see also Barth's contribution in the present volume.

Newspapers

The Bengalee
Bombay Gazette
The Englishman
Times of India

Published Primary and Secondary Sources

Administration Reports of the Indian Telegraph Department, 1855 – 1900. [BL: OIOC]

Appadurai, Arjun, *Modernity at Large: Cultural Dimensions of Globalization*, Minneapolis, MN 2003.

Arya, Kapil, *The Over-Dependence of Indian English Newspapers on Global Agencies for International News*, M.A. thesis, University of Leeds 2011.

Banerjea, Surendranath, *A Nation in the Making: Being the Reminiscences of Fifty Years of Public Life*, 1925, reprint Bombay 1963.

Bentley, L. and Kretchmer, M. (eds.), *Primary Sources on Copyright (1450 – 1900)*, available online, accessed October 26, 2011, www.copyrighthistory.org.

Blondheim, Menahem, "Slender Bridges of Misunderstanding: The Social Legacy of Transatlantic Cable Communications," in: N. Finzsch and Lehmkuhl, U. (eds.), *Atlantic Communications: The Media in American and German History from the Seventeenth to the Twentieth Century*, Oxford/New York 2003, pp. 153–70.

Bonea, Amelia, The Medium and Its Message: Reporting the Austro-Prussian War in the *Times of India*, *Historical Social Research*, vol. 35, no.1 (2010), pp. 167–87.

Boyd-Barrett, Oliver, Media Imperialism: Towards an International Framework for the Analysis of Media Systems, in: J. Curran/M. Gurevitch and Wollacott, J. (eds.), *Mass Communication and Society*, London 1977, pp. 116–135.

Brake, Laurel and Demoor, Marysa (eds.), *Dictionary of Nineteenth-Century Journalism in Great Britain and Ireland*, Gent/London 2009.

Brown, Lucy, The Treatment of News in Mid-Victorian Newspapers, *Transactions of the Royal Historical Society*, vol. 5, no. 27 (1977), pp. 23–39.

Chadha, Kalyani and Kavoori Anandam, Media Imperialism Revisited: Some Findings from the Asian Case, *Media, Culture and Society*, vol. 22, no. 4 (2000), pp. 415–432.

Chesney, G. M., The Native Press in India, *Nineteenth Century: A Monthly Review*, vol. 43, no. 252 (February 1898), pp. 266–276.

Collins, Henry, *From Pigeon Post to Wireless*, London 1925.

Galtung, Johan, A Structural Theory of Imperialism, *Journal of Peace Research*, vol. 8, no. 2 (1971), pp. 81–117.

The Gazetteer of Bombay City and Island, vol. 1, Bombay 1909.

Half-Yearly Report of the Bengal Chamber of Commerce, November 1860 – April 1861. [BL: OIOC]

Hampton, Mark, *Visions of the Press in Britain, 1850–1950*, Chicago, ILL 2003.

Harcourt, Freda, *Flagships of Imperialism: The P&O Company and the Politics of Empire from Its Origins to 1867*, Manchester/New York 2006.

Hirschmann, Edwin, *Robert Knight: Reforming Editor in Victorian India*, New Delhi 2008.

Jeffrey, Robin, *India's Newspaper Revolution: Capitalism, Politics and the Indian Language Press, 1977–99*, London 2000.

Kaul, Chandrika, Imperial Communications, Fleet Street and the Indian Empire, c. 1850s – 1920s, in: M. Bromley and T. O'Malley (eds.), *A Journalism Reader*, London 1997, pp. 58–86.

—, *Reporting the Raj: The British Press and India, c. 1880 – 1922*, Manchester 2003.

Kperogi, Farooq A., Guerrillas in Cyberia: The Transnational Alternative Online Journalism of the Nigerian Diasporic Public Sphere, *Journal of Global Mass Communication*, vol. 1, nos. 1/2 (2008), pp. 72–97.

Lethbridge, Roper, The Vernacular Press in India, *Contemporary Review*, vol. 37 (Jan./June 1880), pp. 459–473.

Low, Sidney J., Newspaper Copyright, *National Review*, vol. 19, no. 113 (July 1892), pp. 648–666.

Parliamentary Papers, *Reports from Committees: East India Communications, 1 February – 10 August 1866*, vol. IX, London.

Raghavan, G. N. S., *PTI Story: Origin and Growth of the Indian Press and News Agency*, Bombay 1987.

Read, Donald, *The Power of News*, New York 1992.

Sapru, Somnath, *The News Merchants: How They Sell News to the Third World*, New Delhi 1986.

Stark, Ulrike, *An Empire of Books: The Naval Kishore Press and the Diffusion of the Printed Word*, New Delhi 2007.

Storey, Graham, *Reuters' Century*, London 1951.

Watson, A. H., Origin and Growth of Journalism among Europeans, *Annals of the American Academy of Political and Social Science*, vol. 145, part 2 (1929), pp. 169–174.

Wenzlhuemer, Roland, London in the Global Telecommunication Network of the Nineteenth Century, *New Global Studies*, vol. 3, no. 1 (2009), Art. 2.

Wiener, Joel H., *The Americanization of the British Press, 1830s – 1914: Speed in the Age of Transatlantic Journalism*, Basingstoke 2011.

Wright, Arnold, *Baboo English as 'Tis Writ: Being Curiosities of Indian Journalism*, London 1892.

Making the Wire Speak: Transnational Techniques of Journalism, 1860–1930

Volker Barth

Recent research on the press and the media has proven to be particularly interested in the transnational and global aspects of media networks.[1] National boundaries do not provide convincing frameworks of analysis as one of the principal characteristics of media is their capacity to reach beyond established political, social and economic frontiers. This is especially apparent in the period from the middle of the nineteenth century up to the 1930s, which can be described as the formative years of the modern media landscape and a time of unprecedented intensification of global interconnectedness.

This chapter takes the establishment of a global telegraph network as a starting point to investigate the evolution in the production, transmission and verification of news items. As this network was mostly run by private enterprises such as early multinational telegraph companies and extensively used by private news businesses such as news agencies and internationally operating newspapers, commercial requirements formed a constitutive element for the development of modern journalism. Within the time span under consideration journalistic techniques changed significantly. In many aspects the production and transmission of news as it was practiced in the 1860s was no longer considered adequate in the 1930s. In dealing with new means of communication, news businesses and media enterprises had to adapt their daily routines to the requirements of an increasingly well-established world-wide network.

From its inception, telegraph lines and submarine cables, just like decades later wireless telegraphy and the telephone, were far from being mere channels to pass on information. Instead, these new tools of transmission heavily influenced the nature of news. Reciprocally, new ways of defining and evaluating news proved to be decisive for the choice of adequate trans-

1 Cf. for example Bösch, *Mediengeschichte*; McMillin, *International Media Studies*; Golan et al. (eds.), *International Media*; Kishan Thussu (ed.), *International Communication*; Kishan Thussu (ed.), *Internationalizing Media Studies*.

mission channels. The question is to what extent available technical infrastructures shaped not only the work of journalists, editors and foreign correspondents but the very definition of news. How did new technological devices influence standards of professional journalism? What special skills did news producers have to develop in order to be commercially successful all while assuring journalistic reputation? Given the expanding and ultimately global nature of the network, this chapter focuses on the emergence of transnationally applicable techniques of professional journalism and their reliance upon, on the one hand, diverse types of technical infrastructure and, on the other hand, evolving concepts of accuracy, veracity and objectivity.

While accuracy referred mainly to "keeping the paper free of the easily identifiable errors and excesses," veracity pointed towards a civic and moral obligation for every journalist.[2] The term objectivity not only comprised these two meanings but went a significant step further. Over the period under consideration, it slowly developed into the main ideal of the simultaneously evolving journalistic profession. It was not until the 1920s, however, that leading journalists, authors of handbooks for future journalists and especially members of the increasing number of schools of journalism tried to formulate comprehensible definitions as well as a mandatory set of rules in order to put them into practice. Ultimately, objectivity aimed at a clear-cut distinction between facts and values. By the early 1930s, the term had established itself as "at once a moral idea, a set of reporting and editing practices, and an observable pattern of news writing."[3]

Therefore, firstly, I address the institutionalization and operation of global information networks. Secondly, I analyze how and why journalists used these networks. Finally, the article questions evolving journalistic standards of assuring the accuracy, veracity and objectivity of transmitted information within media enterprises operating on a global scale.

Transnational Communication Networks

The four decades between 1860 and 1900 saw the establishment of a global network of telegraph lines connecting the five continents and their major

2 Schudson, *Origins of the Ideal of Objectivity*, p. 177.
3 Schudson, The Objectivity Norm in American Journalism, p. 149.

cities. After the invention of this revolutionary means of communication in the 1830s, many, mostly national lines were built in the following years. In 1851, the first submarine cable linked England to France via the channel and after a first, unsuccessful initiative, an intercontinental cable was laid across the Atlantic ocean in 1866. With the completion of the Pacific cable in 1902, the initial stage of the wiring of the world was completed.

While nation-states operated the majority of national cables, private, multinational companies ran the bulk of the international lines. In order to cover the excessive costs of intercontinental cables, whether land or submarine connections, and to shut out possible competitors, the biggest cable companies rapidly set up business corporations controlling large parts of the global telegraph network. Thus, in 1879 the so-called cable pool consisting of the *Anglo-American Telegraph Company*, the *Direct US Cable Company*, and the French *Societé du cable transatlantique français* operated all three existing transatlantic cables. In 1884, the *Commercial Cable Company*, which had an arrangement with the New York based *Western Union* telegraph company, established an agreement with the pool thus assuring a "decisive access for transatlantic transmission to the US news market."[4]

Even more than for private telegrams or official government messages, telegraph lines were used for the transmission of news items. In Germany, the *Kölnische Zeitung* published its first press telegram as early as October 5, 1849.[5] In England, the first news item was cabled from Portsmouth to Vauxhall in 1845 for the *Morning Chronicle*. News from abroad arriving by ship in English ports had to be transmitted as quickly as possible to the London newspapers, which were eager to satisfy their readers and seize their share of the rapidly expanding news market. Nearly all telegraph companies had their headquarters in London, one of the world's leading newspaper capitals, and were thus equally aware of this business opportunity. By as early as 1852, several English telegraph companies had founded their own news service and delivered news items "relating mainly to sporting and Stock Exchange" to the English press.[6] Thus, from the outset, business considerations played a decisive role not only for the modes of transmission but for the content of news items as well.

4 [„für den Atlantikverkehr entscheidenden Zugang zu den US-amerikanischen Nachrichtenmärkten."] Wobring, *Die Globalisierung der Telekommunikation*, p. 194.
5 Morf, *Die Drahtberichterstattung*, p. 12.
6 Hunt, *Then and Now*, p. 111.

However, subscribing newspapers soon began to complain about the telegraph companies' news service. On the one hand, they found the uniformity of a single news service problematic as it hindered newspapers from distinguishing themselves from competing publications. When in 1856, the *Manchester Guardian* wanted to cable an independent report of a parliament debate, the telegraph companies simply refused. "For some time the country newspapers managers protested in vain against the formidable monopoly which the telegraph companies were acquiring." On the other hand, their complaints regarded the often exorbitant rates that the telegraph companies charged for transmission. Finally, newspapers especially denounced the "late and untrustworthy information" they obtained via the cables.[7]

Almost simultaneously to the English telegraph companies, private entrepreneurs began to enter the emerging market of news telegraphy. As early as 1835, Charles Havas had opened the first modern news agency in Paris. In 1848, he was followed by his former employee Bernhard Wolff, who started *Wolff's Telegraphisches Bureau* in Berlin. In the same year, six New York newspapers convened to form the *Associated Press*, the first American news agency. In London, several weeks after the completion of the submarine cable across the channel and shortly before the English telegraph companies started their news service, Julius Reuter, like Wolff a former Havas co-worker, set up the *Reuters* news agency. After several years of concentrating exclusively on stock exchange rates *Reuters* cabled its first "newspaper telegram for general news" on October 8, 1858.[8]

These first four news agencies quickly managed to work out special tariffs with the telegraph companies and successfully obtained cable privileges within state run telegraph networks. The *Associated Press* for example signed a contract with *Western Union* obtaining very advantageous financial conditions in exchange for the promise to rely exclusively on their cable network. Additionally, the news agencies – similar to the telegraph companies they so heavily depended upon – formed a news cartel. In 1859, the three European agencies signed their first cooperation treaty. The *Associated Press* joined the cartel in 1870 as a junior and in 1893 as a senior partner.[9] The cartel treaties

7 Bourne, *English Newspapers*, p. 278.

8 Hohenberg, *Foreign Correspondence*, p. 9.

9 Diverging dates are given concerning the exact year the AP joined the cartel. Most recently Evans upheld the year 1900. Evans, The Path to Freedom?, p. 213. The year 1893 is given by Kent Cooper, the General Manager of *Associated Press* after World War I. Cooper, *Barriers Down*, p. 16. For the history of AP in general see Blondheim, *News over the Wires*.

were renewed every decade or so up to the 1930s and divided the world into four geographically defined news monopolies. *Reuters* dominated the British Empire, the Far East, and the Netherlands, while sharing Turkey, Egypt and Belgium with *Havas*. In addition to these three countries, *Havas'* zone of influence included France and the French colonies, as well as Italy, Spain, Switzerland, large parts of Eastern Europe, and the entire South American continent. Germany, Austria, Northern Europe, and Russia, were *Wolff's* domain. News from the United States was under the exclusive purview of *Associated Press*. The treaty was based on the "principle of commercial exclusivity" and stipulated an "absolute interdiction to act in reserved territory," which meant that an agency could not produce or sell news reports within a foreign zone of influence, but must instead refer to the relevant competitor.[10]

The boundaries of reserved territories evolved according to international politics. Thus, in 1890, the cartel treaty was only renewed after significant modifications in exchange for substantial payments: *Havas* was driven out of Serbia, Montenegro, Romania, Bulgaria, and Turkey; the Balkans became the territory of an Austrian news agency connected with *Wolff*; most of Eastern Europe came under the influence of *Reuters*. Thanks to the growing influence of the United States, *Associated Press* was able to significantly improve its position through renewal of the cartel treaty in 1903. It absorbed Central America, Cuba, and the Philippines into its exclusive area of coverage.[11]

Being the only provider of a global news service – as they referred to themselves —, the cartel's commercial success was almost guaranteed. Its news-wire service constituted a "spider web, fanning out from a central office to connect with bureaus and beats and also to join diverse news media to a central location."[12] *Reuters*, for instance, collected news items in six regional centers (Cape Town, Alexandria, Bombay, Hong Kong, Teheran, Yokohama) before cabling it to their London headquarters.[13] Foreign news was by far the agencies' most valuable product. In 1919, director Hermann Dietz stated that

10 Heerdegen, *Der Nachrichtendienst der Presse*, p. 43; treaty of 1870 quoted in Basse, *Wolff's Telegraphisches Bureau*, p. 255.

11 For the history of the cartel treaties and global news agencies, cf. Boyd-Barrett, *The International News Agencies*; Boyd-Barrett and Palmer, *Le trafic des nouvelles*; Boyd-Barrett and Rantanen (eds.), *The Globalization of News*; Rantanen, *When News was New*; Read, *The Power of News*.

12 Tuchman, *Making news*, p. 20. Cf. as well Winder, London's Global Reach?.

13 Heerdegen, *Der Nachrichtendienst der Presse*, p. 18.

Wolff gained almost all its revenues from news items from abroad.[14] Once obtained, the agency passed on the news via its domestic branches to its German customers.[15] In accordance with the central importance of foreign news, *Associated Press* opened its first foreign branch as early as 1849 in Halifax, Nova Scotia, the closest ship link to Europe.[16]

The news agencies' cartel and its diverse links to the telegraph businesses practically granted them a monopoly over the global news market for almost seven decades. Until 1907, when competing agencies were founded, *Associated Press* was the only agency able to supply a foreign news service within the United States. Until well into the twentieth century, *Reuters* was virtually the only source for news from abroad for the entire Indian subcontinent.[17] In 1915, the English *Press Association*, an agency specialized in English domestic news and collaborating with *Reuters*, possessed a "practical monopoly of the collection of news" and controlled over 88 percent of the domestic market.[18]

For newspapers, however, depending on the news services of the cartel was as problematic as their earlier reliance on the British telegraph companies' news service. Newspapers continued to worry about issues of trustworthiness and the uniformity of dispatches. In order to cope with these difficulties, they first formed alliances to exchange independently produced news items. In May 1874, the London *Times* laid a direct cable to its Parisian partner newspaper *Le Matin*, transmitting messages independently from the *Havas* and *Reuters* services.[19] Secondly, in 1890, backed up by several newspapers from other countries, *Le Matin* and *The Times* subscribed to the recently established *Dalziel* agency in New York. *Dalziel* specialized in human interest stories and *fait divers*. This allowed its subscribers to address a different readership than the cartel customers. As a response, *Reuters* and *Associated Press* started their own *fait divers* service and *Dalziel* went broke in 1893.[20]

The news agencies' cartel and its cooperation with the telegraph companies revolutionized the news market. Thanks to the telegraph, London newspapers were no longer the only news source for English provincial papers. In

14 ["fast ausschließlich im Auslandsgeschäft erzielt"]. Dietz, *Das Zeitungswesen*, p. 76.

15 ["Es laufen den ganzen Tag Meldungen ein, die sofort strahlenförmig verbreitet werden."] Morf, *Die Drahtberichterstattung*, p. 20. Cf. as well ibid., 53.

16 Fortner, *International Communication*, p. 90.

17 Bonea, The Medium and Its Message, p. 180. See also Amelia Bonea's chapter in this book.

18 Silberstein-Loeb, The Structure of the News Market in Britain, p. 769.

19 Blowitz, *Memoirs of M. de Blowitz*, p. 37.

20 Mathien and Conso, *Les Agences de Presse Internationales*, p. 51.

the 1880s, English journalist Fox R. Bourne observed not only a "complete revolution of provincial journalism," but also considered the service of news agencies "more comprehensive and accurate as regards facts than much that might be put together by ill-informed or careless editors."[21] For the agencies, quick access to the telegraph was decisive. In 1898, *Associated Press* correspondent J. Hilgert was the first journalist to transmit the news of the outbreak of the Spanish-American war. This scoop was only possible because the agency was allowed to use the US-Navy's cable network. Competitors had to travel from Cuba to Kingston, Jamaica, to reach a telegraph line.[22]

The speed of the telegraph was another crucial factor, of course. However, it was not always the fastest way to get news into the newspapers. The agencies relied on a range of different media to provide the most effective service. At the beginning of the twentieth century, *Havas* produced several daily newsletters consisting of six pages and approximately 30,000 characters, "every single one of which forms a complete newspaper."[23] The agency sent ready-to-print stereotypes by train from Paris all over the country where they could be printed and distributed half an hour after arrival.[24] German newspapers also obtained the bulk of the *Wolff* service by train. The telegraph – as it helped to gain less time than the stereotypes – was only used to transmit late incoming news.[25]

With regard to short distance communications, the telephone fulfilled a similar function but posed the additional problem of leaving no tangible trace of transmissions. Furthermore, it contradicted the agencies' principal rule of delivering news simultaneously to all its clients.[26] The abundance of news items rendered it impossible to transmit them entirely over the phone.[27] Nevertheless, in the Paris headquarters of *Havas*, 20 telephone operators took stenographic notes of all incoming calls and transmitted them by telegraph.[28]

21 Bourne, *English Newspapers*, p. 257, 259. He was referring to the *Central Press*.

22 Hohenberg, *Foreign Correspondence*, p. 49.

23 ["deren jede für sich eine vollständige Zeitung darstellt"]. Bücher, *Unsere Sache und die Tagespresse*, p. 27.

24 The English *Central Press* used the same system and additionally send news items arriving too late to be put into printing form as "duplicate flimsy" with the same train. Hunt, *Then and Now*, p. 72.

25 Wrede, *Handbuch der Journalistik*, p. 253.

26 Morf, *Die Drahtberichterstattung*, p. 20.

27 Heerdegen, *Der Nachrichtendienst der Presse*, p. 31.

28 Morf, *Die Drahtberichterstattung*, p. 49.

An agency's choice of the adequate media for the transmission of news depended on several factors, the most important of which were efficiency, reliability and the cost-benefit-relation. New, faster media were not automatically the best transmission tools within a global and rather fragile network.[29] After all, long distance telegraphy demanded a heavy technical and human infrastructure. The 1884 terrestrial line between Europe and India required transmitting posts every 30 to 80 miles, which were operated and maintained by "European employee[s]." These were "mostly old R.E. sergeants or corporals who had been originally lent by the Government of India to built the line." Not the least of their concerns were Persian nomads who time and again stole the wires and used them to fabricate weapons.[30] Decades later, wireless technology did not really overcome such problems either as it depended heavily on meteorological conditions. Up to 1920, more than ten percent of all transmitted messages never reached their intended recipients.[31]

Besides adequate transmission channels, the careful selection of news was another big issue for the agencies. The choice and the amount of delivered news determined quality, reputation and profitability alike. Consequently, a correspondent from every agency working in the headquarters of every partner agency preselected relevant news items in order to avoid transmission costs for irrelevant ones.[32] Keeping their clients in mind, these correspondents had to add "obviously necessary explanations" in case messages were judged unintelligible.[33]

Uses of Networks

In order to obtain information, newspapers relied above all on other newspapers. Reading the press and screening it for useable news was a primary task for every journalist. Smaller newspapers especially could not afford to hire special correspondents to produce their own news. They depended on specialized agencies to obtain anything but strictly local news. Thus, only a handful of major newspapers had permanent correspondents in some of the

29 See Jonas Harvard's chapter in this book.
30 Chirol, *Fifty Years in a Changing World*, pp.77, 145.
31 Evans, The Path to Freedom?, p. 219.
32 Baumert, *Die Entstehung des deutschen Journalismus*, p. 70.
33 ["nötig erscheinende Erklärungen"]. Morf, *Die Drahtberichterstattung*, p. 51.

world's most important political centers. But for news from outside Washington, London or Paris, major newspapers too were forced to subscribe to a news agency. Up to the 1930s, the agencies' cartel was the only provider of a supposedly global service.

In 1938, Henry Wickham Steed, the Paris correspondent of Joseph Pulitzer's *New York World*, described the tasks of a newspaper correspondent: "They take current or normal news from the News Agencies and supplement or explain it by news and comments which their own correspondents at home and abroad collect and supply."[34] But even if a newspaper had local collaborators, they were not always able to deliver a service equivalent to the big agencies. During the Boer War (1899–1902), only three German newspapers sent correspondents to South Africa. However, those correspondents were not allowed to use British telegraph lines. The British army controlled telegraph lines and only made them available to *Reuters* correspondents who cabled their news items to London. From there, they were distributed via *Wolff* within Germany. The German special correspondents sent news by mail, which traveled far too slowly for publication in daily news bulletins. Therefore the reporters wrote extensive descriptions and detailed reportages rather than short, factual messages. When advertising those articles to their readers, newspapers stressed correspondents' German nationality as opposed to British origin and pointed out the "more objective picture" they painted.[35]

A correspondent's ability to report particular types of news depended above all on access to and availability of technical infrastructure. As commercial merchandise, news items had to be saleable to customers. Valentin Chirol, head of the London *Times*' Berlin office in the years preceding World War I, instructed his staff: "A correspondent, while studying every question with all the diligence of a specialist, should seek to place the results of his studies before the public in such a shape as to render them easily accessible and attractive to the largest numbers of readers possible."[36] William T. Stead, editor of the *Pall Mall Gazette* and one of the founding fathers of investigative journalism, explained the work of an effective foreign correspondent to Henry Wickham Steed as follows:

> When you have found something to say and have written it down, imagine that it has to be telegraphed to Australia at your own expense. Telegrams cost, I think, five

34 Steed, *The Press*, p. 193.

35 ["objectiveres Bild"]. *Kölnische Zeitung* 15.4.1900, quoted in Bender, *Der Burenkrieg und die deutschsprachige Presse*, p. 43.

36 Chirol to Morrison 18.5.1896. Quoted in Fritzinger, *Diplomat Without Portfolio*, p. 494.

shillings a word. Then, when you have cut out every superfluous word and spoiled all your favourite phrases, copy it out legibly and send it to an editor.[37]

Every news item had to be evaluated in relation to its transmission and human labor costs. This was not only true for telegraphy's early years when rates were exuberantly high. Even when telephone, wireless and radio had significantly reduced expenses for communications, according to Steed, the fundamental rule of the editor of every newspaper or agency still was: "More messages out than in."[38] It remained more efficient to tell reporters what news to produce rather than await randomly incoming items.

Obviously, the reporter's task was to obtain and transmit the requested information as soon as possible. In order to do so, journalists invented various investigation techniques. Erich Gentz, special correspondent for the *Tägliche Rundschau* during the Boer War, joined the Boer army in order to come closer to the front line.[39] Henri Georges Opper de Blowitz, the *Times* correspondent in Paris at the turn of the century and one of the star journalists of his times, referred to methods of interviewing informants in his memoirs. Blowitz, too, used techniques of disguising his professional identity for obtaining authentic facts:

When a man gives a correspondent an important piece of news, the latter should continue to remain with him for some time, but change the conversation and not leave him until it has turned to something quite insignificant. If the correspondent takes his departure abruptly, a flash of caution will burst upon his informant.[40]

At the end of his career, George Seldes, foreign correspondent for several newspapers and agencies all over the world from the 1910s to the 1930s, characterized his profession's principal investigation techniques without any nostalgia. They mainly consisted of "ear-biting," i.e. "pumping dry your colleagues," and a careful screening of the local press. Furthermore, one had to read the agencies' dispatches and official press releases as well as hire collaborators. "Original work by Native Assistants" who – as news items were usually not signed got "no recognition" – was equally important. For Seldes, think pieces, which journalists had to write "when there is no news or when

37 Steed, *Through Thirty Years*, pp. 6–7.

38 Steed, *The Press*, p. 226.

39 However, he considered himself far less an undercover journalist than a fighter for his political convictions, which he equally advertised in his articles and on the battle field. Bender, *Der Burenkrieg und die deutschsprachige Presse*, pp. 44–5.

40 Blowitz, *Memoirs of M. de Blowitz*, p. 46.

the situation is in such flux that no definite bulletin news can be made out of it" were no more than mere exceptions.[41]

Once obtained, the mode of transmission of news items proved to be decisive for their final form of publication. To begin with, for many correspondents, news transmission formed a large part of their daily routine. Philip Gibbs, Paris correspondent of the *Daily Telegraph* in the 1920s, had to receive calls from his colleague in Rome and telephone the news to London on a daily basis.[42] In 1928, Hans Tröbst telephoned local news summaries from Hamburg to the Berlin headquarters of *Telegraphen Union* news agency.[43] The increasing quantities of transmitted news created new journalistic tasks like "non-official supply of material (in particular by late-night telephone calls) by a professional journalist."[44]

Even though transmission networks were set up to work as quickly and reliably as possible, successful journalists needed special skills to maximize the network's potential. Heroes of early journalism like Archibald Forbes, the *Daily News* correspondent in Paris in 1871, did not hesitate to ride a horse to Karlsruhe – a journey of more than 500 kilometers – in order to send his description of the Paris Commune via the telegraph to London, thereby beating all his journalistic competitors.[45] When Henri Blowitz sensationally obtained the contracts of the Berlin Congress in 1878 before their release, he immediately took the next train to Brussels as he knew that no German telegraph operator would agree to cable the text to London.[46] In order to end the dependency on German cables, Charles Lowe, Berlin correspondent of the London *Times*, was given his own "special wire" in the 1880s, with which he could send his news via Paris to London.[47]

However, news transmission was far trickier than getting access to a working telegraph line. Even in the early twentieth century, a dispatch from Paris to Berlin could take as long as two hours. Consequently, a reporter eager to meet a deadline or outscore his colleagues had to be creative. Hans Morf described how his knowledge of the earliest vending points for morn-

41 Seldes, *Lords of the Press*, pp. 286ff.

42 Gibbs, *Adventures in Journalism*, pp. 123–4.

43 Gebhardt and Tröbst, *Leben und Arbeit des Auslandskorrespondenten Hans Tröbst*, p. 65.

44 [„die nebenamtliche Materialversorgung (namentlich in Form des Nachttelephonates) durch einen Berufsjournalisten."] Baumert, *Die Entstehung des deutschen Journalismus*, p. 71.

45 Bourne, *English Newspapers*, p. 282.

46 Blowitz, *Memoirs of M. de Blowitz*, p. 132.

47 Lowe, *The Tale of a "Times" Correspondent*, p. 114.

ing newspapers gave him just as much of a decisive advantage as knowing what telegraph operators were most likely to master foreign languages and thus to avoid spelling mistakes.[48] Hans Tröbst in contrast had quite different problems on his correspondent post in Albania in 1927 where foreign mail left only twice a week.[49]

To get his news telegram from the German colony Cameroon to Berlin in 1885, Hugo Zöller depended on the German navy to ship it to the closest telegraph station. Because of the expensive transmission costs, the military not only obliged him to shorten several passages but also to describe the events "somehow more peacefully and mildly." Referring to that softened version of Zöller's initial description, Bismarck later responded to critical questions in the Reichstag. Dependency on a transmission network and military censorship thus transformed a colonial conquest into a "'peaceful' campaign."[50]

Nevertheless, agencies and journalists presented sophisticated technical transmissions as a means of guaranteeing the accuracy of given information. *Wolff's* director Hermann Dietz advertised his agency by pointing to the private wires at its disposal. For Dietz, telegraph lines were decisive advantages over competitors especially when messages contained very detailed information. In his 1919 study of the German newspaper business, he noted that *Wolff* possessed "numerous rented telegraph lines which were specifically used to transmit difficult or extensive information that was strictly bound to the wording."[51] Henri Blowitz mobilized an important technical infrastructure to make sure that the *Times* would inform its readers faster and better than other London newspapers. During the Paris municipal elections of 1877, he hired special wires connecting every *arrondissement* directly to his office.[52]

But the increasing speed of the telegraph could just as easily trigger inexactitudes. Blowitz' memoirs also illustrate that cutting-edge technology required special journalistic skills for efficient use. Writing about the "method of improvising upon the most recent events that were in progress as one wrote an account of them," Blowitz explained how "letters logically con-

48 Morf, *Die Drahtberichterstattung*, pp. 34–6.

49 Gebhardt and Tröbst, *Leben und Arbeit des Auslandskorrespondenten Hans Tröbst*, p. 51.

50 ["etwas friedlicher und milder"], ["'friedlicher' Eroberungszug"]. Zöller, *Als Jurnalist* [sic] *und Forscher*, pp. 176–7.

51 ["zahlreich gemietete Telegraphenleitungen, die namentlich für die Wiedergabe textlich schwieriger oder umfangreicher, streng an den Wortlaut gebundener Informationen benutzt werden."] Dietz, *Das Zeitungswesen*, p. 76.

52 Giles, *The Life and Times of Henri Stefan Opper de Blowitz*, p. 122.

ceived throughout had to be written while the telegraph waited, without opportunity for revision."[53]

Almost 60 years later, foreign correspondent Hans Tröbst argued similarly with regard to the replacement of the telegraph by the telephone: greater speed required adequate techniques. Thus, his employer, *Telegraphen Union*, gave its reporters detailed instructions of how to accurately transmit information by phone. Correspondents were told to wait for the editor to signal his attention before beginning to speak; they had to systematically spell out proper names and to document every transmission.[54] These instructions were also meant to solve telephone reports' many further problems that were not mentioned in the leaflet. During the day connections along the principal lines were difficult to obtain, often disturbed by meteorological interferences and strictly limited to six minutes of talking time.[55]

The choice of transmission had a decisive impact on the information the editor received. When Charles Lowe learned that news of a ceremony in Berlin was already known in London, he obtained an interview with one of the protagonists and only then "hastened to the mail" – and not the telegraph – to get it to his English readers.[56] Furthermore, the powerful competition of the news agencies changed the correspondents' production and transmission of news:

> The creation of the press associations had enabled editors to instruct their staffs to minimize spot news and use the slower, cheaper mails for all but the most pressing stories. The application of this politic brought problems, however, since reporters and editors differed in their assessment of important news.[57]

Newspaper editors, however, were very aware that readers judged the importance of news by the speed of its transmission. The "latest news received by cable" became a prominent feature of nearly every newspaper. The breaking stories of major American dailies like the *New York Times* or the *Chicago Daily News* consisted "simply of a pile of Associated Press (AP) bulletins arranged in reverse chronological order."[58]

53 Blowitz, *Memoirs of M. de Blowitz*, p. 72.

54 Gebhardt and Tröbst, *Leben und Arbeit des Auslandskorrespondenten Hans Tröbst*, pp. 145–7.

55 Morf, *Die Drahtberichterstattung*, p. 34.

56 Lowe, *The Tale of a "Times" Correspondent*, p. 103.

57 Heald, *Transatlantic vistas*, p. 110.

58 Barnhurst and Nerone, *The Form of News*, p. 144.

Thus, within the news business, the relationship between speed and accuracy was manifold and problematic. Between 1865 and 1870, telegrams sent over the landline from India to Europe often arrived later than simultaneously posted letters or, if they arrived more quickly, were completely distorted by their handling along the multiple transmission posts.[59] Henry Collins, the first director of *Reuters'* Bombay office, remembered many occasions when the speed of telegraph transmission led to incredulity concerning the veracity of the information: "So unusual was it for telegrams to arrive in India within less than twenty-four hours that doubts were expressed whether the Battle of Lissa [in 1866 – VB] was but a fragment of Reuter's inventive powers."[60] At the turn of the century, German journalistic manuals referred to the speed of the telegraph and the telephone as a direct threat to the style and correctness of news items.[61]

The imperatives of speed and profitability contradicted careful verification and thoughtful formulation. To reduce the numbers of words and thus the cost of telegraphy, nearly every message was sent in special code. Coding, consisting of a "sender and receiver of a telegram agreeing on a series of random words not transgressing the allowed volume," quickly became daily routine in the news business.[62] *Havas*, *Reuters* and newspapers like the *Daily Mail* exclusively received coded bulletins, which further speeded up transmission processes.[63] As an additional advantage coded telegrams raised the security of a message as stolen items were useless without the key to the code. Therefore editors usually preferred coded cablegrams over wireless transmissions which could be tapped.[64] But codes also rendered messages totally unintelligible for every person in the chain between a journalist's account and its reception by the editor. As a necessary reaction to existing transmission networks, coding threatened the reliability and the accuracy of telegraphed news items.

This problem was further intensified by the "packing" of messages, another technique invented to save time and money. Sometimes telegraph rates were not charged according to the number of words but rather by the time

59 Wobring, *Die Globalisierung der Telekommunikation*, p. 273.

60 Collins, *From Pigeon Post to Wireless*, p. 78.

61 Frizenschaf, *Die Praxis des Journalisten*, p. 91.

62 ["Absender und Empfänger des Telegramms vorher eine Reihe beliebiger, den zulässigen Umfang nicht überschreitender Wörter verabreden"] Lenschau, *Das Weltkabelnetz*, p. 60.

63 Lindau, *Nur Erinnerungen*, p. 238; Heerdegen, *Der Nachrichtendienst der Presse*, p. 20; Morf, *Die Drahtberichterstattung*, p. 46; Jones, *Fleet Street & Downing Street*, p. 164.

64 Thurn, *Die Funktelegraphie*, pp. 78f.

the line was actually occupied. Consequently, journalists developed telegraph languages designed to transmit various news items through a single message. The resulting series of incomprehensible words were especially hard to cable accurately.[65]

Contrary to current language, coded and packed messages did not allow telegraph operators to check for possible garbling during transmission. In his memoirs, Paul Lindau, editor of the *Kölnische Zeitung*, described how "the meaning of these pieces of language was utterly misunderstood, and that even more as the unavoidable mutilations by the cable transmissions abetted misunderstandings."[66] For many journalists and editors, transcribing the "cablese" became an important and time consuming part of their job, which – depending on the code – varied from "filling in the omitted articles and expanding the abbreviations" to looking up every single word in specially edited codebooks, all the while guessing what abbreviations and nicknames actually stood for.[67] For Lindau and his colleagues, compressing entire sentences into single words became a "funny sport" regardless of the fact that "some talent and practice are necessary to decode them for reproduction in a newspaper."[68] As late as the 1940s, Josef Hofmann, journalist for the *Kölnische Zeitung*, experienced difficulties with the correct transcription of coded messages.[69]

Availability, access and reliability of telegraph lines in conjunction with codes, packing and transcriptions necessarily engendered multiple mistakes.[70] Henri Collins mentioned several examples of inaccurate *Reuters* bulletins in India. In 1866, a spelling mistake almost led to the bankruptcy of

---

65 Ibid., pp. 135–6.

66 ["dieses Sprachextrakts dunkler Sinn oft gröblich mißdeutet wurde, um so mehr als die unausbleiblichen Verstümmelungen durch die Drahtbeförderung dem Mißverstehen noch Vorschub leisteten."] Lindau, *Nur Erinnerungen*, p. 238.

67 Heald, *Transatlantic Vistas*, p. 49.

68 ["einige Gewandtheit und Uebung in Auflösung derselben zur Wiedergabe in der Zeitung erforderlich ist." "Von uns allen wurde die Kunst des Zusammenpressens von längeren Sätzen in einige wenige längere Worte als heiterer Sport betrieben"] Lindau, *Nur Erinnerungen*, p. 237; Frizenschaf, *Die Praxis des Journalisten*, p. 50.

69 Hofmann and Morsey, *Erinnerungen*, p.106.

70 "Excessive reliance upon skeletonized cables caused difficulties, too, requiring reconstruction of the message at the receiving end, filling in missing words, and elaborating truncated messages into a coherent story. Even with experienced operatives this meant virtually limitless opportunities for misunderstanding. Complaints, explanations, reprimands, and rationalizations on both sides filled the correspondence between Chicago and the field." Heald, *Transatlantic vistas*, p. 111.

an Indian bank as investors hastily withdrew their money. Around the same time, the subcontinent's cotton market almost collapsed due to an additional "0" in a *Reuters* message from the stock exchange.[71] Here again, speed and the desire to beat competitors increased the risk of false reports. When Henri Blowitz cabled the Berlin Congress' act out of a hurry to London, his message contained only 57 out of the 64 original paragraphs.[72]

But such mistakes were not always the fault of correspondents. Usually they only saw the final version of "their" news bulletins once they had been published in a newspaper. Although a Bohemian Jew, Blowitz wrote his articles for the London *Times* in French, which his assistants translated into English. General Elber, the native German Vienna correspondent of the same newspaper, systematically complained when he found out that the *Times'* office had polished the English of his dispatches.[73] Quite often, correspondents were not even sure if their messages made it into the newspapers. While following a story in Denmark, Philip Gibbs only "knew next day, that my dispatch had been published by my paper, for great extracts from it were cabled back to the Danish press."[74] When in the 1930s, Herbert Matthews tried to get his story of the Abyssinian War to the headquarters of the *New York Times*, he had no chance of staying in control over it. Radio transmissions were limited to 10,000 words a day and Matthews had no impact at all on military censorship, possible garbling during transmission or his New York editor's style to "de-skeletonize" what was left of the story.[75] Other than the message's accuracy and the journalist's reputation, consequences could be quite dramatic. Thus, Hans Tröbst was expelled from Greece because of the heading *Telegraphen Union* placed without his knowledge over one of his published dispatches.[76]

Networking the Truth

News items were not simply locally produced and neutrally handled through transmission channels until their publication. Available transmission net-

71 Collins, *From Pigeon Post to Wireless*, pp. 70–2.

72 Giles, *The Life and Times of Henri Stefan Opper de Blowitz*, p. 108.

73 Lowe, *The Tale of a ,"Times"* Correspondent, p. 62.

74 Gibbs, *Adventures in Journalism*, p. 45.

75 Matthews, *Eyewitness in Abyssinia*, p. 170. Cf. as well p. 36.

76 Gebhardt and Tröbst, *Leben und Arbeit des Auslandskorrespondenten Hans Tröbst*, p. 77.

works and techniques played a "constitutive role in defining what news is and means."[77] For editors and journalists alike, news had two principal characteristics: speed and accuracy. As news needed to be new, the former was essential. From a journalistic perspective, the term "news" only made sense in relation to professional competitors. Accuracy rather served as a basic concept for the profession but lacked – at least up to the second half of the twentieth century – a common definition and standardized applications.

The speed of the news powered by an increasingly dense transmission network revolutionized journalism to the same extent as the network's global scale. As a consequence, inventing and developing techniques to assure the accuracy of news became a common task. In this regard, journalists' opinions about the technical infrastructure turned out to be rather ambivalent. From the beginning reporters were skeptical about the supposed "truth" of the telegraph. As early as 1862, a British journalist warned his colleagues that the telegraph "do[es] not invariably convey news, but sometimes merely the counterfeit of news."[78]

Fierce competition and complex transmission channels increased the risk of publishing inaccurate news, which could have dramatic consequences not only in terms of reputation and profit but also politically. In September 1890, the London *Times* caused a diplomatic conflict with Germany when it reported ongoing slavery by German settlers on the recently British island of Zanzibar. The local *Times* correspondent had charged a "very reliable messenger" to inquire and cabled his reports to London without personal verification.[79] When the story turned out to be false, the *Times* faced the protest of the German ambassador and the British government alike as the newspaper had produced the news independently instead of relying on the government-backed *Reuters* service.

For news agencies, it was especially problematic to deliver inaccurate news. In 1896, Hofrat de Grahl, a *Wolff* correspondent within the German court, cabled a false report of a meeting between the German Kaiser and the Russian Czar, which forced the agency to publish a correction. Otto Hammann, head of the German government's press office, understood that de Grahl wanted to meet the deadline of the morning editions but was surprised that the correspondent had not counterchecked his information: "It remained obscure and striking that the court correspondent could pass on

77 Evans, The Path to Freedom?, p. 211.
78 Quoted in Bonea, The Medium and Its Message, p. 168.
79 Lowe, *The Tale of a "Times" Correspondent*, p. 300.

the message that he had heard immediately despite its importance without ascertaining the content."[80] Sometimes the transmission network itself was used as a form of confirmation. When the *Daily Mail* heard rumors about a possible illness of Queen Victoria, the editors turned to the telegraph office to inquire whether such news had actually arrived.[81] As the information had not been transmitted by telegraph, the *Daily Mail* was unable to publish it as news.

According to the circumstances, the telegraph could also inspire mistrust. During the Algeciras conference following the Morocco crises in 1905–06, German newspapers obtained all their information in accordance to the cartel treaties from *Havas*. The French agency delivered it to *Wolff*, which signed and published it. As the German public was well aware that *Wolff* operated under the guidance of the German government, newspaper readers were confused that the official news did not match the official diplomacy.[82]

Prior to World War I, correspondents became more and more reluctant to transmit unchecked information. When in 1910 Philip Gibbs heard about the death of King Edward VII, he decided to obtain confirmation from two independent sources.[83] In the 1930s, counterchecking news items had become a professional obligation and a journalistic practice. In his Abyssinian war account, Herbert Matthews affirmed for his readers the correctness of all reported facts: "I saw some of them with my own eyes, and when I did not see them, I was on the spot immediately afterwards to check the official versions."[84]

Often, however, it was impossible to personally confirm received news. Within a global transmission network, reporters were forced to rely on somebody else. Thus, naming the source of obtained information became an additional technique of assuring accuracy. Back in the nineteenth century, star journalists like Henri Blowitz felt no need to provide the name or any other information regarding the sources of their news. Instead, when he wrote of a "well informed individual," his readers could guess that what followed was nothing else than Blowitz' own opinion.[85]

80 ["Dunkel und auffällig blieb aber, wie der Hofberichterstatter das von ihm gehörte trotz der Wichtigkeit seines Inhalts sofort weitergeben konnte, ohne sich vorher zu vergewissern."] Hammann, *Der neue Kurs*, p. 115.

81 Jones, *Fleet Street & Downing Street*, p. 196.

82 ["Die öffentliche Meinung, von fremden Leuten bedient, war irreführend, aus diesem Grund hatte die Diplomatie sie nicht hinter sich."] Morf, *Die Drahtberichterstattung*, p. 75.

83 Gibbs, *Adventures in Journalism*, p. 24.

84 Matthews, *Eyewitness in Abyssinia*, p. 16.

85 Giles, *The Life and Times of Henri Stefan Opper to Blowitz*, p. 137.

Identifying the sources could also serve as criticism of information. During the Boer War, German newspapers warned their readers that all the news was delivered through *Reuters* and therefore not trustworthy.[86] Meanwhile, a British communication officer in South Africa used *Reuters'* name to render his information more trustworthy in the eyes of his superiors.[87] One way or the other, naming a source reduced responsibilities. At the beginning of the twentieth century, the news profession knew that sources were a tool to build up trustworthiness rather than a guarantee to obtain it.[88] In 1917, the US government gave *Associated Press* permission to publish news on the beginning of Germany's unrestricted submarine warfare only under the condition that its source – the famous Zimmermann telegram – would not be acknowledged.[89] In the 1920s, Alfred Lückenhaus, a Berlin correspondent of *Telegraphen Union*, learned that the German government's daily press conference provided three types of information: "Messages that can be published with indication of the source," "messages that can only be published without indication of the source," and "messages that due to their confidential character may absolutely not be published in a newspaper."[90] Accordingly, some journalists had nothing but suspicion for the connection between naming the sources and the accuracy of given information. For George Seldes, "it has been an unwritten rule among American newspapermen to hang the news on someone else."[91]

Finally, the responsibility for published news items lay in the hands of the editors, not the journalists. Therefore, they also needed to come up with techniques of checking information while remaining in their newspaper's headquarters. When in 1870, *Reuters* was accused of transmitting inaccurate messages, its director could simply point to the received telegram: "I have published the dispatch in question exactly as it came, without adding, omit-

86 Bender, *Der Burenkrieg und die deutschsprachige Presse*, p. 38.

87 Collins, *From Pigeon Post to Wireless*, p. 284.

88 ["Die Agenturen verschmähen es auch nicht, Meldungen der Blätter, die ihnen selbst nicht zugegangen sind, mit Quellenangabe weiterzugeben."] Morf, *Die Drahtberichterstattung*, p. 22.

89 Hohenberg, *Foreign Correspondence*, p. 100.

90 ["Mitteilungen, die unter Quellenangabe veröffentlicht werden konnten," "Mitteilungen, die nur ohne Quellenangabe veröffentlicht werden konnten" and "Mitteilungen, die wegen ihres vertraulichen Charakters in keiner Form in die Zeitung durften"]. Lückenhaus, *Von Draussen gesehen*, p. 57.

91 Seldes, *Lords of the Press*, p. 288.

ting, or altering a single word."[92] But for editors too, laying the entire responsibility on the transmission network soon proved to be insufficient. During the Russian-Ottoman war of 1877–78, New York newspapers like the *Times*, the *Tribune* and the *Herald* sent correspondents to both sides of the front line to produce not only more, but also less biased news.[93] In the early 1880s, the London *Times* telephoned Parliament debates to its headquarters, where they were first set into type and then compared to correspondents' original notes which meanwhile had arrived by messenger. Using a multitude of transmission means maintained the speed of the incoming information while increasing its accuracy.[94] In 1917 Georg Bernhard, the director of the Ullstein publishing house in Berlin, explained in a guide for future journalists: "In the interest of his own safety, a conscientious editor thus checks the veracity of even apparently insignificant messages."[95]

But newspapers and agencies also instructed their correspondents about what kind of news they wished to obtain. Sometimes, as in the summer of 1914, that was not very helpful, as illustrated by a *United Press* directive for its Rome correspondent: "War interest diminishing. Hold down. World Series begins Monday."[96] However, reporters were told how to secure accuracy as well. In 1920, the *Chicago Tribune* advised its correspondents to look for "inside" instead of "official" informants to get confirmation for interesting stories.[97] In 1927, the *Washington Times* explicitly instructed correspondents to transmit news containing at least one and if possible all three of the following elements: "1. Self-preservation, 2. Love and reproduction, 3. Ambition."[98] The question of how to check the accuracy of this kind of news was not addressed.

Such instructions were triggered by newspapers' dependency on the cartel agencies, which delivered the same news to all their clients. For journalist Paul Baumert, the *Wolff* service of the late 1920s was not at all sufficient for German newspapers as it contained "only official and officious news, sec-

92 Collins, *From Pigeon Post to Wireless*, p. 100.

93 Hatton, *Journalistic London*, p. 249.

94 Ibid. 70.

95 ["Schon im Interesse seiner eigenen Sicherheit läßt deshalb ein gewissenhafter Redakteur die Wahrheit auch der unscheinbarsten Meldung nachprüfen."] Bernhard, Stellungen im Zeitungswesen, p. 167.

96 Heald, *Transatlantic Vistas*, p. 31.

97 Ibid. 135.

98 ["1. Selbsterhaltung, 2. Liebe und Fortpflanzung, 3. Ehrgeiz"] Dovifat, *Auswüchse der Sensationsberichterstattung*, p. 10.

ondly no independently obtained foreign news and thirdly – and this aspect is especially decisive for the small papers in the provinces – no commented news."[99] As commercial problems, the uniformity and veracity of agencies' news reports obliged newspapers to produce their own news. After a lifetime as a foreign correspondent, Henry Wickham Steed summed up the problem: "The contest lies between what I may call the standardisation and the individualisation of news."[100]

For newspapers, individualization, in the sense of distinguishing themselves from competitors, became a question of survival. In the 1900s, Edward Price Bell, the newly appointed chief editor of the *Chicago Record*, instructed Victor Lawson, the head of the foreign news department, to deliver "sidelights" in opposition to the bare facts *Associated Press* promised its clients. "It is your business to gather matter that explains, that illuminates, the bare facts."[101] When in 1919 the *Philadelphia Public Ledger* organized an independent foreign service, editor Ackermann criticized his correspondents for "sending news bulletins instead of special articles." In fact, some of them were former *Associated Press* reporters and Ackermann had doubts about their usefulness: "An AP man cannot be remade, I fear."[102] In 1912, foreign correspondent Henry Lucy remembered how newspaperman Henry Labouchere explained his job. For the founder of a daily newspaper entitled *The Truth*, his description was rather surprising. The task of a newspaper, Labouchere told Lucy, is "to create sensation" by interpreting the news in the most unexpected manner. "The great difference between a good and a bad editor is this: the former looks at things subjectively, the latter objectively."[103]

World War I turned out to be a watershed for the quest for subjectivity in objective information. Correspondents and editors felt an increasing obligation to rely on eyewitnesses assuring the accuracy of reported events. In order to control information about the war, the British army excluded all journalists from the battlefield. In response to the persistent protests of English newspapers, the army finally hired Edward Swinton, as an "official eyewitness" charged with the collection of relevant news. Newspapers did not

99 ["nur offiziöse und offizielle Nachrichten, zweitens keine selbstbeschafften Auslandsnachrichten und drittens - und dieses Moment war besonders für die kleine Provinzpresse maßgebend - keine kommentierten Nachrichten."] Baumert, *Die Enstehung des deutschen Journalismus*, p. 70.

100 Steed, *The Press*, p. 206.

101 Heald, *Transatlantic vistas*, p. 24.

102 Ibid. 106.

103 Lucy, *Sixty Years in the Wilderness*, p. 86.

appreciate Swinton's work, however, as they found publishing unchecked information highly problematic. In July 1915 after 103 official eye witness reports, he was dismissed from his functions.[104]

Due to the cartel treaties, the news agencies faced a similar problem. After all, their mutual dependency hindered them from checking news items produced outside their zone of influence. Paul Eltzbacher complained in 1918 that *Wolff* received only "second hand information" and thus could not live up to Germany's international ambitions.[105] Three years earlier, Rudolf Rotheit had pointed out that *Havas* and *Reuters* deliberately transmitted false information as long as they could blame it on other sources.[106] Thus, Eltzbacher and Rotheit joined in the choir of many other journalists pleading for an abolition of the cartel and the creation of genuinely independent news agencies.

For many of them, *Associated Press* could serve as a role model. Run exclusively by professional journalists and as independent as possible from the national government, such an organization would serve German interests much better than the semiofficial *Wolff* agency.[107] However, with regard to *Associated Press*, American journalists had concerns surprisingly similar to the ones of their German colleagues:

> America controlled no cables, manipulated no press agency, operated no propaganda machinery of any kind. We were, and always had been, dependent on foreign press agencies for intercourse with the world. The volume of information that went from our shores was comparatively small, and after it had been filtered in London, or Paris, it grew smaller and smaller until it amounted to mere 'flashes' when it reached a far country.[108]

Here, dependency on a transmission network and on fellow agencies were described as two sides of the same coin: the problem of assuring news' accuracy, while increasing its commercial and political impact. When in 1906 Fritz Walz laid out his vision of accurate, German news, he proposed providing an individual submarine cable to every correspondent.[109]

104 Swinton, *Augenzeuge*, p. 49.

105 ["fast nur Nachrichten aus zweiter Hand"] Eltzbacher, *Die Presse als Werkzeug der auswärtigen Politik*, p. 52.

106 Rotheit, *Friedensbedingungen der deutschen Presse*, p. 10.

107 Bücher, *Unsere Sache und die Tagespresse*, p. 61.

108 Creel, *How We Advertised America*, p. 238.

109 Walz, *Die Presse und die deutsche Weltpolitik*, p. 8.

Conclusion

From the middle of the nineteenth century up to the interwar period, an expanding market for international news and an equally expanding communication infrastructure based on new devices like terrestrial, submarine, and wireless telegraphy as well as the telephone led in a mutual reinforcing relationship to the emergence of a global news network. On the one hand, this network opened up unprecedented possibilities for media businesses. Thanks to the speed of the telegraph, they could deliver their news exponentially faster; thanks to its transmission capacities, they could sell more news to more clients; and thanks to transnational telegraph links, media enterprises could build cooperative exchanges with news producers and customers all over the world. On the other hand, the telegraph networks engendered new problems and concerns. From a business perspective, the high charges increased the difficulties to keep the cost-benefit-relation for the transmission of news items in balance. Financially and in regard to journalistic standards, the insufficient reliability of available lines often hindered a smooth news exchange between collaborators. Additionally, it became especially problematic to assure the accuracy of information received via the telegraph.

Thus, business opportunities engendered by the speed of the telegraph network conflicted with evolving journalistic ideals of accuracy and objectivity. As the style of news changed in accordance with available means of communication and the increasingly global network they were linked to, journalists and media businesses alike were obliged to develop new techniques to establish the veracity of given information. Objectivity, as a technique of guaranteeing the truth of the news, could be codified partly because the interplay of cartels of news agencies, cooperation of cable companies and newspapers served to standardize international markets. As an important technical shift, the invention of the telegraph had a huge impact on evolving concepts of objectivity. In many instances, new technologies allowed news producers, i.e. private, profit-driven bodies, to argue that they established objectivity, which proved to be a key element for the success of their businesses.

Bibliography

Barnhurst, Kevin G./John C. Nerone, *The Form of News. A History*, New York 2001.

Basse, Dieter, *Wolff's Telegraphisches Bureau 1849 bis 1933. Agenturpublizistik zwischen Politik und Wirtschaft*, München 1991.

Baumert, Paul, *Die Entstehung des deutschen Journalismus in sozialgeschichtlicher Betrachtung*, Altenburg 1928.

Bender, Steffen, *Der Burenkrieg und die deutschsprachige Presse. Wahrnehmung und Deutung zwischen Bureneuphorie und Anglophobie, 1899 – 1902*, Paderborn 2009.

Bernhard, Georg, "Stellungen im Zeitungswesen," in: Paul Eltzbacher (ed.), *Berufswahl. Ein Ratgeber für kriegsbeschädigte Offiziere und andere gebildete Kriegsbeschädigte sowie für Schüler höherer Lehranstalten*, Berlin 1917.

Blondheim, Menahem, *News over the Wires. The Telegraph and the Flow of Public Information in America, 1844–1897*, Cambridge, MA 1994.

Blowitz, Henri George, *Memoirs of M. de Blowitz*. New York 1903.

Bonea, Amelia, "The Medium and Its Message. Reporting the Austro-Prussian War in the Times of India," in: Roland Wenzlhuemer (ed.), *Global Communications. Telecommunication and Global Flow of Information in the Late 19th and Early 20th Century*, Köln 2010.

Bösch, Frank, *Mediengeschichte. Vom asiatischen Buchdruck zum Fernsehen*, Frankfurt am Main 2011.

Bourne, Fox R., *English Newspapers. Chapters in the History of Journalism*. London 1887.

Boyd-Barrett, Oliver, *The International News Agencies*, London 1980.

Boyd-Barrett, Oliver and Michael B. Palmer, *Le trafic des nouvelles. Les agences mondiales d'information*, Paris 1981.

Boyd-Barrett, Oliver and Terhi Rantanen (eds.), *The Globalization of News*, London/Thousand Oaks, CA 1998.

Bücher, Karl, *Unsere Sache und die Tagespresse*, Tübingen 1915.

Chirol, Valentine, *Fifty Years in a Changing World*, London 1927.

Collins, H. M., *From Pigeon Post to Wireless*, London 1925.

Cooper, Kent, *Barriers Down. The Story of the News Agency Epoch*, Port Washington 1942.

Creel, George, *How We Advertised America. The First Telling of the Amazing Story of the Committee on Public Information that Carried the Gospel of Americans to Every Corner of the Globe*, New York/London 1920.

Dietz, Hermann, *Das Zeitungswesen*. Berlin 1919.

Dovifat, Emil, *Auswüchse der Sensationsberichterstattung*. Stuttgart 1930.

Eltzbacher, Paul, *Die Presse als Werkzeug der auswärtigen Politik*. Jena 1918.

Evans, Heidi J., "'The Path to Freedom?'. Transocean and German Wireless Telegraphy, 1914–1922," in: Roland Wenzlhuemer (ed.): *Global Communications. Telecommunication and Global Flow of Information in the Late 19th and Early 20th Century*, Köln 2010.

Fortner, Robert S., *International Communication. History, Conflict, and Control of the Global Metropolis.* Belmont, CA 1994.

Fritzinger, Linda B., *Diplomat Without Portfolio. Valentine Chirol, his life and the Times*, London/New York 2006.

Frizenschaf, Johannes, *Die Praxis des Journalisten. Ein Lehr- und Handbuch für Journalisten, Redakteure und Schriftsteller.* Leipzig 1901

Gebhardt, Hartwig/Hans Tröbst, *Mir fehlt eben ein anständiger Beruf. Leben und Arbeit des Auslandskorrespondenten Hans Tröbst (1891–1939)*, Bremen 2007.

Gibbs, Philip, *Adventures in Journalism*, London 1923.

Giles, Frank, *A Prince of Journalists. The Life and Times of Henri Stefan Opper de Blowitz.* London 1962.

Golan, Guy J. Ü./Thomas J. Johnson/Wayne Wanta (eds.), *International Media. Communication in a Global Age*, London, New York 2010.

Hammann, Otto, *Der neue Kurs. Erinnerungen*, Berlin 1918.

Hatton, Joseph, *Journalistic London. Being a Series of Sketches of Famous Pens and Papers of the Day*, London 1882.

Heald, Morrell, *Transatlantic vistas. American journalists in Europe, 1900–1940*, Kent 1988.

Heerdegen, Ernst, *Der Nachrichtendienst der Presse*, Leipzig 1920.

Hofmann, Josef/Rudolf Morsey, *Journalist in Republik, Diktatur und Besatzungszeit. Erinnerungen 1916–1947*, Mainz 1977.

Hohenberg, John, *Foreign Correspondence. The Great Reporters and Their Times*, Syracuse, NY 1995.

Hunt, William, *Then and Now. Or, Fifty Years of Newspaper Work.* London 1887.

Jones, Kennedy, *Fleet Street & Downing Street.* London 1920.

Kishan Thussu, Daya (ed.), *Internationalizing Media Studies.* London/New York 2009.

— (Ed.), *International Communication. A Reader*, London 2010.

Lenschau, Thomas, *Das Weltkabelnetz*, Frankfurt am Main 1908.

Lindau, Paul, *Nur Erinnerungen*, vol. 1, Stuttgart/Berlin 1916.

Lowe, Charles, *The Tale of a "Times" Correspondent (Berlin 1878–1891)*, London 1927.

Lückenhaus, Alfred, *Von Draussen gesehen. Berichte eines Auslandskorrespondenten aus Großbritannien, den Vereinigten Staaten von Amerika, Japan, China. 1924 bis 1945*, Düsseldorf 1955.

Lucy, Henry, *Sixty Years in the Wilderness. More Passages by the Way*, London 1912.

Mathien, Michel/Catherine Conso, *Les Agences de Presse Internationales.* Paris 1997.

Matthews, Herbert, *Eyewitness in Abyssinia. With Marshal Badoglio's Forces to Addis Ababa*, London 1937.

McMillin, Divya C., *International Media Studies*, Oxford 2007.

Morf, Hans E., *Die Drahtberichterstattung im modernen Zeitungswesen*, Bern 1912.

Rantanen, Terhi, *When News was New*, Malden, MA 2009.

Read, Donald, *The Power of News. The History of Reuters*, Oxford [2]1999.

Rotheit, Rudolf, *Friedensbedingungen der deutschen Presse. Los von Reuter und Havas!*, Berlin 1915.

Schudson, Michael, *Origins of the Ideal of Objectivity in the Professions. Studies in the History of American Journalism and American Law, 1830–1940*, New York/London 1990.

—, "The Objectivity Norm in American Journalism," *Journalism*, vol. 2 (2001), pp. 149–170.

Seldes, George, *Lords of the Press*. London 1938.

Silberstein-Loeb, Jonathan (2009), "The Structure of the News Market in Britain, 1870–1914," *Business History Review*, vol. 83, (2009), 759–788.

Steed, Henry Wickham, *Through Thirty Years 1892–1922. A Personal Narrative*, London 1924.

—, *The Press*, London 1938.

Swinton, Ernest D., *Augenzeuge. Hinter den Kulissen der britischen Weltkriegsführung*, Berlin 1940.

Thurn, H., *Die Funktelegraphie*, Leipzig 1913.

Tuchman, Gaye, *Making News. A Study in the Construction of Reality*, New York 1980.

Walz, Fritz, *Die Presse und die deutsche Weltpolitik*, Zürich 1906.

Winder, Gordon M., "London's Global Reach? Reuters News and Network, 1865, 1881, and 1914," *Journal of World History*, vol. 21 (2010), pp. 271–96.

Wobring, Michael, *Die Globalisierung der Telekommunikation im 19. Jahrhundert. Pläne, Projekte und Kapazitätsausbauten zwischen Wirtschaft und Politik*, Frankfurt am Main 2005.

Wrede, Richard, *Handbuch der Journalistik*, Berlin 1902.

Zöller, Hugo, *Als Jurnalist* [sic] *und Forscher in Deutschlands großer Kolonialzeit*, Leipzig 1930.

Space|Time

Telecommunications Technology and News of Disaster: Earthquake Reporting in *The Los Angeles Times*, 1917–1939

Gordon M. Winder

Telegraph and radio transmission of the news of earthquakes made fascinating press copy, and had far-ranging implications for the ways that such disasters were understood. This chapter analyses *The Los Angeles Times'* (*LAT*) reporting of major Pacific Rim earthquake disasters from 1917 to 1939 in order to reveal how telecommunications technology shaped disaster coverage and was reported in it.[1] It compares and contrasts the roles of telecommunications and rapid transport in the reporting of sixteen devastating earthquakes that shook parts of the Pacific Rim (Table 1). These disasters were selected from the lengthy list of earthquakes compiled by the US Geological Survey on the grounds that each earthquake extinguished the lives of at least 1,000 people. Significantly, where Japan and many other parts of the earthquake-prone Pacific Rim lay within the telegraphic world accessible from Los Angeles during the inter-war period, large parts of China remained only poorly connected, because of inadequate infrastructure or political and military obstacles to journalistic access. In these circumstances, how was the fact that China lay on the margins of the 'telegraphic world' made manifest in the *LAT*?

Globalization was already in progress under the force of electric telegraphy at the beginning of the twentieth century. The electric telegraph helped to produce a global sphere comprising spaces of flows themselves detached from spaces governed by material movement.[2] This was a distinctive form of globalization, and one not to be confused with the globalization of the

1 The emergence in 1906 of in-house earthquake disaster reporting in the *LAT*, featuring coverage of charity initiatives in Los Angeles, and the newspaper's subsequent application of this form of disaster journalism to earthquakes in foreign lands is explored in Winder, Mediating Foreign Disasters, pp. 195–226. The changing reporting of Japanese earthquakes during the twentieth century in the *LAT* are the subject of a second paper: Winder, *The Los Angeles Times* Reports Japanese Earthquakes, 1923–1995 .

2 Wenzlhuemer, Telecommunication and Globalization in the Nineteenth Century, pp. 7–18, and Wenzlhuemer, Globalization, Communication and the Concept of Space, pp. 19–50.

late twentieth century in which communication and computer technology are coupled as information technology to dramatically change the scale, intensity and effectivity of communication.[3] We should expect an identifiable development of the earthquake disaster story within the pages of newspapers like the *LAT* as the new information technologies of the late twentieth century were brought to bear.[4] Nevertheless, even in the 1930s, the *LAT* claimed to deliver "all the news all the time," a message carried in its masthead. Even earlier, on Monday, July 31, 1917, when the *LAT* trumpeted its publication of a giant Sunday Edition of 138 pages, including 560 columns of reading material, the Edition was said to be comprehensive:

> The general telegraphic news and features, with illustrations, occupied thirty-three columns, exclusive of market and sporting items. Every happening of importance in the world to which news agencies of America had access was recorded. The resources of the greatest news-gathering agency, of special correspondents in world centers, of leading American newspapers and of representatives in every sizeable community in the Southwest were all drained to present to the readers of *The Times* an accurate record of Saturday's events and a sensible forecast of important things the future has in store.[5]

This was a reasonable portrait of the extent and organization of the newspaper's content and telegraphic news production network: this newspaper could deliver the news of the world to southern Californians; it aimed to generate a community in Los Angeles based upon American news agency.[6] Nevertheless, from her reading of the *LAT*'s Sunday and Monday editions in mid-1917, the discerning reader in Los Angeles might have added two further observations.

First, this "globalization" of the news was the product of an ensemble of technologies, not simply electric telegraphy but also telephony, the typewriter, print and photography, railway, steamship and automobile transport, of course assisted by boys delivering messages and newspapers in the built environments of modern cities. In subsequent decades radio transmissions and aircraft transport played additional roles. In trumpeting its Sunday Edition of July 1917, the *LAT* hid this technological ensemble behind the modern news gathering organization that made use of it, but both the organization

3 Castells, *The Rise of the Network Society*. See also Appadurai, *Modernity at Large* and Winder, London's Global Reach?.

4 Wilke, Historical Perspectives on Media Events.

5 Sunday "Times" Greatest Yet, *LAT*, July 30, 1917.

6 Anderson, *Imagined Communities*.

and the technologies were readily understandable to readers, since both were repeatedly portrayed in the pages of each edition. For example, datelines and names of news agencies appeared under each headline, and an entire section was devoted to news and advertising of automobiles.

Second, this modern ensemble of technologies was not available everywhere, but only in privileged places, like Los Angeles. The technological ensemble and the news gathering organization combined to produce specific "digital divides" between those in Los Angeles with access to the press and those without, as well as imagined geographies of other places in the world conformed to the ways of knowing made possible through the modern newspaper.[7] In 1917 the *LAT* represented its news services by outlining the geography of its news production network: its "global" coverage was sourced from its contracts with Associated Press (AP) and other American newspapers, its own journalists located in Los Angeles, "world centers" and throughout the South West of the USA. Such a network was at once metropolitan, regional, national and international in scope, but also "modern," partly because its geography adhered to the infrastructure of telecommunication and news gathering. This also meant that the modernity attributed to the telecommunications technology helped to code news from distant places in terms of civilization discourse.[8] Was news from other places immediate, comprehensive and trustworthy, or delayed, incomplete and suspect? The absence of modern telecommunications technology indicated backwardness within a shifting geography of "modernity."

This paper explores how the *LAT* coped with the geographic asymmetries produced by the interaction of these components of the news. It investigates the geographies associated with telecommunications technology in modern disaster reporting. Copies of the *LAT* (Table 1) were collected for the week following each event. Earthquake coverage in these weeks was then analysed using spreadsheets of authors, agencies, datelines, headlines, word counts and images. The changing geographies of news production were identified along with the ways in which telecommunications technology and transport

7 See Mann, The Deep Digital Divide, pp. 188–208 for one effective use of this concept in an historical setting. The geographical imagination is explored by Harvey, Cosmopolitanism and the Banality of Geographical Evils, Gregory, *The Colonial Present,* and Hannah, Torture and the Ticking Bomb. Regarding telegraph services and news see Bonea, The Medium and its Message, Winder, Imagining Citizenship in the Networked Newspaper, and Winder, Mediating Foreign Disasters.

8 See Anderson, White Natures.

systems were written into these news stories. The rest of the newspaper was analysed in order to gain an appreciation of the broader context for disaster news: for example the reporting of other, less devastating, earthquakes in the same week, the placement of the stories within the newspaper, and the relationships of the earthquake reporting to other important news of the day. *The Times* of London and *The New York Times* were consulted to help make sense of the *LAT*'s coverage of one of the earthquakes. However, tracing the life history of each disaster, which is a feature of research on the political ecologies of disaster, is not attempted here.[9] Similarly, the roles of gatekeepers in making decisions about what becomes news cannot be discerned in this approach.[10] Instead, the aim is to identify the role of telecommunications in the framing of the news.[11]

The Edge of the Modern World: Bali and Yunnan 1917; Nan'ao 1918

The earthquake in Bali, Dutch East Indies, January 20, 1917 claimed 1,500 lives but, in the week after the disaster, it did not make front page news in the *LAT*. In fact it warranted only one, 116-word bulletin sourced from AP (Table 1). [12] This bulletin contained several different estimates of the death toll, and was bolstered by general information on Bali, which was said to be an island of 5,439 square kilometres, located in the Dutch East Indies, and with a population of about 700,000 inhabitants. Further, the bulletin is placed on page 12 and buried under the headlines "Buildings in Montreal Shaken by Earthquake" (36 words), and "Recorded at Ottawa" (12 words). Despite the much greater destruction caused by the Bali earthquake, this placement effectively made the news from Bali far less important than that from Montreal. Together, these bulletins attested to the geographic scope

9 Oliver-Smith and Hoffman (eds.), *The Angry Earth*, p. 30.

10 Whitney and Becker, Keeping the Gates for Gatekeepers, and Singer, More Than Ink-stained Wretches.

11 Benthall, *Disasters, Relief and the Media*, Boyle and Hoeschen, Theorizing the Form of Media Coverage Over Time, Adams, The September 11 Attacks as Viewed From Quebec, and Christensen, Pocketbooks or Prayer Beads?

12 The earthquake contended for news space with the Battle of Verdun and tensions in Mexico. "Hundreds Killed in Bali," *LAT*, January 27, 1917.

of the *LAT*'s news services, but they also revealed the delays involved in getting news to Los Angeles from "distant" parts of the world: seven days from Bali, one from Montreal. Indeed, the *LAT* took care to elaborate precisely which route the bulletin had taken from origin to the news page. It had received a news bulletin by Atlantic Cable and AP from London, on January 26, at 9.10 pm. This cable cited an official Dutch report received at The Hague and transmitted to London by Reuters. This was official news and the newspaper thus emphasized both the telecommunications technology at work and the intricate set of translations between national institutions that made this international news available. The bulletin thus highlighted the circuitous route, both geographically and politically, that this news had taken. From the telegraphic news viewpoint of Los Angeles in 1917, which itself owed to economic and political biases in communication, the Dutch East Indies lay on the edge of the modern world. Telegraphic bulletins could arrive from there only via London. In terms of steamship travel time but not telegraphic communication time, Bali was further away from Los Angeles via London than via Manila. Nonetheless, Bali was on the margin of the *LAT*'s telegraphic world and a disaster there warranted little more than a bulletin attesting to an event at a distant time and location. This bulletin was primarily a statement about the telecommunication and news gathering capabilities of the *LAT* rather than an account of the disaster.

If the Indonesian archipelago was on the edge of the *LAT*'s telegraphic news world in 1917, then China's interior provinces lay outside it. There were no reports in the *LAT* of the earthquakes in Yunnan, July 30, 1917 or Nan'ao, Guangdong, February 13, 1918. Despite sixteen references to "China" between July 1 and August 31, 1917 there were no reports of the disaster in Yunnan (Table 1).[13] In the week after the Yunnan event, reports of earthquakes streamed into Los Angeles from various points on the telegraphic network, but not from central China. Officials at the Chilean Embassy in Washington denied reports of an earthquake in their country saying instead that the quake had occurred in Argentina.[14] While residents of Mexico City raised funds to donate to victims of an earthquake in Salvador, both Argentina and New Zealand felt earthquakes, and another broke the Los Angeles

13 "China's President at Peking," *LAT*, August 3, 1917, "Few Joy Rides Taken in China," *LAT*, August 5 1917.

14 "Earthquake Report Denied in Chile," *LAT*, July 31, 1917.

aqueduct, though water services were unaffected.[15] Similarly, in the week after the disaster in Guangdong in February 1918, a thirty-six-word AP bulletin from Amoy, China, was published. It announced that no estimate of casualties from the quake that shook Amoy had yet been made. This news from Amoy took more than a week to reach Los Angeles, and Amoy, a treaty port with a telegraph and cable station, was in the telegraphic world.[16] The small amount of news from China that did arrive in Los Angeles during the First World War was sourced from Shanghai, Beijing, Washington and Tokyo, not directly from China's interior provinces.

Simon Winchester contends that the events associated with the Krakatoa eruptions and tsunami of August 26–27, 1883 established a pattern of modern coverage of a natural disaster with global implications and reporting.[17] But neither Guangdong nor Yunnan was in Los Angeles' telegraphic world during the First World War, so it was impossible to know in Los Angeles of the terrible events there and the Krakatoa model was thus not followed. As the *LAT* commemorated the life of its owner, General Harrison Gray Otis, whose passing was announced on July 31, 1917, and the paper's engagement with the war in Europe increased into 1918, readers were probably only vaguely aware of the boundaries of the telegraphic world, and certainly could not have known of the terrible disasters in central China.

The Seismograph vs. the Telegraph: Haiyuan, China, 1920

During the 1920s, the telegraph appeared to be no match for the seismograph, an outcome made plain in the *LAT*'s coverage of the Haiyuan earthquake. On December 17, 1920, readers learned that a great earthquake had been recorded the day before on seismological devices at universities in Washington DC, Chicago, San Francisco and Seattle, but that neither the location of the tremors nor the extent of the damage could be pinpoint-

15 "Aid Salvador," *LAT*, July 13, 1917, "Argentina is Shaken," *LAT*, July 28, 1917. "New Zealand is Shaken," *LAT*, August 7, 1917. "Mother Earth is Vanquished," *LAT*, July 11, 1917, and "Temblor Blamed," *LAT*, July 24, 1917.

16 "Earthquake's Deaths Not Yet Estimated," *LAT*, February 18, 1918. Amoy, now Xiamen, was a treaty port and heavily engaged in trade. Sugiyama and Grove (eds.), *Commercial Networks in Modern Asia*.

17 Winchester, *Krakatoa*.

ed.[18] Four days later, the editor of the *LAT* quipped that "[i]f there must be great quakes, that (the middle of the ocean) is the best place for them. A lot of it can be passed off as seasickness."[19] That week readers of the *LAT* found reports of earthquakes in Argentina and Japan,[20] and their eyes would have been drawn to news of war-ravaged Europe, and particularly Hungary, as grain prices soared and Edgar Hoover led a relief campaign.[21] The great earthquake, which actually occurred in the loess terrain of Haiyuan, China and not the middle of the ocean, was deadly but in the circumstances, the *LAT*'s readers learned nothing of the disaster.[22] Instead they read about the detection of seismographic waves.

Coverage in *The New York Times* (*NYT*) was better. A short, forty-six-word bulletin printed on December 25, 1920 reported a devastating earthquake with 2,000 deaths on December 16. The dispatch was recorded as originating from Tokyo, where the *Asahi Shimbun* sourced its report from Shanghai. [23] A second bulletin, this one of sixty words and datelined Tirana, Albania, December 22 appeared beneath this notice in the same column. It reported forty-two deaths and 500 made homeless by an earthquake in Albania. The next day special reports from Washington DC and Chicago, in all 2,023 words, noted that scientists had been unable to agree on the location of the epicenter, but that China was very likely.[24] The placing and recording of the earthquake took over as the main news story in the *NYT*, so that here too, the horrible consequences of the quake in Haiyuan were poorly reported.

18 "Earthquake a Mystery," *LAT*, December 17, 1920.

19 "Wet Quakes," *LAT*, December 21, 1920.

20 "Temblor in Argentina," *LAT*, December 19, 1920, "Argentine Earthquakes Raze Towns," *LAT*, December 20, 1920, "Andes Sink Sixty Feet," *LAT*, December 23, 1920. "More Than Four Hundred Killed by Earth Shock," and "Yap Island Rocked by Earthquake," *LAT*, December 21, 1920. "Looks That Way," "Jap Village is Entirely Destroyed," "Two Thousand Slain in Japanese Temblor," and "Quake Victims Live in Terror," *LAT*, December 15, 1920.

21 In one cartoon Edgar Hoover says "Merry Christmas" to "one of the 3,500,000 starving children in Central and Eastern Europe whose life you can save by sending Hoover $10.00." "Hungary, in Dire Need, Seeks Aid of America," *LAT*, December 16, 1920, "Europe's Needs and Ours," "Need Help to Provide Gifts," *LAT*, December 17, 1920, "Ask Funds to Save Children," *LAT*, December 18, 1920.

22 Zhong and Wang, Geological Disasters in Loess Areas During the 1920s.

23 "Earthquake Claimed 2,000 Victims in China," *NYT*, December 25, 1920.

24 "Scientists Puzzle Over Quake Records," *NYT*, December 26, 1920.

In London on December 17, *The Times* also reported the signs of a great earthquake left on seismographs on December 16.[25] A map showed a radius of 8,000 kilometres around the recording seismograph in London and thus the areas of the world in which the quake could have occurred. Over the next few days *The Times* informed readers that the epicenter of the quake could not yet be located. This drew Professor H. H. Turner of Oxford's Observatory to speculate on the location of the earthquake, first in the Gulf of Mexico and then off the Alaskan coast. On December 28 a further letter to the editor from him was published, this time locating the quake in China.[26] As reports in the *NYT* suggested that there were two, widely separated, earthquakes (December 26), British scientists puzzled publicly over their quake records. Their estimates were over 9,600 kilometers off target. British scientists blamed erroneous information emanating from the USA for their error.

Simon Winchester has highlighted the ways in which scientists used telegraphic news to aid them in locating the epicenters of earthquakes,[27] but coverage of the Haiyuan disaster in these three newspapers reveals that both the extent of coverage and the story line were determined by the cable and telegraph network. In the centers of scientific calculation like Oxford, Toronto, Seattle, Washington and Chicago, seismographs recorded disturbances but scientists could not accurately place the epicenters of these tremors without access to accurate news bulletins. Since central China lay beyond the extent of their cable network, the press could not supply satisfactory reports. That meant that both the way distant earthquakes were understood and the ways that they could be reported were determined by their location relative to the cable network. Different time-spaces operated inside and outside the telegraphic world.

The Full Ensemble: Kanto Plain, 1923

The *LAT* brought a sophisticated array of telecommunications technologies to the task of reporting the earthquake and fire in Tokyo and Yokohama 1923. It published 77,000 words on the Kanto Plain earthquake in the first

25 "Great Earthquake Reported," *Times*, December 17, 1920.

26 "Earthquake Not Yet Located," *Times*, December 18, 1920, "The Great Earthquake," *Times*, December 21, 22 and 28, 1920.

27 Winchester, *Krakato*.

week after the event (Table 1), and Japan's disaster remained a leading story for weeks. The *LAT* had access to photographic images cabled from the disaster zone, and it boasted news correspondents of its own, located in New York, Los Angeles, Washington and Kobe, who helped to report the disaster. Short bulletins from widespread datelines reported high tides in Hawai'i and seismic readings at US and European observatories.[28] Reports emphasized the terrible devastation wrought by the earthquake using photographs, lists of damaged buildings, reports of the dead, and eye witness reports.[29] Photographs of Angelinos who were known to be in Tokyo were published.[30] Their fates and the concerns of their relatives were a further substantial component of the news.[31] The *LAT* published lists of Californians known to have been in Japan, and announced that ships and individuals were safe or missing. Readers learned that "America will rebuild Japan" as a demonstration of its civilizing power,[32] and that the US government planned a massive relief effort.[33] The US Navy delivered aid to the other side of the Pacific.[34] Through-

28 "Japan Records Four Temblors Daily Average," and "California Needles Record Earthshock," *LAT*, September 2, 1923. "New Tremors Recorded," and "Felt in Guam," *LAT*, September 3, 1923. "Ground Swells Hit Coast," and "Temblors Felt Often in Japan," *LAT*, September 4, 1923.

29 For example, see: "All Tokio in Flames: Death Toll Staggering," "Yokohama in Ruins as Temblor Rocks East Coast of Nation," and "Scenes in Japanese Cities Stricken by Earthquake and Fire," *LAT*, September 2, 1923; "Summary of Japan's Disaster," *LAT*, September 3, 1923; "Historic Spots in Tokio Ruined," and "Imperial Hotel in Path of Fire," *LAT*, September 4, 1923; or "Flaming River Consumes City," *LAT*, September 8, 1923.

30 "Angelinos in Nippon; Views of Razed Cities" *LAT*, September 4, 1923.

31 For example, see. "Californians in Fire Zone," "American Officials in Earthquake Zone," and "Americans in Yokohama," *LAT*, September 3, 1923; "Church Schools in Tokio Feared Lost in Tremor," "Many From State are in Japan," "Mrs. Hughes Not Injured in Temblor," "Film Official Feared Lost in Tokio Tremor," "No Fear for Mrs. Edith Hagin," *LAT*, September 4, 1923; "Americans Listed in Japan," and "Tennis Star's Father Killed in Earthquake," *LAT*, September 7, 1923; or "Missionaries of Southland Now in Far East Safe," and "Fear Expressed for Dr. Moore and His Wife," *LAT*, September 8, 1923.

32 "Hands Across the Sea," and "America Will Rebuild Japan," *LAT*, September 4, 1923.

33 "Federal Aid is Granted Japan," "Rebuilding Aid Offered Japan," and "America Has Started Help for Japanese," *LAT*, September 6, 1923. "America to Hasten Aid," *LAT*, September 7, 1923. "Cabinet Maps Relief Plans," *LAT*, September 8, 1923.

34 For example, see: "American Ships Ordered to Speed to Yokohama to Help Temblor Sufferers," *LAT*, September 2, 1923. "Americans Landed at Yokohama," "Naval Ships to Rush Aid," *LAT*, September 3, 1923. "Relief Vessels Speed to Tokio," "*Times* Relief Fund for Japanese," and "Transport Held for Japan Relief Voyage," *LAT*, September 7, 1923.

out southern California, relief committees sprang into action.[35] Los Angeles' Japanese community alone aimed to raise a half a million dollars, and Japanese Americans were also raising funds in San Francisco, Chicago and Honolulu.[36] The editor invited readers to participate in the relief campaign. Together, these individual themes combined to present a mass mediated ritual of the USA's aid programe, international networks caught in the disaster, and a foreign government restoring order.[37] The *LAT* reported the Kanto Plain earthquake as a matter of transnational sympathy, of Californians at risk in the disaster zone, and of a large-scale US relief effort. The extent and the intimate character of these transnational relations are perhaps surprising.

The *LAT* signalled its superior communications which could overcome the interrupted service in the quake zone. Notably, it celebrated the potential of radio technology.[38] It ran its own radio station and advertised its content in a weekly section of the newspaper.[39] It had the latest radio bulletins from British and American shipping and from Shanghai and Kobe.[40] There may have been only one radio operator left in Tokyo, but the *LAT* had his account.[41] In these ways the *LAT* compensated for the obvious failure of communications from Tokyo and Yokohama. The *LAT* wrote its wider scope of operations and its mastery of communications into its reports of the Tokyo earthquake.

35 "Relief Work is Japanese Plan in Long Beach," *LAT*, September 3, 1923. "For Earthquake Victims," "Fresno Japanese Raise relief Fund," *LAT*, September 4, 1923. "Southland Opens Hearts and Purses to Nippon," "All Southland Aids Stricken," "Fresno Red Cross will Aid Japanese," *LAT*, September 5, 1923. "Long Beach Council Aids Japanese Relief," *LAT*, September 7, 1923. "Southern Pacific Gives Large Sum to Aid Japanese," and "Long Beach to Do Bit Toward Aiding Japanese," *LAT*, September 8, 1923.

36 "Japanese Here Raise Stricken Area Fund," "Bay City Mayor Aids in Relief of Japanese," "Chicago's Japanese Plan Relief Efforts," *LAT*, September 4, 1923. "Open Campaign to Obtain Aid for Devastated Nippon," "New York Help Offered Japan," and "Honolulu Opens Fund for Japanese Relief," *LAT*, September 5, 1923.

37 Couldry, *Media Rituals*, Couldry, Hepp, and Krotz (eds.), *Media Events in a Global Age*, and Cottle, Mediatised Rituals.

38 Winder, *The Los Angeles Times* Reports Japanese Earthquakes, 1923–1995. "Radio Tells it to the Worlds, Wireless Waves Infinite in Time and Space, KHJ Impulses Transmitted to Farthest Star," *LAT*, September 2, 1923.

39 "*Noticias Mundialles de Ultima Hora*," and "Radioland KHJ Section," *LAT*, September 2, 1923.

40 "Messages accepted For Japan," *LAT*, September 4, 1923. "Japanese Toll 240,000 Says Fleet Wireless," *LAT*, September 5, 1923.

41 "Lone Operator Tells of Quake," *LAT*, September 4, 1923.

Between September 16 and the end of October readers were treated to almost 40,000 more words relating to the Tokyo disaster, and in this reporting telecommunication and transport technologies were important themes. Among its regular shipping reports the *LAT* published news of US relief ships en route to Yokohama.[42] As refugees from Tokyo arrived at Canadian and US ports, they brought with them eye-witness reports and photographs.[43] Their reports were published in the newspaper and the *LAT* tracked the onward journeys of the images as they were flown on to eastern cities, and driven by car to Los Angeles.[44] Not only had the first news of the disaster arrived over the radio waves, but now news photos arrived in Los Angeles by wireless, and the *LAT* claimed yet another beat.[45] The Pantages Vaudeville Theater advertised its showing of the "First and Only Authentic Japanese Earthquake Pictures," on September 30 and throughout the first week of October. In turn, the editor related the costs of the disaster to the movies: "The Japanese earthquake and fire were not so costly in a monetary way as might be imagined. It has been figured out at $13 per capita. Shucks, we spend that much at the movies."[46]

This relativity trivialised the disaster by distorting the geography of the impacts (the costs were distributed equally among the entire Japanese population), but it also placed the Tokyo earthquake within a modern calculus of disasters, which used movie-going as the modern standard, and, in the process, implied a contrast in modernity between Tokyo and Los Angeles. In these ways, the *LAT* again used reports of telecommunications and transport technology to produce geographies of difference between Los Angeles and other cities.

Finally, despite the technologies that made reports of the Tokyo disaster possible, and the recognition given to them in the press, analysis of the *LAT*'s

42 For example, see "News of Ships and Sailings at Pacific Ports," *LAT* September 19, 1923, "San Francisco Relief Ship Sails for Japan," *LAT* September 20, 1923.

43 "First View in America of Tokio Fire Started by Earthquake," and "First Quake refugees Landed at Victoria," *LAT* September 16, 1923. "First Pictures of Japan's Earthquake Horror," "President Jefferson Arrives with Quake Refugees," and "Quake Inferno Yields up Cargo of Human Salvage," *LAT*, September 17, 1923. "Photos Come From Japan," *LAT*, September 18, 1923.

44 "Drive From Bay City in Eight Hours," *LAT*, September 25, 1923. "Flyer With Japanese Films Speeds East," *LAT* September 26, 1923. "Plane at New York with Quake Films," *LAT*, September 27, 1923.

45 "First News Photo Ever Sent by Wireless," *LAT*, September 21, 1923. "Radio First in Earthquake," *LAT*, September 30, 1923.

46 "Pen Points," *LAT*, October 6, 1923.

coverage reveals that, in fact, most of the news generated by this modern technological ensemble did not originate in Japan. The editor largely reported the quake from the USA (including Hawai'i, Guam and Manila), and paid scant attention to Japanese news agencies. Only four percent of the first week of disaster news was actually sourced in Japan. Together, Hong Kong, Shanghai and European capitals supplied more copy than Japan. Even AP supplied less than 29 percent of the copy. Instead, Los Angeles, with 43 percent of copy, was the single most important source point, and the main story, amounting to 41 percent of copy, was the American relief effort. This geography of news production indicates both biases in the news agency system and constraints in the news services infrastructure. The *LAT*'s disaster coverage turns out to have been largely composed of stories about Americans caught up in the disaster, about American ways of knowing and responding to disasters, and about the technologies used to produce the news. It was news produced by a national press system and so really was neither "global news" nor news of Japan.

Underreporting of Disasters in China in the Early Radio Years

No news bulletins arrived from central China along the cables, and so the Luhuo earthquake of March 1923 went unannounced in the *LAT*. Tensions between Japan and China, including anti-Japan boycotts in China, meant that "China" was often in the news: it was mentioned in eighteen separate news stories in the week March 24–31, 1923. At best, this reported "China" was located in Nanking, Shanghai or Beijing and was actually reported from Honolulu, Tokyo or Los Angeles. Radio had arrived in Los Angeles and the technical and entertainment capabilities of radio were celebrated by the *LAT*. The newspaper advertised and then commented favorably upon the KHJ radio station's program, which that week featured the precocious talent of the boy reader Emerson Frome Livingston, who read "The Earthquake," a bed-time story penned by Marshall Stedman, and a talk by Julian Arnold, US Department of Commerce, Beijing.[47] But, unlike in the Kanto Plain

47 "KHJ *The Times* Radiophone," *LAT*, March 28 and 29, 1923, "Easter Music Broadcast," *LAT*, March 30, 1923.

disaster just five months later, radio and telegraph operators failed to pass on news of the disaster from central China.

In 1925, one news bulletin of sixty-eight words announced that thousands were homeless in Yunnan Province following a severe earthquake, but this was the only news of this disaster in the *LAT* in the week after the event.[48] Datelined Shanghai, March 20, 1925, four days after the quake, the exclusive cable dispatch informed that Shanghai officials had learned via telegraph that the city of Talifu in West Yunnan was destroyed, that further details were not known but that foreign missionaries residing in the city were reported safe. The *LAT* paid more attention to China in this week than it had in March 1923. Further troubles were expected in China following the death of San Yat Sen. The US government was reorganising its diplomatic representation in the region. [49] Even past US relief aid to China proved controversial, with the *LAT* reporting a dispute between Shanghai and Beijing relief bodies over the future of US relief monies left over from a famine in 1919, and sought by the national relief organization for use in North China.[50] Nevertheless, datelines in China were responsible for only seven of the 18 items. The rest hailed from the USA or Manila.[51] This was news about "China," opinion about what Americans might learn from Chinese "civilization," and how they might understand China's geo-political and commercial situation.

Comprising over 4,000 words, *LAT* coverage of the Tango quake of March 7, 1927 (Table 1) shows a more intense interaction between Los Angeles and Kobe, Japan, than with Yunnan, China, but there is nothing like the reporting that went with the Kanto Plain disaster of 1923. AP bulletins reported the devastation around Kobe, the rising death toll, and that a winter storm had impeded Japanese relief efforts.[52] President Coolidge offered aid,

48 "Temblor Devastates City in China," *LAT*, March 21, 1925.

49 "Memorial for San Yat Sen is Held at Canton," *LAT*, March 18, 1925. "China Has Peace Interval," *LAT*, March 22, 1925, "New Chinese Puzzle," and "China Situation is Grave," *LAT*, March 23, 1925. "Schurman is Confirmed," *LAT*, March 18, 1925. "More Envoys Confirmed," *LAT*, March 19, 1925.

50 "China Fighting Over Charity Fund Surplus," *LAT*, March 23, 1925.

51 Shanghai (4), Peking (1), Canton (1) and Kunshan (1) were the Chinese datelines. Editorials, letters a sermon and various reports originated in Los Angeles (4), Washington DC (3), New York (1), Chicago (1), Seattle (1) and Manila (1).

52 "Sweep of Fires Adds to Quake Horror," *LAT*, March 8, 1927. "Refugees in Earthquake Region Suffer Agonies," and "New Earthquake Lasts for Two Hours," *LAT*, March 10, 1927. "Storm Hits Refugees," and "Quake Survivors in New Peril," *LAT*, March 11, 1927. "Many Still Perishing in Quake Area," *LAT*, March 12, 1927. "The Osaka Temblor," *LAT*,

and encouraged the American Red Cross to send relief.[53] Few Americans were in the disaster zone.[54] The full range of communications technologies was used to deliver the latest news to Los Angeles and AP again got a beat on the Japanese embassy in Washington.[55] US scientific credibility was asserted with respect to earthquakes. The seismic activity was too much for the needle of one English seismograph,[56] and the US would study earthquakes from a new facility in the University of Hawai'i.[57] In these ways, the *LAT* confirmed that the USA had a modern understanding of earthquakes, that Americans were in the danger zone, that the newspaper's superior communications network allowed the distant disaster to be reported, and that it was America's moral duty to be charitable.[58]

Significantly, both the efforts of Californians to raise relief funds and the rhetoric of the US rebuilding Japan were absent in reports of this earthquake. The limited coverage was partly a matter of the smaller scale of this disaster (Table 1). In addition, the Japanese ambassador in Washington declared that the Japan Government was already active in relief, expressed thanks for the US President's kind offer, but declined to accept it.[59] This reluctance on the part of the Japanese government was the result of its humiliating experience of 1923, when its acceptance of foreign aid signalled that it was not capable of dealing with the disaster. Moreover, there was a growing US antipathy for Japan following that country's invasion of China. The *LAT* explained to readers that America, Japan and Europe were entangled in China,[60] and

March 13, 1927. "Heavy Snows in Japan Cut Off Communication," *LAT*, March 14, 1927.

53 "War Secretary Wires Offer of Aid," *LAT*, March 9, 1927. "Coolidge Asks Red Cross Offer Aid," *LAT*, March 10, 1927.

54 The *LAT* mentions American tourists, a businessman from New York and American missionaries. "American Tourists Hurt as Quake Strikes Japan," *LAT*, March 8, 1927. "Missionaries in Japan Reported Safe," and "New York Businessman Killed in Quake," *LAT*, March 10, 1927.

55 "Embassy Hears Earthquake Details," and "Steamer Line Gets More Reports," *LAT*, March 9, 1927.

56 "Tremors Ruin English Seismograph," *LAT*, March 9, 1927.

57 "Quakes Will Be Studied," *LAT*, March 10, 1927.

58 "Easter Giving Season Opens Here," *LAT*, March 13, 1927.

59 "Government Sends Aid to Stricken Sections," *LAT*, March 9, 1927. "Japanese Envoy Expresses Thanks," *LAT*, March 10, 1927. "Japan Intends to do All of Relief," *LAT*, March 11, 1927.

60 This message is illustrated in two cartoons. In "The China Egg," *LAT*, March 11, 1927, a "Dove of Peace" sits on an egg labelled "China" atop a nest of swords, and wonders "... if this thing is *ever* going to hatch?" In "An Asiatic Laocoon," *LAT*, March 13, 1927, a

that peace in China was long delayed. Readers learned that same week of the march of Japanese forces on Shanghai,[61] and that Britain and Japan had agreed to President Coolidge's invitation to negotiate limits to naval build up and to sign a non-aggression pact.[62] Thus the earthquake news confirmed the general theme about Japan: relations were difficult.

Improved scientific instrumentation and detection of earthquakes were persistent features of *LAT* news. AP reports of seismic activity streamed into the editor's office from diverse datelines.[63] Automatic cameras were set near fault lines in California to help scientists better understand earthquakes.[64] And, on May 23, 1927 AP bulletins reported that scientists at Georgetown University, Washington DC and Loyola University, New Orleans had recorded big and distant earth tremors the day before, but with no way of knowing the direction from which the tremors came.[65] These two bulletins, in all just 146 words, were the only impression that readers of the *LAT* got of the Tsinghai earthquake. Ironically, the latest telecommunications and seismic detection technology were newsworthy while the disaster went unreported.

Thus the *LAT*'s extensive coverage of the Tokyo disaster of 1923 proved to be exceptional not only in relation to the coverage devoted to other twentieth century Japanese earthquakes but also in the context of 1920s disasters from around the world.[66] In the week after each of four other major Pacific Rim disasters in the 1920s, the *LAT* carried either no report or bulletins announcing the detection of earthquakes using seismographic equipment. These earthquakes – Luhuo, 1923, Yunnan, 1925, Tango and Tsinghai, 1927 – coincided with the rise of radio but, in these years, radio did not measurably alter patterns of reporting. While the *LAT* promoted its own radio program in the pages of its newspaper, it remained silent over the

dragon labelled "China" encircles and squeezes figures representing the "USA," "Europe" and "Japan."

61 "Peril Grips Soochow," and "Eyes of the World are on Shanghai," *LAT*, March 7, 1927. "New Light on China's Plight," *LAT*, March 13, 1927.

62 "Japanese Accept Bid," *LAT*, March12, 1927.

63 "Earth Shock Felt in Southern Italy," *LAT*, May 26, 1927, "Ecuador Earthquake Strongest in Years," and "Virginians Shaken by Earthquake in Night," *LAT*, June 11, 1927, "Earthquake Shakes Town in Crimea," *LAT*, June 28, 1927.

64 "Films to Catch Quakes," *LAT*, June 22, 1927.

65 "Temblors of Gigantic Force Felt," *LAT*, May 23, 1927.

66 See Winder, *The Los Angeles Times* Reports Japanese Earthquakes, 1923–1995, which deals with the effects of changing US-Japan relations on reports of Japanese earthquakes in the *LAT*.

terrible disturbances in China, and even the Tango disaster in Japan, 1927 warranted only a fraction of the coverage given to the Kanto Plain disaster.

Beachhead of Modernity: Managua, 1931

LAT coverage of the earthquake in Managua, Nicaragua, March 31, 1931 amounted to 7,441 words in the week after the event. This extensive coverage was sourced from Washington, Los Angeles, New York, Philadelphia, Seattle, San Francisco, New Orleans, the SS *City of Panama*, Managua, Mexico City, London and Vatican City, all of which were points on the modern telecommunications network. Managua was reported to have been "one of the most modern, enterprising and active cities in Central America."[67] Now razed by quake and fire, it was also portrayed as a beachhead of modernity surrounded by jungles and insurgents. Thus, a photographic essay juxtaposed the National Palace and the American Legation in Managua, two aerial photographs (the US Marine Aviation Field and a general view of Managua), with photographs of a "typical cantina" and an oxcart. The essay quotes "late radio dispatches," and reports that martial law has been declared.[68] In addition to reports from the scene of the disaster, the newspaper featured a list of destructive earthquakes of the last 25 years, a list of Americans evacuated from the disaster zone, and accounts of relief efforts. With Nicaragua's President Moncada homeless and fighting an insurrection, and with water supplies in Managua disrupted, the *LAT* and AP concentrated their reports of the Managua earthquake on the roles of Americans: US Marines, who were stationed in the city at the time; the US Navy, which delivered relief and evacuated refugees; and American reactions to the disaster.[69]

Managua's modernity was in peril, an interpretation highlighted in stories of the city's quake-disrupted communications. Radio and air services were now significant because the railroad was destroyed and with it the telegraph

67 "Angelino Concerned Over Safety of Brothers," *LAT*, April 2, 1931.

68 "Hundreds Perish as Nicaraguan Capital is Razed by Quake and Flames," *LAT*, April 1, 1931.

69 US Marines were conducting a canal survey and ensuring security against the insurgency of Augustino Sandino. According to the *LAT* their time in Nicaragua would now be prolonged. "Marines Face Extended Stay," *LAT*, April 1, 1931.

lines.[70] Initial reports from Managua pay special attention to the United Fruit Company's radio station. Back in service just two hours after the quake, it sent "the first direct news of the disaster."[71] While two army transport planes flew medics from the Panama Canal Zone to Managua, the US Navy scrambled its ships to aid the relief effort, dispatched amphibian planes to Lake Managua, and ferried evacuated Americans to Panama.[72] Pan-American Airways Company cabled to Matthew Hanna, the United States Minister, that it would be pleased to place its planes at the service of the relief effort, and alerted him to the fact that Earnest Swift of the Red Cross would take a direct flight from Miami to Managua.[73] Californians were among the first US citizens evacuated by airplane to Corinto, Nicaragua, by Pan-Am.[74] "Serum, medical supplies and physicians converged on the city by every available means of transport, with the vanguard expected to arrive by air by nightfall tonight," the *LAT* reported.[75] Five days later, with regular air mail services restored, the German Consul in Managua was reportedly concerned over the circulation of a cable to one insurance representative that stated that insurance on damage from the earthquake will not be paid.[76]

Comings and goings from Managua continued to be reported in the weeks after the quake. The *LAT*'s special correspondent, Will Rogers, arrived in Managua and filed a report.[77] A news reel was released in southern California theatres which, in part, showed the disaster zone in Managua. This was possible because "Graham McNamee, star of the National Broadcasting Company and pioneer talking reporter" had flown to the scene and could give an eyewitness account.[78] A photograph taken in San Diego showed refugees from Managua reaching port.[79] Then a San Francisco importer was reported to be a "harbinger of earthquakes," who had been "hounded by seismological disturbances" over his fifty-four years of travel in Central Ameri-

70 "Eyewitness Account Given By Marine Flyer," *LAT*, April 1, 1931.

71 "Radio Staff Quickly Rigs Up New Station," and "Thousands Die in Quake at Nicaraguan Capital," *LAT*, April 1, 1931.

72 "Quake Area Aid Rushed," *LAT*, April 1, 1931, "Americans Depict Scenes of Suffering," *LAT*, April 8, 1931.

73 "New Severe Earth Shocks Hit Stricken Managua," *LAT*, April 2, 1931.

74 "Californians Among Evacuated Refugees," *LAT*, April 4, 1931.

75 "Relief Work Directed at Preventing Disease," *LAT*, April 2, 1931.

76 "Managua Takes Stock of Loss," *LAT*, April 7, 1931.

77 "Mr. Rogers Remarks," *LAT*, April 9, 1931.

78 "Great Nicaragua Disaster Shown by News Films," *LAT*, April 11, 1931.

79 "American Survivors Tell of Nicaragua Horror," *LAT*, April 19, 1931.

ca.[80] These reports indicated not only that, together, radio telecommunication and mobility by automobile, railway, airplane and ship, provided the substance of the modern disaster story from Central America, but also that Managua's continued modernity depended on these services.

Edges of the News World: Xinjian, 1931, Sanriku and Sichuan, 1933, Formosa, 1935

There was no appreciable improvement in telecommunications services with China's interior in the *LAT*'s reporting of the Xinjiang disaster of October 1931, or that in Sichuan in August 1933. But when scientists at the Pasadena observatory reported another distant earthquake in August 1931, AP signalled a scientific break-through: Japanese seismologist Akitsune Imamura located the epicenter of the quakes in Sinkiang.[81] This was the news. The bulletin continued: "As Sinkiang lacks telegraph facilities, weeks probably will pass before direct reports can be received from that area."[82] News of an earthquake in Sichuan, China, August 25, 1933 arrived with further confirmation of the infrastructure shortcomings in China. The *LAT* published the news from Chengdu of the Sichuan earthquake disaster, in an AP bulletin with a Shanghai dateline in its August 31, 1933 edition.[83]

For different reasons, the *LAT*'s coverage of the Sanriku, Japan earthquake of March 1933 amounted to little more than a 1,000 words (Table 1). All of the news came from news agencies and from Japan, and there were few signs of Los Angeles engagement with the disaster. Readers learned that thousands had died but that the victims were being relieved by the Japanese Government.[84] It was not expected that Americans would give aid to Japan. Readers were informed of where, Japan was "battling for new territory" in

80 "Traveler in Americas Hounded By Temblors," *LAT*, April 25, 1931.

81 For Imamura's contribution to seismic research see Clancy, *Earthquake Nation*.

82 "Violent Quake Great Distance Away Recorded," *LAT*, August 11, 1931, "Great Quake's Center Located in Turkestan," *LAT*, August 14, 1931.

83 "Chinese Quake Fatal to 100," *LAT*, August 31, 1933.

84 "Quake and Tidal Wave Kill Hundreds in Japan," *LAT*, March 3, 1933. "Japanese Quake Death List Climbs to 1535," *LAT*, March 4, 1933. "Quake Victim's Needs Relieved," *LAT*, March 5, 1933.

China.[85] Readers learned of Filipinos objecting to Japan's war in China, and of their determination to boycott Japanese goods.[86] They also read of Japan's troubled financial state – trading was suspended by Japanese banks, and on exchanges in Tokyo, Osaka and Kobe.[87] Like the USA, Japan faced daunting troubles in the Depression, but, with President Roosevelt sworn in, and taking immediate action to stop runs on banks, Americans could hope that the US economy was turning the corner on the depression.[88] The Japanese government sought to know what the US government would do about the financial crisis, but the *LAT* turned this situation into a moral lesson: for Japan, financial trouble followed from its aggression in China.[89] So, even though the radio waves and telegraph cables carried news to Los Angeles of the disaster across the Pacific Ocean, the *LAT* did not report the earthquake in any depth, largely because Japan had already placed itself outside the reach of American charity.

Two major earthquake disasters struck Taiwan, then administered as Formosa by Japan, in 1935 (Table 1). The first of these quakes was reported by the Franklin Institute on April 20.[90] The time difference between Taiwan and Los Angeles made it possible to publish this news the day that the quake occurred. An editorial appeared the next day and, over the following days, nine further bulletins, in all 1,613 words, one map and one photograph all from AP, and all datelined Taihoku, Formosa. These announced the destruction of thousands of houses in central Taiwan which left perhaps 250,000 people homeless. The death toll was attributed to collapsing Chinese-style houses made of sun-dried brick, examples of which were shown in an AP photo. But there was also news of destruction of the province's communications infrastructure. Refugees clogged the roads in central Taiwan as the Japanese authorities rushed soldiers, doctors and water to the disaster zone. Military carrier pigeons, "the only means of quick communication available," were

85 "Japanese Grab Jehol Centers," and "Invaders Smash Jehol Defense," *LAT*, March 3, 1933. "Jehol City Captured," *LAT*, March 4, 1933.

86 "Philippines Boycott Japanese," *LAT*, March 6, 1933.

87 "Japanese Banks Suspend Exchange Trade," *LAT*, March 6, 1933.

88 In the cartoon "There's Always a Bottom to 'em," *LAT*, March 8, 1933, Uncle Sam hits "Business Bed Rock" after a long slide and declares "Ah ha! The turning point has been reached at last!"

89 A cartoon "The Camp Follower," *LAT*, March 2, 1933, shows a ragged figure labelled "National Bankruptcy" pursuing advancing Japanese tanks.

90 "Foreign Yacht Held by Japan in Spy Inquiry," *LAT*, April 27, 1935.

reported to have carried the news that relief was on the way.[91] Meanwhile, an aerial survey revealed that the disaster zone extended over 5,000 square kilometers and an AP bulletin from Taiwan highlighted the tensions in the region between Japan and the other Great Powers. It reported the detention of a yacht by Japanese authorities who suspected the French-German crew of spying.[92] United Press bulletins reported that the second disaster in July 1935 leveled 305 dwellings, and destroyed communications with the coast and an oil storage tank. For this earthquake, the bulletin from Formosa was published without any scientists' seismographic report, although the delivery of a seismograph to Griffith Observatory was reported, along with an explanation of how it worked and a photograph.[93] In both cases, the Japanese government's refusal to accept foreign offers of relief assistance made news of these disasters of little interest to American news agencies and newspapers. In the absence of a US relief campaign, they focused on the struggle to modernize Taiwan's housing and communications, and on the ways of knowing the earthquake made possible by air surveillance.

Air Power: Chillan, Chile, 1939

The *LAT* reported the Chillan earthquake by paying special attention to aviation and radio. True, an initial AP bulletin reporting seismic activity that shook Santiago and Valparaiso, Chile and Mendoza, Argentina arrived by cable from Santiago, and the quake was front page news the next day.[94] Messages whizzed over telegraph wires and telephone lines carrying estimates of the death toll and damage, but, in the face of disrupted communications, United Press (UP) noted that the best estimates came from an aerial survey of the 100,000 square kilometer area of central Chile that had been devastated. UP's correspondent, Charles Serby, wrote that "there was no room for passengers in the Junkers bomber" and that he was "obliged to lie on the

91 "Thirst Adds Torture in Quake-stricken Isle," *LAT*, April 23, 1935.
92 "Foreign Yacht Held by Japan in Spy Inquiry," *LAT*, April 27, 1935.
93 "Quake Kills Two Score in Formosa," *LAT*, July 17, 1935, "New Temblors Jolt Formosa," *LAT*, July 19, 1935, "Japan Suffers Another Shake," *LAT*, July 20, 1935, "Earthquake's Autograph Shown," and "Temblors Reproduced by Device," *LAT*, July 22, 1935.
94 "Heavy Quake Rocks Chile," *LAT*, January 25, 1939.

floor," wear a parachute and view the disaster "through a slit in the plane."[95] After landing in Chillan, he toured the destroyed city in a car. On the return flight he noted the tragic loss of modern institutions like the new Roxy picture theatre and the city's newspaper office. Other aviators supplied copy to UP's correspondent and to the Buenos Aires newspaper, *La Nación*.[96] As the Chilean President left by "special train" to direct relief efforts, the *LAT* was already publishing AP wire photos of Santiago and Concepcion from before the earthquake, and a seismographic chart from Carnegie Institute of Seismology, Pasadena.[97] As ships were made ready in Valparaiso Harbor to carry supplies along the coast, military and private planes ferried medical staff into the disaster zone, emergency landing fields were prepared, and a US Air Force "Flying Fortress" headed south from the Panama Canal Zone on a mission that criss-crossed Latin America. It carried medical supplies, powdered milk and a Red Cross director.[98] From Chillan radio broadcasts appealed for chlorine and sanitary equipment to disinfect water. In these circumstances, aircraft and radios supplanted the telegraph as the communications technology of the disaster. While cables from datelines including Santiago, Chillan, Concepcion, Buenos Aires, Cristobal (Santa Cruz), Lima, Washington, Pasadena, and San Francisco supplied the copy, altogether 6,437 words in 29 articles, accompanied by seven photographs, one map and one reading from a seismograph, aviators and radio operators supplied the eye witness reports.

In fact, radio, television and aviation received considerable attention in the week following the Chillan earthquake. Mention of radio and television communication was less frequent than for air travel technologies but provided arresting news.[99] At the end of January 1939, Senator Key Pittman was pictured in Washington DC listening to Adolf Hitler's speech to the German Reichstag, the full text of which was published in the *LAT*.[100] Brit-

95 "Quake Toll Mounts to Thousands," *LAT*, January 26, 1939.

96 "Flyer Reports 5,000 Dead in Chillan," *LAT*, January 26, 1939, "Official Describes Quake Scene Horrors," *LAT*, January 27, 1939.

97 "Thousands Reported Killed as Severe Earthquake Rocks Central Chile," and "Intensity of Chile and Long Beach Temblors Compared," *LAT*, January 26, 1939.

98 "Disease Perils Quake Areas," *LAT*, January 27, 1939, and "Chile Quake Toll Set at 30,000," *LAT*, January 28, 1939, "Army Planes Leave With Relief Supplies," *LAT*, January 29, 1939.

99 "Plane Radios It's Down in Bering Sea," *LAT*, January 31, 1939, "Capital Views Television Test," *LAT*, January 28, 1939.

100 "Senator Pittman Views Hitler Speech as Hopeful," "Hitler Gives Warning in Demand for Colonies," and "Reich Will Go Onward Hitler Tells Powers," *LAT*, January 31, 1939.

ish Prime Minister Neville Chamberlain's Birmingham address defending his appeasement of Hitler in the freshly signed Munich pact was also printed.[101] However, reports of the use of civilian and military aircraft streamed into Los Angeles from all around the globe. Two German pilots reached Manila on a goodwill mission, while Guatemalans announced plans for a goodwill flight to Mexico.[102] Not only were the British planning a round-the-world air service, but, as the Royal Air Force practiced bombing Buckingham Palace as the King watched, a British expert expressed doubts over Germany's air superiority.[103] As the United States government invited bids from companies on new warplanes and aircraft carriers, promised an aircraft deal to France and urged an air and naval base to be built on Guam, a US airman set a world speed record in a dive, and another was ready to test planes at high altitude.[104] Simultaneously, Chinese planes bombed a Japanese fleet, while General Franco's fighter-bombers filled the roads around recently captured Barcelona with the bodies of Republican soldiers who had fled the city.[105] Readers were asked to consider the vital question "Is the air-raid bugaboo the bunk?"[106] The great powers were each preparing for war, and ascendancy in the air would be vital, so was the USA ready?

With this question mark planted in the heads of its readers, admittedly five days after the first news of the quake, the *LAT* signalled the importance of air power for both war and disaster. Photographs of Catalonian houses, towns and roads ruined by General Franco's bombardment appeared in the *LAT* on January 28 and 29.[107] So, when the front page carried a photograph of a "deserted, debris-choked street in Concepcion, Chile," readers could not but compare the damage of war and earthquake. The photograph had been

---

101 "Text of Chamberlain's Birmingham Address," *LAT*, January 29, 1939.

102 "Germans on Goodwill Flight Reach Manila," *LAT*, January 28, 1939, "Guatemalans Plan Goodwill Flight," *LAT*, January 29, 1939.

103 "British Air Expert Doubts Nazi's Great Superiority," *LAT*, January 26, 1939, "British Eye Pacific in Plan For Round-the-world Air Service," and "British Air Force 'Bombs' King's Palace as Royal Family Looks On," *LAT*, January 29, 1939.

104 "Pilot Sets World Speed Mark in Dive," *LAT*, January 25, 1939, "New Plane Bids Opened By Army," *LAT*, January 26, 1939, "Plane Carrier Bids Invited," and "Guam Naval and Air Base Urged by Admiral Hepburn," *LAT*, January 27, 1939, "French Given Air Force Aid by Roosevelt," *LAT*, January 28, 1939, "High Flight Tester Ready," *LAT*, January 29, 1939.

105 "Chinese Bomb Japanese Fleet," *LAT*, January 26, 1939, and "Rebels Bomb Refugees," *LAT*, January 28, 1939.

106 "Is the air-raid bugaboo the bunk? Some Say It Is," *LAT*, April 29, 1931.

107 "Typical Scene in Wake of Gen. Franco's Conquest," *LAT*, January 28, 1939.

"flown to Buenos Aires and flashed to New York by wireless."[108] Similarly, Chilean military authorities took charge of the disaster, not only imposing martial law but also commandeering "all long and short-wave radio stations." Interior Ministry bulletins were soon the only reports available to the press. As war fever mounted in Europe, the *LAT* was preparing Californians for the new war technologies. Radio and air power would have center stage.

Conclusion

Reporting of distant earthquakes changed in the 1920s and 1930s, as modern newspapers used radio and aircraft technology as well as the telegraph in news production. The new technologies enhanced the news gathering capabilities of the newspaper and thus facilitated and authenticated stories about globalization of the news. Newspapers related readers to the earthquake using stories about communication technology and scientific knowledge, and the two were interrelated. How knowledge of earthquakes was recorded and transmitted was a main story when a terrible earthquake shook a distant land, and the *LAT* celebrated its superior access to the world's latest stories whenever it reported an earthquake. The combination of seismographic and telegraphic work offered opportunities for newspapers to celebrate superior access, notably in the form of "beats:" being the first to publish the news. The work of producing the story was made into a narrative in space as the newspaper informed readers of where the story was compiled, under what conditions, and how reliable or authoritative it was, all of which were matters shaped by telecommunications technology.

Additionally, telecommunications technology shaped the news of earthquakes in the following specific ways. First, disasters in different regions were reported in different ways because of their different relation to the geography of the network of news production. While the transmission speeds of telegraphic and radio telecommunication were faster than with transport, this advantage was compromised by other aspects of modern disaster reporting, notably the detection of seismic disturbances at long range, the desire for eye-witness accounts and photographs, and the slow expansion of the telecommunications infrastructure. This meant that the telecommunications

108 "First Earthquake Picture – Chile City Laid in Ruins," *LAT*, January 29, 1939.

network available in Los Angeles shaped the earthquake story into one or other of several basic forms: no news at all; simple news bulletins reporting the earthquake and disaster; news of communications technology and seismic disturbances rather than news of disaster; and disaster reporting from diverse locations including coverage of events in the disaster zone, charitable acts at home, and relief efforts that might be reported from the distant disaster zone, from home and anywhere in between.

Second, the new technologies also helped to demarcate and distinguish between those regions that were "modern" and those that were not. Readers could compare the length of time that it took for news to arrive from different places, and would be reminded of why it took so long to arrive from central China rather than from Japan. Readers were told about where the story was compiled, under what conditions, and how reliable and authoritative it was: they could rely on the *LAT*. Of course, what they could rely on was the ensemble of technologies and the news gathering networks that made the modern newspaper possible. In turn, those networks confined and constrained the news that could be known. Thus, reports of only one of the disasters studied here, the Kanto Plain earthquake and fire of 1923, amounted to full media coverage of disaster, relief, science stories, survivors' tales and rebuilding, and that was partly because of the ensemble of technologies and infrastructure connecting Los Angeles and Tokyo. As the technologies supporting news production were developed, so too could the disaster story be developed from the limited knowledge and accuracy associated with early news wires to fuller accounts featuring photographic essays, editorials and eye-witness accounts. But this was a matter of geography as well as "progress:" after the Kanto Plain earthquake coverage of other disasters were not so comprehensive partly because of the more limited ensemble of network connections to places other than Tokyo.

Third, particular ideas about space were promoted using juxtapositions and resonances within the pages of the newspaper. For example, in 1939 readers in Los Angeles could compare the damage done by an earthquake in Managua with that from bombing or artillery barrage in Barcelona. Since almost every page carried at least one story related to aviation, and most of these were about war-time use of aircraft, the short accounts of the roles of aircraft in the reports of earthquake disasters acquired additional meaning: relief would not reach victims in time without them; no combatant was safe from them; air power was vital. Moreover, the remote sensing of earthquakes using aerial photography, seismographic equipment and radio transmission interception became an important theme in the coverage.[109] The *LAT* used an ensemble of war technologies to cover Depression era earthquakes, and so the *LAT* became a promoter of war technologies and narratives.

Finally, interwar telecommunications and transport technology produced a pattern of geographical asymmetries in the *LAT*. By shrinking time-space between places already located on the network infrastructure, it simultaneously highlighted the time-space delays between those places in the telegraph and radio worlds and those that lay outside the orbit of such networks. In the hands of the newspaper, these disparities became emblematic of the distance, in civilization terms, between "modernity" and its other. In this context, and since earthquake detection was a matter of modern seismology and telegraphy, newspaper reports of distant seismic events served to emphasise the disjuncture between places, like Los Angeles, where, it was reported, technologies and practices of cutting edge modernity could be found, those, like central China, apparently dominated by centuries old "tradition," and those, like Managua, that were construed in the press as outposts of "modernity."

Bibliography

Adams, Paul C., "The September 11 Attacks as Viewed From Quebec: The Small Nation Code in Geopolitical Discourse," *Political Geography,* vol. 23, no. 6, (2004), pp. 765–795.

Anderson, Benedict, *Imagined Communities: Reflections on the Origin and Spread of Nationalism,* London 1983.

Anderson, Kay, "White Natures: Sydney's Royal Agricultural Show in Post-humanist Perspective," *Transactions of the Institute of British Geographers* NS, vol. 28, (2003), pp. 422–441.

Appadurai, Arjun, *Modernity at Large: Cultural Dimensions of Globalization,* Minneapolis 1996.

Benthall, J., *Disasters, Relief and the Media*, London 1993.

Bonea, Amelia, "The Medium and its Message: Reporting the Austro-Prussian War in the Times of India," *Historical Social Research/Historische Sozialforschung,* vol. 35, no. 1, (2010), pp. 167–187.

Boyle, E.H./A. Hoeschen, "Theorizing the Form of Media Coverage Over Time," *Sociological Quarterly,* vol. 42, no. 4, (2001), pp. 511–527.

109 Charles Owens mapped distant disaster zones for the *LAT* before World War Two began. Cosgrove and Della Dora, "Mapping Global War," 2005, pp. 373–390.

Christensen, Christian, "Pocketbooks or Prayer Beads? US/UK Newspaper Coverage of the 2002 Turkish Elections," *Harvard International Journal of Press-Politics,* vol. 10, no. 1, (2005), pp. 109–128.

Clancy, Gregory, *Earthquake Nation: The Cultural Politics of Japanese Seismicity, 1868–1930,* Berkeley, CA 2006.

Cosgrove, Denis/ Veronica Della Dora, "Mapping Global War: Los Angeles, the Pacific and Charles Owens' Pictorial Cartography," *Annals of the Association of American Geographers,* vol. 95, no. 2, (2005), pp. 373–390.

Cottle, Simon, "Mediatised Rituals: Beyond Manufacturing Consent," *Media, Culture and Society,* vol. 28, no. 3, (2006), pp. 411–432.

Couldry, Nick, *Media Rituals: A Critical Approach*, London/New York 2003.

Couldry, Nick/Andreas Hepp/Friedrich Krotz (eds.), *Media Events in a Global Age,* London/New York 2010.

Gregory, Derek, *The Colonial Present: Afghanistan, Palestine, Iraq*, Malden, MA 2007.

Hannah, Matthew, "Torture and the Ticking Bomb: The War on Terrorism as a Geographical Imagination of Power/knowledge," *Annals of the Association of American Geographers,* vol. 96, no. 3, (2006), pp. 622–640.

Harvey, David, "Cosmopolitanism and the Banality of Geographical Evils," *Public Culture,* vol. 12, no. 2, (2000), pp. 529–564.

Hewitt, K., "The Idea of Calamity in a Technocratic Age," In K. Hewitt (ed.), *Interpretations of Calamity: From the Viewpoint of Human Ecology*, Boston, MA 1983, pp. 3–32.

Mauch, Christof/Christian Pfister, (eds.), *Natural Disasters, Cultural Responses: Case Studies Toward a Global Environmental History*, Lanham, MD 2009.

Oliver-Smith, A./S.M. Hoffman (eds.), *The Angry Earth: Disasters in Anthropological Perspective,* New York/London 1999.

Singer, J. B., "More Than Ink-stained Wretches: The Re-socialization of Print Journalists in Converged Newsrooms," *Journalism and Mass Communication Quarterly* Winter, vol. 81 no. 4, (2004), pp. 838–856.

Sugiyama, Shin'ya/Linda Grove, (eds.), *Commercial Networks in Modern Asia,* Richmond, Surrey 2001.

Wenzlhuemer, Roland, "Editorial – Telecommunication and Globalization in the Nineteenth Century," *Historical Social Research/Historische Sozialforschung,* vol. 35, no. 1, (2010), pp. 7–18.

—, "Globalization, Communication and the Concept of Space in Global History," *Historical Social Research/Historische Sozialforschung,* vol. 35, no. 1, (2010), pp. 19–50.

Whitney, D.C./L. Becker, "Keeping the Gates for Gatekeepers: The Effects of Wire News," *Journalism Quarterly,* vol. 59 (Spring) (1982), pp. 60–65.

Wilke, Jürgen. "Historical Perspectives on Media Events: A Comparison of the Lisbon Earthquake in 1755 and the Tsunami Catastrophe in 2004." In Nick Couldry/Andreas Hepp/Friedrich Krotz (eds.), *Media Events in a Global Age,* London 2010, pp. 45–60.

Winchester, Simon, *Krakatoa: The Day the World Exploded 27th August 1883*, London 2003.
Winder, Gordon M., "London's Global Reach? Reuters News and Network, 1865, 1881, and 1914," *Journal of World History*, vol. 21, no. 2, (2010), pp. 271–296.
—, "Imagining Citizenship in the Networked Newspaper: *La Nación* Reports the Assassination at Sarajevo 1914," *Historical Social Research – Historische Sozialforschung*, vol. 35, no. 1, (2010), pp. 140–166.
—, "Mediating Foreign Disasters: *The Los Angeles Times* and International Relief, 1891–1914," in A. Janku/G. Schenk/F. Mauelshagen, (eds.), *Historical Disasters in Context: Science, Religion, and Politics*, Oxford 2012, pp. 195–226.
—, "*The Los Angeles Times* Reports Japanese Earthquakes, 1923–1995," In C. Mauch/S. Mayer (eds.), *American Environments: Climate,-Cultures-Catastrophe*, Heidelberg 2012, pp. 133–157.
Zhong, Zenzhong/Lanmin Wang, "Geological Disasters in Loess Areas During the 1920, Haiyuan Earthquake, China," *Geojournal*, vol. 36, no. 2–3, (1995), pp. 209–274.

Table 1. Coverage of Severe Pacific Rim Earthquakes in the LAT, 1917–1939

Year	Date	Location	Magnitude (R)	Deaths (No.)	Words (No.)	Delay (Days)
1917	Jan 20	Bali, Dutch E. Indies	Nd	1,500	116	7
1917	Jul 30	Yunnan, China	7.5	1,800	No report	No report
1918	Feb 13	Nan'ao, Guangdong	7.3	1,000	No report	No report
1920	Dec 16	Haiyuan, China	7.8	200,000	538	1
1923	Mar 24	Luhuo, Sichuan, China	7.3	3,500	No report	No report
1923	Sep 1	Kanto, Japan	7.9	143,000	76,876	1
1925	Mar 16	Yunnan, China	7.1	5,800	68	5
1927	Mar 7	Tango, Japan	7.6	3,020	4,222	1
1927	May 22	Tsinghai, China	7.6	40,900	146	1
1931	Mar 31	Managua, Nicaragua	6.0	2,500	7,441	1
1931	Oct 10	Xinjiang, China	8.0	10,000	176	1
1933	Mar 2	Sanriku, Japan	8.4	2,990	1,025	1
1933	Aug 10	Sichuan, China	7.4	9,300	40	6
1935	Apr 20	Formosa, Japan	7.1	3,270	1,613	0
1935	July 16	Formosa, Japan	6.5	2,740	141	3
1939	Jan 25	Chillan, Chile	7.8	28,000	6,437	0

Sources: US Geological Survey and *The Los Angeles Times*
Note: The place names used in this table are those used by the USGS.

Northern Experiences of Global Telegraphy: Materiality and Technology in the Scandinavian Periphery

Jonas Harvard

For sparsely populated areas with adverse geographical conditions, the possibility of bridging distances through technology would seem a godsend. So when telegraph lines began to appear in the countries of Scandinavia in the 1850s, provincial authorities petitioned their respective governments to construct lines in their regions. In many cases the petitions were successful and the network of telegraph lines grew steadily in Sweden, Norway and Finland during the period from 1850 to 1870.

The successive extensions of lines may be fitted into an over-arching narrative of globalization, in which the "tentacles of progress" reached farther and farther, as northern Europe became more and more integrated into the global network.[1] Such histories of globalization focus on the systemic level and imply universalizing perspectives. The scope of the perspective – looking at the entire world – often breeds, or rather necessitates far-reaching generalizations. In these generalizations, the electric telegraph is routinely presented as a symbol of the immense possibilities created by technological development, heralding the dawn of a global exchange of ideas and information.[2] Such arguments tend to look at the telegraph as an invention with particular qualities, rather than as a technology interwoven in a complex web of contexts and relationships. From a strictly technological point of view, communication could indeed take place with the speed of electricity. In theory, as James Carey noted, the electric telegraph separated communication from transportation.[3] In the fields of finance and diplomacy, the changes were immense, as actors with large resources appropriated the tech-

1 The term comes from Daniel Headrick. *The tentacles of progress: Technology transfer in the age of imperialism, 1850–1940.*

2 See for example Chanda, *Bound Together*, 63. For general histories of globalization see Bayly, *The birth of the modern world*; Bordo, Taylor, and Williamson (eds.), *Globalization in historical perspective.*

3 See his seminal essay "Technology and ideology: the case of the telegraph" from 1983, reprinted in Carey, *Communication as culture.*

nology.[4] However, we know very little of how the electric telegraph operated in the peripheries of Europe, its actual impact and the advantages it brought when conditions were adverse.

In alignment with the general thesis of this book, this chapter attempts to present an alternative to the grand narratives of steady progress and examines the realities of the electric telegraph at the local level. It seeks to move beyond ideas about technology *as such* and focus more on the stories of technologies in use.[5] Applying this perspective reveals other images of the electric telegraph, such as how material conditions, sometimes hostile attitudes among the local population, as well as institutional tensions and other external factors, affected the operation of the telegraph in northern Europe. In relation to the general theories and large-scale histories of globalization, the chapter thus argues that specific local conditions determined the impact of new communication technologies to such an extent that the grand narratives become less valuable as tools for understanding the development.

The paper seeks to narrow in on the role of place, by juxtaposing the role of the telegraph in the mid-nineteenth century, in particular for newspaper publishing, in two Swedish cities: Helsingborg on the southern coast, close to the Danish border, and Piteå far up in the North. In both these cities, newspapers were founded in 1847: *Öresunds-Posten* in Helsingborg and *Norrbottens-Posten* in Piteå.

Newspapers were typically important users of the new technology, eager to embrace the opportunity of a speedier delivery of the latest information. In this respect, the cities were in very different situations, separated by around 1500 kilometres. Piteå was a small coastal city in the far North, near the Russian border, with little trade. Helsingborg in the South was situated close to the Danish border, with lively trade, and experienced strong population growth in this period, multiplying its size in few decades. However, they shared an important factor: they were both connected to the electric telegraph in the 1850s, in 1854 and 1857 respectively. Focussing on the first decades of the use of the electric telegraph also means focussing on the process of adopting the new technology and how it was integrated into existing communication structures. Sometimes it is not the existence of entirely different processes of adaption and implementation that reveal differences in

4 Nickles, *Under the wire*; Chandler, *The visible hand.*

5 Edgerton, *The shock of the old*; Edgerton, From Innovation to Use: ten (eclectic) theses on the history of technology.

how new technologies become embedded in societal structures, but rather that similar processes occur on different time-scales.

We can learn a great deal about the importance of context by comparing two Swedish cities, which were part of the same national network, which in turn was connected to the growing international net of telegraphs. In theory these cities held the same possibilities for long-distance communication. What impact did the access to new technology really have in these two different settings? Was distance in fact "annihilated" through the electric telegraph, as recent historical writing claims, recycling the words of nineteenth-century telegraph proponents?[6] Or were local conditions the determining factor, so that the telegraph, instead of shrinking communicative distances, actually increased the relative difference between the cities, concerning the time delay of news dissemination?

My source material consists of newspaper articles, documents from the different archives of the National Telegraph Agency in Sweden (Telegrafverket) and the archives of telegraph stations, private letters, as well as secondary sources.

The chapter begins with a general presentation of the expansion of the electric telegraph in Sweden, followed by a discussion of the situation in Helsingborg, which is compared to the developments in Piteå. I then compare the cities through two main themes: the reliability of telegraph technology and institutional tensions within the national telegraph network. The chapter concludes with an analysis of how the case of the northern periphery relates to the general history of the development of global interconnectivity in the nineteenth century.

The Swift Expansion of the Swedish Telegraph

In Sweden, electric telegraphs were built by the state and the matter was the subject of several debates debate in Parliament from 1851 onwards. In 1852, the Swedish king asked the head of the optical telegraphs, General Carl Akrell, to investigate the possibility of setting up electric telegraphs in the country. Based on his proposal, the Swedish government in February 1853 took the decision to establish a system of electric telegraphs. Once the decision was taken and funding in place, the actual construction of telegraph

6 Roland Wenzlhuemer, The Dematerialization of Telecommunication.

lines across Sweden was an uncomplicated matter. Following the general European pattern of construction in cooperation with the military, a first line of wire stretched across some 2000 poles was built from Stockholm, the Swedish capital, to Uppsala during the summer of 1853.[7]

The first electrical tests of the equipment were successful, and in 1854, new lines were built all the way down to Malmö, the most southern Swedish city of any prominence. The density of the network increased rapidly in the southern part of the country, and beginning in 1856, lines also found their way north. In 1857, nearly the whole length of the country was covered. When a submarine cable was laid to Denmark in 1854, Sweden became connected to Hamburg and the rest of the European continent. The route east was established in the north, when Haparanda was linked to the Finnish and thus also the Russian telegraph network.[8]

As historical telegraph statistics show, the period under consideration in this chapter saw relatively little international use of the telegraph. Traffic increased greatly in the following decades until the turn of the century, and reached a peak during World War I.[9] What did the new technology mean for news reporting in the Swedish papers in the decades after the establishment of electric telegraphs in Sweden in the 1850s? What ideas were associated with it and how were the papers' usage of the telegraph determined by the practicalities of its operation? We now turn to the city Helsingborg in the Swedish south.

Between 1850 and 1880, the population of the city nearly tripled, growing from around 4000 inhabitants to close to 12,000. The circulation of the local paper, *Öresunds-Posten*, also multiplied many times over, increasing from around 500 daily copies in the 1850s to 1800 copies in the 1870s. Although placed in Sweden, in European terms a northern country, in relation to other towns in Scandinavia, this was not a small town in the periphery, but rather an important border town close to Denmark and the continent.

7 Heimbürger, 37–42; Kaijser, *I fädrens spår.*

8 Heimbürger.

9 On the dynamics of development inherent in the successive phases establishment, expansion and stagnation, see Kaijser, 75–96; Brian Mitchell, *International Historical Statistics: Europe, 1750–2005,* Pulgrave Macmillan, New York, 2007.

Lines Reaching Helsingborg and *Öresunds-Posten* in 1854

Helsingborg was one of the first Swedish cities to be connected to the telegraph. The local telegraph was set up in 1854. On the occasion of the inauguration, the local paper, *Öresunds-Posten*, published celebratory texts, which hailed the blessings of the new technology. The editor of *Öresunds-Posten* was personally present at the opening ceremony and gave speeches and proposed toasts at the festivities celebrating the inauguration of the Helsingborg station. In fact, as new stations opened throughout the country, the celebrations soon became ritualised to such a point that the general director of the National Telegraph Agency became fed up with all the festivities. In a letter, he specifically complained about the never-ending celebrations in Helsingborg.[10]

The editor of the paper at the time, Oscar Patric Sturzenbecker, embraced the technological novelty of the telegraph with great enthusiasm. He later left the paper to become its foreign correspondent, first in Copenhagen, then in Paris and elsewhere.

In a number of articles, the paper commented on the new technology and its possibilities. An article from 1855, most likely written by the new editor F. T. Borg, who took over the paper in 1856, explained how the telegraphic service in Sweden might be improved and praised the innovation, effectively echoing Saint-Simonian visions of universal communication:[11]

> The great importance of the electric telegraphs lie in the fact that they seem to speed up the development of mankind through the ever more rapid distribution of the word and thereby through the command of action.[12]

The local paper's attachment of great importance to the telegraph is apparent from the paper's annual presentations for the coming year. Most papers published such declarations around December or January each year, in which they told current and prospective readers about the ambitions and directions of the paper for the year to come. In the first announcement after the opening of the telegraph, *Öresunds-Posten* emphasized its extensive use of the

10 Heimbürger, *Svenska telegrafverket*, 51.

11 On the Enlightenment and later visions of communication as a tool for peace see Mattelart, *Networking the world*. On early visions of different technologies see Marvin, *When old technologies were new*.

12 *Öresunds-Posten*, January 12, 1855.

telegraph to set itself apart from other regional newspapers. The paper took pride in delivering news faster than any comparable Swedish paper:

> With regard to *telegraph news*, it should already be obvious that we in this respect have an advantage over most, if not all, of our colleagues, and for the future, we expect to become even more certain of having a definitive lead in this respect.[13]

Already during 1854, the paper published a range of telegraphic news. An example from December 29 displays an extraordinary number of dispatches with a wide geographic distribution. The headline "The telegraph dispatches of Öresundsposten" gave the telegram section a distinct identity in the paper. And then, just as in numerous other papers around Europe, the world according to the telegraph unfolded: *Paris the 26th*. Napoleon confirming a number of alliances. *London the 25th*. Latest news from the House of Commons. *Madrid the 18th*. Cortes opposing the sale of Cuba. *Petersburg the 25th*. Weak English firing at Sebastopol. *Bombay November 26*. The English planning actions Teheran to prevent alliance with Russia. *London the 27th*. Very little trading in the grains. *Paris the 27th*. The emperor offering subscription on 500 million loan. *Petersburg the 26th*. The empress is sick. *Vienna the 28th*. Envoys from all the great powers gathered.[14]

The dispatches came all the way from London in the west to St. Petersburg in the east, and even from as far away as Bombay. Of course, *Öresunds-Posten* did not have its own correspondents in all these places, so how did it get the telegrams? This is an interesting question, but difficult to answer with precision.

The telegrams may have been provided by a person collecting these news items from foreign papers available in Copenhagen, and then sending them by telegraph to *Öresunds-Posten.* A similar practice is known to have been used by Norwegian papers in 1855, whose correspondents copied news from the printed Stockholm papers and relayed them by telegraph to Christiania in Norway.[15]

The "telegraph news" of *Öresunds-Posten* could also have been actual telegrams that had been printed in papers arriving by regular mail. The use of

13 *Öresunds-Posten*, December 18, 1854. There was fierce competition among local papers in the south, with many claiming to present fast news. This might have contributed to the eagerness to make the most out of the possibilities offered by the telegraph.

14 *Öresunds-Posten*, December 29, 1854. The actual wording in the paper was slightly more verbose than in the preceding summary, and the quantity of text attributed to the cities mentioned spanned between three and sixteen lines of text in a single column.

15 Giverholt, *Nyhetsformidling i Norge, 20.*

the particular headline "The telegraph dispatches of Öresunds-Posten" seems to indicate that they had been relayed by telegraph to the paper. It is not known if the paper at this point subscribed to any news agency service. It later announced such subscriptions, but did not mention any such arrangements at this point.

Although Agence Havas had been operating since 1832, and Reuters and Wolff began operations in 1851 and 1849 respectively, by 1855 no organized telegraph service existed in Scandinavia. Through an agreement between the three in 1859, Scandinavia became designated as part the interest sphere of Wolff's bureau, but it was not until 1867 it opened a branch in Stockholm, called Svenska telegrambyrån [The Swedish Telegram Agency]. In 1866, the Ritzau bureau opened in Copenhagen and the following year, Norsk telegrambureau [The Norwegian Telegram Agency] opened in Christiania. A general assessment is that the Franco–Prussian war of 1870 provided a similar boost for the Scandinavian agencies that the Crimean War gave their continental counterparts.[16]

The attribution of news quoted from other papers to "the telegraph" was a pattern, which proved persistent in the in the early years of the telegraph. Any time a message had been passed on by means of the telegraph, even for a short distance, it was thereafter labelled as telegraphic news, even if it was taken from a paper printing the information. "Telegraph" and "telegrams" soon came to symbolise something like "the latest" or "short rapid messages." Researchers in the history of journalistic form, point out that in many cases, the telegram category simply replaced the category "By the latest mails."[17] This shows that the use of the telegraph in terms of news transmittance was actually not as frequent as implied by the newspapers in its first decades of use.

Later in 1855, *Öresunds-Posten* announced that through special "telegraph arrangements," it would now publish foreign news sooner than any other paper which was not published daily. It did not specify what these special arrangements were. The paper foresaw that foreign news would soon interest its readers the most. Therefore, the correspondence articles from Copenhagen would henceforth concentrate on "European politics in general" rather than the political issues of Denmark and Scandinavia.[18] The perception that the temporal horizon was shrinking was thus coupled with an increased em-

16 Giverholt, 9–26.

17 Barnhurst and Nerone, *The Form of News.*

18 *Öresunds-Posten*, April 2, 1855.

phasis on long-distance news. Irrespective of whether news arrived quicker or not, the idea that it did warranted focusing on European news rather than on regional items.

On numerous occasions in the following years, *Öresunds-Posten* boasted about its extensive use of telegrams. It published some acerbic remarks about the high costs, but made it clear that the benefits and continued support of its readers outweighed the costs . In December 1857, by which time a new editor, F. T. Borg, had taken over the paper, the paper gave its presentation for 1858, wherein the readers were told that the paper had acquired "foreign telegraph dispatches" which arrived on "the very same day that they can be read in the evening papers of Copenhagen" and that a "reliable telegraph correspondent" had been hired in Stockholm.[19] On December 30, 1857, under the headline "Telegraphic dispatches to Öresunds-Posten," the latest parliamentary news was presented under the dateline "Stockholm, today, 11.15 a.m." News from Paris in the same message was dated the 26th and news from London was from the 28th of that month and hence several days "outdated."[20]

The telegraph figured prominently in the paper in many ways in these early years. Firstly, every edition mentions whether items had arrived by telegram or referenced telegrams from other papers. Secondly, numerous reports dealt with technological development. Here of course, the establishment of a connection with the Danish coast via submarine cable was considered a big step forward as were subsequent developments on the Danish side.[21] News about the worldwide construction of new telegraph lines was also a recurring type of news item.

Through the many reports of new telegraphs constructed around the world, the paper created an image of a worldwide network spanning the globe. The idea of a world telegraph also already figured in 1855 when early ideas of Taliaferro Preston Shaffner's world telegraph project were announced. The American T. P. Shaffner was a colonel entrepreneur who attempted to create a unified telegraph network around the globe, entering

19 *Öresunds-Posten*, December 11, 1857.

20 *Öresunds-Posten*, December 30, 1857.

21 The cable was placed by the ship Christopher Hage, beginning in November 1854. Heimbürger 54–56. The successful completion was reported in *Öresunds-Posten* on January 2, 1855. The report praised the quality of the submarine cable and emphasised how easy the exchange of messages between Stockholm and Copenhagen had been, considering the large distance between the two cities.

into negotiations with the governments of countries which were potential hosts for the lines. Although the project of a North Atlantic Telegraph line received a lot of attention, it never came to fruition.[22] However, the temporary success of a transatlantic cable in 1858 following the route of the Atlantic Plateau in mid-Ocean showed that submarine cables were technologically realistic.[23] As the plans were revealed in 1855, *Öresunds-Posten* presented all the details of the "world telegraph," this "giant enterprise," in a longer article filled with the names of all the distant cities, countries, rivers, straits and mountains the cable would cross on its way around the globe. Taking a route between Europe and North America via the Shetland Islands, Greenland and Iceland, such a project was also important to Swedes. The article even concluded that every step towards the perfection of the telegraph system, every addition to the system of "world-communications" was "to our advantage." The editor waxed lyrical about the idea and expected it to bring advantages beyond imagination:

> This enormous project will, once completed, have consequences that the human imagination cannot even begin to grasp. It gives us the possibility of communicating daily with every civilised nation on earth, and also with many less civilised peoples.[24]

The article envisioned speedy communication around the globe. And although Shaffner's original plan for a transatlantic cable along the northern hemisphere never became a reality, history was to prove him right in the following decades, when submarine cables and land wires eventually came to connect the major cities across the earth.[25]

Upon the first successful, albeit short-lived completion of a cable across the Atlantic in 1858, *Öresunds-Posten* published a longer article translated from *The Times*, celebrating the "chain of electric communications." The time had now come, the article claimed, when "distance as a reason for uncertainty" would be excluded from the calculations of merchants and statesmen. The author called the telegraph "an instrument of practical omnipresence" and the article promised that in the close future, it would be possible

22 A historical overview arguing for the "feasibility and practicability" of such a route was presented in *Great North Atlantic telegraph route.* London: William Cloves and Sons, 1866, 2.

23 Gordon, *A thread across the ocean.*

24 *Öresunds-Posten*, March 19, 1855.

25 The first permanently functioning transatlantic cable was opened in 1866, in 1880 nine cables crossed the Atlantic but not until 1902 (Pacific Telegraph) did the telegraph lines reach all the way around the globe. Gordon, 210–212.

to every morning publish reports from all around the world about what had happened the day before.[26]

Once the telegraph had become an option for communication in the region, speed became so important, that a local correspondent writing from the countryside had to excuse the fact that due to poor communications, his article had not arrived "with the speed of the telegraph."[27] The birth of a Russian prince in St. Petersburg in April 1857 was called "remarkable," not because of the consequences of the event itself, but because of the speed with which the news was delivered to Paris from Petersburg. The event caused "the fastest telegraphic dispatch, which has ever been exchanged between Petersburg and Paris," and the prince was only 3 hours old when officials in Paris learned about his birth, explained the paper.[28]

An Editor Obsessed with Speed

In the mid-nineteenth century, the editor of a newspaper had a tremendous influence over its development. A driven entrepreneur and writer named Oscar Patric Sturzenbecker started *Öresunds-Posten* in 1847. Although he sold the paper in 1856–1857, he continued to contribute articles and for an extended period, served as the paper's correspondent in Paris. An avid traveller with many ties to Denmark, Sturzenbecker took a keen interest in technological developments in communications. Sometimes he expressed frustration when things took longer than expected. In an article reflecting on the historical development of communications, he used the example of a "spiritual letter" by Madame Marquise de Sevigné, written in 1672. In it, she had complained that a journey from Marseille to Paris took one month, even using the fastest mode of transportation available. Sturzenbecker doubted that in those times a person would have been deemed sane had he prophesied that a day would come when using three days to travel between London and the Mediterranean would be considered, as he wrote, "a scandalously uneconomical use of valuable time."[29] Noting that Mme. De Sevigné wrote

26 *Öresunds-Posten*, August 20, 1858.

27 *Öresunds-Posten,* April 30, 1855.

28 *Öresunds-Posten*, May 22, 1857.

29 Undated article in capsule 10, "Anteckningar och utkast" O. P. Sturzen-Beckers arkiv, Byrån för enskilda arkiv, Riksarkivet, [National archives of Sweden].

some 200 years before, he pointed out that the real revolution had come only in the past few years:

> We tend to forget, that the ability to shrink distances – what am I saying? To *nullify* distances – as a matter of fact is completely new, so new that to almost all of us, it is a contemporary occurrence.[30]

His general interest in communication also surfaced in a series of articles concerning the historical development of steamship traffic in Sweden before 1840.[31] Impatience was his trademark and difficulties in communication clearly caused him frustration. In an expressive letter from 1847, written to a friend in Copenhagen as he was about to start his paper, he expressed his concerns:

> I have today not yet received anything at all from Copenhagen. --- Have you gotten the Prussian constitution which was expected? --- The lousy 20 kilometres which separate Copenhagen from Skåne seem like 200. We have here reports from China, but *none*, not a single one from Copenhagen. This is unbearable, and I am telling you frankly that in the end I can't stand it.[32]

Despite such general enthusiasm for new technology by its editor, in *Öresunds–posten* it was also noted around this time that speed sometimes could come before substance. Occasional articles lambasted the pseudo-importance of all news arriving by the telegraph, even if the subject was only Napoleon's umbrella.[33] Despite this, the key way of reporting about the electric telegraph, focused on the accelerations in communication. One essay retold an anecdote about a large purchase of bonds in this new communication era, which showed how the tables could turn in a heartbeat. If the investor slept a little too long, when he woke up, the investment could be worthless because of news that had arrived by the telegraph. On the basis of such observations, the piece even presented "a rule to follow before all others: Never sleep through a telegraphic dispatch!" The moral of the story was that

30 Undated article in capsule 10, "Anteckningar och utkast," O. P. Sturzen-Beckers arkiv, Byrån för enskilda arkiv, Riksarkivet, [National archives of Sweden].

31 "Some notes on the steam-ship traffic in Sweden before 1840." Undated article in O. P. Sturzen-Beckers arkiv: 10, Anteckningar och utkast, Riksarkivet. Box E 5687, O. P. Sturzen-Beckers arkiv.

32 Letter to Carl Plough, 1847, E 5675, O. P. Sturzen-Beckers arkiv. Also reprinted in Ragnar Sturzen-Becker, *Oskar Patrick Sturzen-Becker (Orvar Odd): Hans liv och gärningar framställda på grundval av bl. a. dagboksanteckningar och korrespondens till och från hans samtida, del 1, 1811–1857.* (Stockholm: Norstedt, 1911), 117.

33 *Öresunds-Posten,* August 17, 1857.

the speed of the telegraph could change conditions very quickly, for one was immediately affected by events occurring far away.[34]

For a newspaper such as *Öresunds-Posten*, the proximity to Denmark made quick journeys to continental Europe easy. Although placed in the periphery from a European point of view, from a Swedish perspective it was at the centre of events. The paper made frequent use of the new communications technology and its columns proclaimed the idea that the telegraph was an invention with great effects on communication across large distances. When links through telegraph cables and wires were established to the city of Helsingborg, it strengthened its ties to the continent. Not only did the use of a correspondent in Copenhagen speed things up, but foreign news from other sources now arrived more quickly as well.

Lines Reaching Piteå and Norrbottens-Posten in 1857

When the steamship *Lennart Torstensson* reached Stockholm harbour on December 8, 1851, the journey from Lübeck had lasted 5 days, including several stops on the way. As the passengers left the ship, eager hands also unloaded the French, German and English papers it carried as cargo. Soon these reached the newspaper offices of the Swedish capital, where editors copied the most important news items to be printed in their next edition. The Stockholm papers, in turn, transported by regular mail, eventually reached the small northern town Piteå, where finally on December 20, the readers of the provincial paper *Norrbottens-Posten* read about events in the French National assembly on December 2, almost three weeks earlier.

Up in the cold North of Sweden, winter was a most trying time for those eager to know what was going on in the world. The steamship traffic no longer ran as far north as in the summer, due to ice in the harbours. The northbound postal service faced regular delays, and once it arrived, the papers it brought were much older than local readers had hoped for. During summer, the area already lagged a number of days behind Stockholm, which in turn learned about events on the continent with a substantial delay. In winter, things got even worse. Far away from the continent and with scarce communications even with the capital Stockholm, in the 1850s this region

34 *Öresunds-Posten*, January 8, 1855.

eagerly awaited the arrival of the electric telegraph. This invention would bring Piteå in closer contact with Stockholm, Europe and the world. Or so it was hoped.

The existing conditions of transportations of post and newspapers were anything but propitious and weather was a popular subject of discussion. The worst time was spring, when sleds could no longer be used as transportation and steamships had not yet begun their traffic. A letter in *Norrbottens-Posten* summarised the situation in the spring of 1851 as follows: "South of the town the roads are now at their worst. The mail arrives two days later than it should."[35]

Besides the weather, the general unreliability of postal services in these northern areas was a matter of great concern for people living there. The editor of *Norrbottens-Posten*, Axel Lundström, complained about the "irregularity with which the so-called exchange papers from some cities arrive." He had been told that the problem lay with the post office of Umeå, where the papers frequently stayed an extra day for no apparent reason. He conceded that "no definitive proof" was at hand, but recapitulated his suspicions. He had back-traced the route of news remediation, and concluded that news items from the papers of Sundsvall and Hudikswall reached Piteå faster by being quoted in the Stockholm newspaper *Aftonbladet*, which arrived north before the quoted papers themselves arrived.[36]

As a telling example, he mentioned the paper of Härnösand, a smaller town down the coast, which some years ago had arrived most regularly. Published on Thursdays, it used to arrive with the Tuesday mail the following week, but now seldom arrived until the Friday mail, and often not until the following day. His comments underscored that southbound communication was unreliable and that physical distances were prohibitive to interaction across space.

As the telegraph lines successively reached farther north along the coast in 1856–1857, *Norrbottens-Posten*, which by then was a weekly newspaper with about 500 copies in circulation, reported on the progress. On September 13, it was announced that the station was operational. On September 26, it published a copy of the obligatory telegram to the king, in which the inhabitants expressed their gratitude for the "incredible advantages offered by the electric telegrap." On October 3, the letter of gratitude from the province

35 *Norrbottens-Posten*, April 26, 1851.

36 *Norrbottens-Posten*, September 20, 1851

governor was quoted, in which he called the telegraph "the most beautiful invention of our times" and on October 10, 1857, the station was finally declared open to the public.[37] Reports about the opening of a telegraph station and its festivities stretched over almost a month.

When the telegraph office opened in Piteå, there was a shortage of personnel. Other stations were asked if anyone could travel there and initiate the operation.[38] As it happened, one of the interested individuals, named Ringius, came from Helsingborg, the town where *Öresunds-Posten* was published. So in the diary of the Helsingborg Telegraph office, an order was kept that Ringius should travel to Stockholm and there await further orders.[39] His tracks lead to Piteå, and in the second note ever in the "Journal" of the station in Piteå, it was noted that assistant Ringius now had arrived.[40] He was to stay on at the Piteå station for a number of years, and in time became a respected and important figure in the local community.[41]

How did the opening of the telegraph affect the reporting of long distance news in the paper of Piteå? Its initial use was rather uneven, but some examples show that the paper did indeed come to make use of the novelty. The message that the mayor of the town had died in Stockholm was received via telegraph, although the news was already some days old when the paper was published.[42]

Despite the existence of this and other examples, it soon became clear that *Norrbottens-Posten* adopted quite a different strategy from that of its southern counterpart. Following the previously mentioned practice in the press at this time, before the actual lines reached Piteå, the paper published "telegrams", albeit telegrams quoted from other papers.[43] An example of the

---

37 *Norrbottens-Posten*, September 13, 1857; September 26, 1857; October 3, 1857; October 10, 1857.

38 Incoming telegram, July 28, 1857, in Televerket, Helsingborgs centralstation, Orderjournaler 1854–1857, Ankomne Tjenstedepescher 1854–1857, E1:1, Landsarkivet i Lund [County archive of Lund].

39 Incoming telegrams. September 3, September 5, 1857, in Televerket, Helsingborgs centralstation, Orderjournaler 1854–1857, Ankomne Tjenstedepescher 1854–1857, E1:1, Landsarkivet i Lund [County archive of Lund]..

40 Note on September 27, 1857, Televerket, VI distriktet, Piteå CSO, Passage-Journaler, 1857–1906, D1:1. Härnösands landsarkiv [County archive of Härnösand.].

41 Stig Jacobsson, *När telegrafen kom till Piteå.*

42 *Norrbottens-Posten*, November 21, 1857. The death of mayor Baudins in Stockholm on the 17 is announced. The news was received by telegram the 17th.

43 From the occasional telegrams in the first years one stands out. It was published on June 19 and was, dated March 16th 10 o'clock recounted news about the birth of a prince. *Norrbottens-Posten*, June 19, 1858.

continued indirect use of telegrams was the news from Stockholm, published on March 27 1858, dated Stockholm March 17, which referred to telegrams dated in Paris the 13th.[44] This habit continued long after the telegraph office had opened. Apparently, the relative advantage in time did not merit the expense associated with hiring a telegraph correspondent in Stockholm. In its yearly presentations of the editorial profile, telegrams were mentioned only once when the paper stated that telegrams from Stockholm would be published "when the course of event calls for it."[45]

The frequency of publishing also played a role. Whereas *Öresunds-Posten* started as a weekly paper, it soon expanded into several editions per week. *Norrbottens-Posten*, on the other hand, continued to be published only once a week. Still, with the aid of the telegraph, the news just preceding the publishing day could be completely up to date. A corresponding paper in another smaller city in the north, also coming out weekly, actually spent considerable sums on telegrams. It was also from this paper – *Sundsvalls-Posten* – that *Norrbottens-Posten* often quoted its telegrams. So the frequency of publication alone is no explanation for the lack of will to begin publishing regular telegraph news.[46]

Although the telegraph opened a new channel of communications, it could not change climate conditions. Winter was still a difficult period. As a letter from a reader in *Norrbottens-Posten* described the situation on November 15, 1862, not even the telegraph could remedy the feeling of being cut off from the world:

> Again we are isolated and practically bound to stay at home, where we may only sit and wait for ice and snow to entirely wrap land and sea in its chains. All communication with the south, which during the summer months is so lively and easy, unfortunately during winter is much more difficult and slow. A few times a week we are of course offered land mails from the capital which are 8–9 days old, and he who wishes to travel does indeed not look forward to the pleasure of bumping along the road after slow wagon-horses; but, except for the telegraph, this is at present all that we are offered.[47]

44 *Norrbottens-Posten*, March 27, 1858.

45 *Norrbottens-Posten*, December 12, 1863.

46 Harvard, Modernitetens depescher? Telegrafen och den norrländska pressens tidshorisonter 1850–1870, *Presshistorisk årsbok*, 2007.

47 *Norrbottens-Posten*, November 15, 1862.

On some occasions, however, the telegraph did provide a previously unknown opportunity to take part in European affairs. A political question of great European interest in 1863 was the fate of Poland, where during the January uprising, insurgents protested the governance of the Russian Empire[48]. Thanks to the electric lines running south, the Piteå community was able to stay updated on events with reasonable speed and accuracy. Although they were close to the Russian border, their news came via Stockholm. In May, people were still looking forward to the time when the steamships would begin their summer traffic and the literature published during the winter months would finally reach those sorry souls who so long had been feeding on "old news."[49] Until that happened, the telegraph provided relief and a temporary means to break the isolation:

> To also mention something about 'the Polish business,' about 'war or peace,' which also among us is spoken about with compassion. Thanks to the telegraph, the only means, which at this time connects us at some kind of speed to the rest of the world, we twice a week receive direct news from the capital and from the scene of the war. On those occasions the matters are discussed and politicised, and the warmest sympathies for Poland then are brought to light. Yes! Also our neighbours 'on the other side' [the Russians] follow the shifting fortunes of war with agitated attention.[50]

In this case, the telegraph indeed integrated the local public temporally with the changes on the European political scene. The citizens were engaged and formed opinions on current matters, and they were able to stay updated on the latest developments despite the harsh winter.

Norrbottens-Posten published a report from the nearby town Hudiksvall in 1862 pointing out that the telegraph had helped to integrate the world. The article admired the incredible speed of a telegram sent to Hudiksvall from a town just outside Constantinople. Just as in Helsingborg, the idea of the telegraph as a harbinger of immediate communication was strong:

> As an example of the swift expedition of a telegram, we call attention to a telegram sent on the 22 this month at 10 hours 10 minutes from Pera, a suburb of Constantinople, which arrived here on the same day at 2 and 4 minutes, thus in less than 4 hours.[51]

48 Davis, Europe, A History, p. 828.
49 *Norrbottens-Posten*, May 23, 1863.
50 *Norrbottens-Posten*, May 23, 1863.
51 *Norrbottens-Posten*, February 8, 1862.

It may be tempting to reconstruct the actual route this telegram may have taken and to use the notice as a measure of the speed development relative the European network. At this point, however, it is more interesting to note the similarity between Helsingborg and Piteå in their fascination with the new technology. The news was about the speed of transmission. Just as in the previously recounted news report about a prince, the substance of the message was less important than its newness.

In some indirect ways, we may also find signs that the telegraph entered into larger debates on modernity. In 1862, *Norrbottens-Posten* published a petition defending the church from accusations that it was attracting fewer and fewer people. The reason for the small number of visitors to the church, the petition said, was the cold. At minus 30 or 40 degrees Celsius, few people in any town would visit a church built of stone. This was only natural, since it was not "as warm in winter as e.g. a Telegraph office." Here, the telegraph office was portrayed as a warm and welcoming place and it was hinted that people would rather frequent this place than go to church.[52] It is worth noting that the personnel working at the station probably wouldn't have agreed, since they wrote to the National Telegraph Board complaining that the office in winter at times was as cold as minus 3 degrees Celsius, despite three steady fires per day. Although decreasing religious piety was not stated as the reason, as it was presented as a warm locality where the community could gather, the telegraph office did draw people from the church.[53]

We may assume that newspapers' use of the telegraph in some ways reflected the general intensity of trade and business in the respective cities. According to public statistics, traffic from the Piteå station was modest, indicating a weak local need for costly but time-sensitive information. A table published in the paper summarised the total number of telegrams sent from the local station the first years. The figures did not reveal how many of those telegrams that were sent by private individuals or by businesses or if the paper itself had made any use at all of the telegraph. As the table shows, the number of telegrams sent from the telegraph office in Piteå did not exceed 220 per month, or about seven per calendar day on average:

52 *Norrbottens-Posten*, June 7, 1862.

53 Utgående Skrivelser 1865, nr. 25, August 31, 1865 Televerket, VI distriktet, Piteå CSO, Utgående skrivelser 1861–1885, B1:1, Concept-Bok 1861–1866. Härnösands landsarkiv [County archive of Härnösand].

Table 1. Number of telegrams sent monthly from Piteå. Local telegram statistics reprinted in Norrbottens-Posten, November 3, 1860.

Month	Year		
	1858	1859	1860
Jan	92	74	37
Feb	60	95	31
Mar	60	130	73
Apr	91	103	82
May	139	130	106
Jun	186	179	143
Jul	188	147	157
Aug	219	213	180
Sep	172	173	172
Oct	123	133	
Nov	44	50	
Dec	68	26	

The traffic to and from foreign telegraph stations was also miniscule, as revealed by the statistics for the first quarter of 1862. During this period, a total of 176 telegrams had arrived, of which 19 were foreign, and in the same period, 222 telegrams were sent, of which 28 went to foreign recipients.[54]

This volume can be compared to that in Helsingborg, and the difference further emphasizes the varying local realities to which the newspapers were connected. The difference between the offices in Piteå and Helsingborg was clear. From the start, the Helsingborg office had been extremely busy, and soon the number of operator tables had to be increased.[55] *Öresunds-Posten* even reported that the personnel were strained due to the large number of

54 *Norrbottens-Posten*, April 26, 1862.

55 Instruction to the Helsingborg Station to immediately acquire a third telegraph table, August 3 1855. Televerket, Helsingborgs centralstation, Orderjournaler 1854–1857, Ankomne Tjenstedepescher 1854–1857, E1:1, Landsarkivet i Lund [County archive of Lund].

telegrams, and that people had become sick from the intensity of the work.[56] The quantitative difference in the volume of traffic was further accentuated by the role of Helsingborg as a gateway for the foreign traffic of both Sweden and Norway. In 1861, the number of telegrams passing through the station amounted to almost 50,000.[57]

Although the colleagues up north had lesser volume to deal with, it was still not easy to run a telegraph station in a deserted area in the mid-nineteenth century. The personnel faced a veritable catalogue of problems. Their success in solving those problems to a large extent determined the possibility of the local users, including the newspaper, to make use of the telegraph. If the wire was broken, it could carry no messages.

Vandals, Thunderstorms and Conflicts

One major problem facing the telegraph lines in the north was vandalism. A small note in *Norrbottens-Posten* on December 6, 1862 reveals this problem. It mentions briefly that a "telegraph watcher" named J Nilson had been awarded the considerable sum of 25 riksdaler, for his vigilance and aid in finding the person responsible for damaging the line to justice.[58]

The problem of sabotage was not unique to this area, and parliament already passed a national law in 1854 to make it possible to fine or imprison those sabotaging wires.[59] But this legal protection was less effective in the north, where the distances were much greater, and the length of the lines made it impossible in practice to protect every inch from vandalism. For a long time there was also only a single wire, which made the service quite vulnerable. If the wire was cut, the service went down. In the south, the main routes were soon expanded into double or triple lines, and where stations were situated near each other, a telegram could easily be rerouted.

56 *Öresunds-Posten,* Dec 2, 1857.

57 Heimbürger, 57.

58 *Norrbottens-Posten,* December 6, 1862. The original reads: "Kongl. Telegrafstyrelsen har den 20 sistl. Nowember såsom uppmuntran för wisad påpasslighet, samt för den brottsliges befordran till laga answar för åwerkan på telegraflinjen tilldelat Telegrafwaktaren J Nilsson i Pitsund en belöning af 25 Rdr Rmt." Compare to examples in Bektas, The Sultan's Messenger.

59 Heimbürger, 68–69.

A variety of actors noted the vulnerability of electric telegraph networks built on land. The French government, for example, remained sceptical of the tiny wires for a long time and instead preferred the older but safer fort-like system provided by the buildings harbouring the optical telegraph.[60] In the American case too, familiar anecdotes exist about how vulnerable the lines were as financial speculators cut the wires to gain an advantage in price information.[61]

But surely the situation in America was not on the mind of telegraph inspector Ringius as he, on July 4, 1859, ventured to mend yet another broken cable. As he described the event in his office diary:

> The telegraph wire south of the city over the river broken due to the fact that the farmers C.O. Wallin and Adolf Lundmark from Kopparnäs ran against it with a raft which had a mast, so that the wire was torn down, and the poles were wrecked. After some difficulty I succeeded in finding out the culprits who tried to hide their deed. The break was mended by Selander.[62]

From the notes in this diary, it is clear that this was not a singular event. On numerous occasions, telegraph personnel had to travel to fix broken cables. Many times, vandalism was the cause, while other times it was falling trees. Apart from regular maintenance of lines which were not actually broken, eleven total breaks were reported between 1858 and 1860.[63] It could take several days before the lines were fixed.[64]

The technical problems with the northern telegraph lines were also noted in the Swedish parliament, where the costly repairs became an argument against further extending the telegraph network in the northern region. When new lines to the north in 1859–1860 were under discussion in the

60 Field, French Optical Telegraphy, 1793–1855.

61 Blondheim, *News over the wires*, 73.

62 Note on July 4, 1859, *Televerket, VI distriktet, Piteå CSO, Passage-Journaler, 1857–1906, D1:1*, Härnösands landsarkiv [County Archive of Härnösand.]

63 In 1858 January 11, June 11 and November 24, in 1859 April 6, July 4, November 3, November 12, and in 1860, breaks were reported on April 7, May 25, December 19 and December 28.

64 Televerket, VI distriktet, Piteå CSO, Passage-Journaler, 1857–1906, D1:1. Härnösands landsarkiv. [County archive of Härnösand.]. It should be noted however, that as late as 1867, the inspector claimed that the station *never* had been inoperable for a period longer than three days. "Utgående skrivelser" 1869, no. 200. Televerket, VI distriktet, Piteå CSO, Utgående skrivelser 1861–1885, B1:1, Concept-Bok 1861–1866. Härnösands landsarkiv [County archive of Härnösand.]

Peasant Estate, Nils Svensson, a representative from Kristianstad complained about the costs:

> When you see how frequently employees at the stations undertake journeys to check if the wires are in order etc., and also know that they travel with two horses and a high daily allowance for expenses, you get a picture of the steep costs created by this institution.[65]

Apart from damage to the wires, signs of another problem appeared on January 15, 1861, when it was reported that someone throwing stones had damaged the insulators.[66] The insulators were a vulnerable point since they were made of porcelain. It became popular across the nation to practice throwing skills by aiming at them. To curb this damage, the National Telegraph Board in 1866 issued a warning to be posted publicly, which strongly discouraged such activities. Other measures included attempts to protect the porcelain "hats" with iron covers or painting them black to make them less visible.[67] As a matter of fact, the person breaking insulators on January 15 later faced prosecution for this act.[68]

A problem with prosecuting the vandals was that the persons contracted as lookouts tended to get into conflict with the rest of the local community. On October 30, the inspector noted that the telegraph watcher Nilsson mentioned above did a good job keeping an eye on his part of the line. After many unsuccessful attempts, through the aid of this telegraph watcher, finally a person had been brought to justice for constant "shooting with marbles" or throwing of stones at the isolators. However, due to the conflict created by this capture between the telegraph watcher and his community, the inspector wrote to the National Telegraph Board, asking them to issue a reward for catching vandals. This would both provide compensation for the social problems telegraph watchers might face and send a deterring signal to

65 Minutes of the Swedish Parliament: Bondeståndets protokoll, 1859–1860, vol. 8, 301.

66 Note for January 15, 1861. Televerket, VI distriktet, Piteå CSO, Passage-Journaler, 1857–1906, D1:1. Härnösands landsarkiv. [County archive of Härnösand.].

67 Heimbürger, *Svenska telegrafverket*, 79, 170, 179. The warning was published in *Norrbottens-Posten* on February 16, 1867, and also stated that those turning in culprits would be rewarded with up to 50 riksdaler. A similar warning was published in 1860. *Norrbottens-Posten*. April 21, 1860. The phenomenon also ocured in Germany. See Geistbeck, Der Weltverkehr, p. 502.

68 Report for the 1st quarter of 1861, Televerket, VI distriktet, Piteå CSO, Utgående skrivelser 1861–1885, B1:1, Concept-Bok 1861–1866. Härnösands landsarkiv [County archive of Härnösand.]

the locals that tampering with the telegraph was serious business.[69] During the following years, more trials took place and the inspector seems to have actively pursued those attacking his lines.[70]

In addition to vandalism, other problems mounted. One was thunderstorms and atmospheric electricity, which could not only create uncontrollable currents and render the apparatus useless, but also hurt the operators. A number of incidents related to thunder and lightning were reported. For example, in 1857, *Norrbottens-Posten* reported that thunder had "done damage to the telegraph" and that near Stockholm 30 poles had been somewhat damaged and that the wire had fallen down.[71] Also in the southern part of Sweden, the problem of thunderstorms was noted. A message received at the telegraph station of Helsingborg in September 1857 mentioned a break on the Stockholm line, "probably due to thunder."[72] While resources for repair were more abundant in the more densely populated south, in the north these events added to the already difficult situation. In February 1862, *Norrbottens-Posten* published reports about thunder and "unusually beautiful beams of light" coming from the instruments at the telegraph station and that the officials of the station had gotten "rather capital electric shocks" as they touched the instruments.[73] However, it seems that bad weather was a national plague causing problems with the electric telegraph. Ordinary cases of bad weather also frequently damaged the lines. In 1858, Helsingborg complained about a "hurricane-like" storm, which had damaged the Danish cable.[74] On January 16 the same year, *Norrbottens-Posten* reported on a storm bringing down poles.[75] Still another cause for worry was faulty material. As

69 Utgående skrivelser 1862, nr 58, 30 October 1862, Televerket, VI distriktet, Piteå CSO, Utgående skrivelser 1861–1885, B1:1, Concept-Bok 1861–1866. Härnösands landsarkiv [County archive of Härnösand.]

70 The report for the 4th quarter of 1863 mentions 2 more ongoing trials. Utgående skrivelser 1864, nr 4. See also Utgående skrivelser 1864 no. 8, Televerket, VI distriktet, Piteå CSO, Utgående skrivelser 1861–1885, B1:1, Concept-Bok 1861–1866. Härnösands landsarkiv [County archive of Härnösand.]

71 *Norrbottens-Posten*, September 26, 1857.

72 Message received 8 September 1857. Televerket, Helsingborgs centralstation, Orderjournaler 1854–1857, Ankomne Tjenstedepescher 1854–1857, E1:1, Landsarkivet i Lund [County archive of Lund.]

73 *Norrbottens-Posten*, February 1, 1862.

74 Message sent to the Danish telegraph station in Helsingör October 2, 1858. Televerket, Helsingborgs centralstation, Konceptböcker 1858–1859, B1:1, Landsarkivet i Lund [County archive of Lund.]

75 *Norrbottens-Posten*, January 16, 1858.

no chemical protection was applied, the original poles began to rot after approximately five years in the soil. During 1863, numerous expeditions to repair falling poles were required.[76]

In addition to a general lack of resources and faced with the general weather problems of the telegraph, which were compounded by the colder climate of the northern part of the country, the local telegraph operators also had to deal with continuous acts of vandalism, which undermined their social role in the community they were set to serve.

Institutional Tensions and Structural Hierarchy

Cooperation within the telegraph institution itself proved a somewhat unexpected source of problems, such as insubordinate or incompetent personnel and lack of cooperation between nearby stations, which were all part of the Swedish national system. The construction of the telegraph system as a network required each node to transmit signals further along. This could be done by repeating the message or by making the signal automatically bypass a station. In some locations, conflicts between stations meant that the superior station would simply assign all telegrams from the disfavoured station a lower priority and transmit them only after all other correspondence was taken care of. If the superior station was a busy one, this caused considerable delays for the telegrams of the subordinated station.

In the case of Piteå, a severe problem seems to have arisen with the station in Haparanda, a bit further north on the line and at the time Sweden's northernmost station. The conflict seems to have begun with a lack of willingness to cooperate. The problems worsened when an employee at the Piteå station insulted his northern counterpart over the telegraph. As the situation was described by inspector Ringius in reports to the head office in Stockholm, one wonders how any telegrams managed to be sent at all. One night an employee woke him up, abruptly declaring, "they are completely out of their minds." He was referring to the personnel in Haparanda, who, according to him, refused to answer or cooperate in repeating telegrams. As Ringius described the story, it soon became clear that this employee had been

76 Notes for 1863. Televerket, VI distriktet, Piteå CSO, Passage-Journaler, 1857–1906, D1:1. Härnösands landsarkiv [County archive of Härnösand.]

involved in conflicts before, such as insulting customers or committing acts of insubordination.[77]

However, the relationships between stations were not only determined by interactions between personnel. The hierarchical structure of the telegraph network assigned every station a role in the chain of priorities. Lines far from the core routes of the network frequently had to wait until all the telegrams of the connecting station were sent. When a minor station opened in the northern town of Östersund in 1863, the inspector there soon realised the problems he faced, and he wrote in his diary that messages from his station would become rather old before "coming through", since they constantly were placed in a waiting line after other messages that were deemed more urgent.[78]

It was clear that both technical problems and difficulties in cooperation between different stations influenced the possibility for newspapers in the north to get a reliable feed of telegrams. Transfer times in the north were a recognised problem, and beginning in 1862, measurements were made to determine the average time that passed from the moment a dispatch was sent until it reached its destination. The average time for dispatches within whole of Sweden was just under two hours, but for to or from the northern section the average was between three and four hours, with times up to seven hours not at all unusual.[79] The reasons were both recurrent technical problems and the large number of middle stations on the way north.

For the Helsingborg office, the hierarchy of telegraph stations had the opposite effect. Due to its role as a gateway to foreign countries, its officials had the ability to suppress incoming messages from nearby stations and make them wait. They seem to have used this ability to such an extent that it resulted in a reprimand from the chief inspector in Stockholm. As the chief inspector wrote in his rather sharp message: "The eastern line complains:

77 See for example memoranda nr. 90, November 17, 1866, regarding the poor behaviour by Winqvist. Televerket, VI distriktet, Piteå CSO, Utgående skrivelser 1861–1885, B1:1, Concept-Bok 1861–1866. Härnösands landsarkiv [County archive of Härnösand.]

78 Harvard, Modernitetens depescher. The hierarchy was also made explicit in telegraph regulations. Smaller stations were ordered not to interrupt traffic between superior stations, and were instructed to wait until special times, when their telegrams would be scheduled. Message to all stations from telegraph director Fahnehjelm on August 30, 1855 regarding the traffic between stations. Televerket, Helsingborgs centralstation, Orderjournaler 1854–1857, Ankomne Tjenstedepescher 1854–1857, E1:1, Landsarkivet i Lund [County archive of Lund.]

79 Heimbürger, *Svenska telegrafverket,* 155.

that the force and perseverance of Helsingborg to push through to Stockholm; that the dispatches thereby become older than necessary; that the station postpones the reception of messages from the southern line in order to use the repeater; and that it sometimes for extended periods lets calls from the southern line go unanswered."[80] The inspector was convinced that there was a natural explanation to these occurrences, and said he was disinclined to believe that the reason was an insufficient sense of duty. He urged the station to do everything possible to avoid similar problems in the future.[81]

When the telegraph convention between Sweden and Russia was signed in 1860, northern cities expressed fears that the increased traffic east would further burden the already poorly functioning northern line. *Norrbottens-Posten* reacted to this and published an article from a correspondent in Stockholm, which discussed the general problems with telegraphy up north. Already without this extra strain, the article said, "telegraphy has always been problematic on the northern line." Both the irritating dependence on a single line and the problems with the "hitherto unexplained forces of nature" were preventing telegrams from being dispatched as fast as they rightly should. Apart from the atmospheric problems, there seemed to be "something rotten" in the Swedish telegraph institution, which could not be explained by the weather. The authorities, hinted the article, simply did not bother with the northern line.[82]

To make its point, the article discussed the situation for a gentleman who had business interests which made him rely on the telegraph. Far too often the telegrams were not sent or were returned due to problems, in which case he still had to pay the "initiating fee." The two alternatives for the sender were either to accept alternative modes of transportation or to request that telegram be returned in case it could not be delivered before a certain time. Again and again the hopes of this "poor gentleman" that the telegrams "would fly like lightning" were put to mockery. In the meantime, the telegram had lost all news value and become "like the Stockholm cookies," which were tasty when fresh, but totally without allure when old. The

80 Message to Director Lagerström, Helsingborg, 19 October 1855. Televerket, Helsingborgs centralstation, Orderjournaler 1854–1857, Ankomne Tjenstedepescher 1854–1857, E1:1, Landsarkivet i Lund [County archive of Lund.]

81 Message dated 19 October 1855, from A. L. Fahnehjelm, Stockholm, to Direktör Lagerström, Helsingborg. 7, Televerket, Helsingborgs centralstation, Orderjournaler 1854–1857, Ankomne Tjenstedepescher 1854–1857, E1:1, Landsarkivet i Lund [County archive of Lund.]

82 *Norrbottens-Posten*, November 3, 1860.

two Riksdaler paid for the telegram were then a total waste. Of course this prevented many presumably "historic" pieces of news which were actually without value from being forwarded. But, asked the paper, "is it really the idea of the telegraph that one should pay dearly for one's telegrams, only to see maybe five out of ten to arrive too late?"[83]

The article concluded that not only was a second telegraph line north necessary, but that reforms were also needed within the telegraph system itself. The article expressed its hope that more people would take interest in the issue so that the problems with the telegraph institution in the northern province would eventually be solved.

Conclusion

The purpose of this chapter was to provide a case study of the electric telegraph's impact in northern Europe. The comparison between the uses of the new technology in two Swedish cities shows several things. First, the symbolic power of the electric telegraph was very strong. In both cities, it was perceived as a sign of progress, even of a coming "global age." Second, the telegraph was not seen as bringing the nation closer together, nor neighbours in more intimate contact. Rather, it appeared to provide a link between the local and an imagined global community, bypassing both nation and region. More importantly, the comparison also shows the big differences in actual operation once telegraphs were working in the respective cities. *Öresunds-Posten* made the most out of its strategic position right at the transit point for Swedish telegraphic communication with the continent. The paper also employed its own correspondent, who wired news specifically to the paper. *Norrbottens-Posten* initially shared the great enthusiasm for the new technology, but after a few symbolic telegrams, reverted to its habit of quoting telegrams from other newspapers. Thus, the telegraph made only an indirect impact. The questions of maintenance and sabotage were important in both cities, but the vulnerability of the northern locality, with larger distances, more difficult climate and lack of alternative routes, made constant breaks a clear factor hampering the use of the telegraph. Furthermore, the institutional structure worked to the advantage of Helsingborg, which had

83 *Norrbottens-Posten*, November 3, 1860.

the right to suppress messages from subordinated stations due to its role as transit hub. By contrast, that same structure worked against the station in Piteå. Along the northern line it was often difficult to get through, and sometimes even personal relations between telegraph operators hindered a smooth flow of dispatches.

In relation to the general theories and large scale histories of globalization, this case study from the periphery thus highlights the degree to which specific local conditions determined the impact of new communication technologies such as the electric telegraph. Rather than diminishing the role of geography or location in space, with regard to communication, the electric telegraph increased the differences between the locations able to make good use of the possibilities of the new technology, and those which could not. On a general level, the electric telegraph, at least in this time period, did not even out the playing field, but instead gave privileged geographic locations the opportunity to leverage their relative advantages. Progress and technological advances as visionary concepts did not help particularly when a telegraph inspector had to travel far through the northern woods to repair a sabotaged wire. The simple realities of frail equipment, harsh winters and a sceptic local population show that globalization was a material process, and as such only partially challenged the limitations of physical space in a globalising world.

Bibliography

Archives and Libraries Consulted

County archive of Härnösand

Televerket, VI distriktet, Piteå CSO, Passage-Journaler, 1857–1906, D1:1.

Televerket, VI distriktet, Piteå CSO, Utgående skrivelser 1861–1885, B1:1, Concept-Bok 1861–1866

County Archive of Lund

Televerket, Helsingborgs centralstation, Orderjournaler 1854–1857, Ankomne Tjenstedepescher 1854–1857, E1:1

Televerket, Helsingborgs centralstation, Konceptböcker 1858–1859, B1:1

National Archives of Sweden, Stockholm

Byrån för enskilda arkiv, O. P. Sturzen-Beckers arkiv: 10, Anteckningar och utkast, Riksarkivet. Box E 5687

Byrån för enskilda arkiv, O. P. Sturzen-Beckers arkiv, Letter to Carl Plough, 1847, E 5675
Parliamentary sources

Newspapers

Norrbottens-Posten, 1851, 1857, 1858, 1860, 1862, 1863, 1867
Öresunds-Posten, 1854, 1855, 1857, 1858, 1865

Published Primary and Secondary Sources

Minutes of the Swedish Parliament: Bondeståndets protokoll, 1859–1860.

Barnhurst, Kevin G. and John Nerone, *The form of news: A history,* New York, 2001.

Bayly, Christopher Alan, *The birth of the modern world, 1780–1914: global connections and comparisons,* Oxford 2004.

Bektas, Yakup, "The Sultan's Messenger: Cultural Constructions of Ottoman Telegraphy", 1847–1880, *Technology and Culture* 41.4 (2000).

Blondheim, Menahem, *News over the wires: the telegraph and the flow of public information in America, 1844–1897,* Cambridge, MA 1994.

Bordo, Michael D., Taylor, Alan M. & Williamson, Jeffrey G. (eds.) *Globalization in historical perspective*, Chicago, ILL 2003.

Carey, James, "Technology and ideology: the case of the telegraph", in James Carey, *Communication as culture: essays on media and society,* Boston, MA 1989.

Chanda, Nayan, *Bound Together: How Traders, Preachers, Adventurers, and Warriors Shaped Globalization,* New Haven, CD 2007.

Chandler, Alfred D. *The visible hand: The managerial revolution in American business,* Cambridge, MA 1977.

Edgerton, David, *The shock of the old: Technology and global history since 1900,* London 2006.

—, "From Innovation to Use: Ten (Eclectic) Theses on the History of Technology", *History and Technology* Vol 16 (1999), pp. 1–26.

Field, Alexander J. "French Optical Telegraphy, 1793–1855: Hardware, Software, Administration", *Technology and Culture* 35:2 (1994), pp. 315–347

Geistbeck, Michael. Der Weltverkehr. Telegraphie und Post, Eisenbahnen und Schiffahrt in ihrer Entwicklung dargestellt. Mit 123 Abbildungen und 33 Karten, Freiburg 1895.

Giverholt, Helge, *Nyhetsformidling i Norge: Norsk telegrambyrå 1867–1967,* Oslo 1967.

Gordon, John Steele, *A thread across the ocean: the heroic story of the transatlantic cable*, London 2002.

Great North Atlantic telegraph route, London 1866.

Harvard, Jonas, „Modernitetens depescher? Telegrafen och den norrländska pressens tidshorisonter 1850–1870", *Presshistorisk årsbok,* 2007.

Headrick, Daniel, *The tentacles of progress: Technology transfer in the age of imperialism, 1850–1940*, New York 1988.

Heimbürger, Hans, *Svenska telegrafverket: Historisk framställning Bd 2, Det elektriska telegrafväsendet 1853–1902,* Stockholm 1938.

Jacobsson, Stig, *När telegrafen kom till Piteå,* Piteå 1989.

Kaijser, Arne, *I fädrens spår: Den svenska infrastrukturens historiska utveckling och framtida utmaningar*, Stockholm 1994.

Marvin, Carolyn. *When old technologies were new: Thinking about electric communication in the late nineteenth century*, New York 1988.

Mattelart, Armand, *Networking the world, 1794–2000*, Minneapolis, MN 2000.

Mitchell, Brian, R. *International historical statistics: Europe, 1750–1988*, Basingstoke, MC 1992.

Nickles, David P., *Under the wire: How the telegraph changed diplomacy,* Cambridge, MA 2003.

Sturzen-Becker, Ragnar, *Oskar Patrick Sturzen-Becker (Orvar Odd): hans liv och gärningar framställda på grundval av bl. a. dagboksanteckningar och korrespondens till och från hans samtida, del 1, 1811–1857,* Stockholm 1911.

Wenzlhuemer, Roland, "The Dematerialization of Telecommunication: Communication Centres and Peripheries in Europe, 1850–1920", *Journal of Global History*, 2 (2007) 3, pp. 345–372.

Power Lines: Arizona Elites, the Telegraph, and the Construction of a Regional Identity, 1870–1910

Torsten Kathke

As a means of quick and reliable communication, in the late nineteenth century telegraphy became a supplement to, and in some cases, a substitution for, the writing of letters. The difference – not just in speed, but in kind – of communication, greatly affected the formation of various levels of both more stable identities, and contextual identifications in the United States, particularly in the West. After the end of the Civil War, the sparsely settled Southwestern territories experienced coalescing forces binding them ever closer to distant points in the nation, essentially replicating processes that had earlier occurred in the East. The availability of the telegraph, however, sped up as well as complicated the spread of federal authority in these regions. The southern part of Arizona Territory, for example, witnessed the construction of distinct local and regional identities at the same time that the notion of national identity took hold during the region's transition from what Thomas E. Sheridan calls its phase of incorporation (into national and transnational economies) to that of extraction (of resources).[1] The years from 1870 on also saw a shift in Arizona's white, or Anglo, settlement patterns, and a significant increase in their numbers over those of the Spanish-descended population who had dominated before.[2] The arrival of the railroad in the territory during the late 1870s and early 1880s brought these Anglo migrants to Arizona. It portended momentous changes in regional culture as well.

This chapter will show how an established Anglo-Mexican elite, rooted in the region and its Hispanic cultural traditions, slowly gave way to a na-

1 Roughly the 1870s through 1890s. Sheridan calls Arizona at the time an "internal extractive colony for Western and Eastern business interests." Sheridan, Arizona, p. xiv–xv.

2 By convention, in the American Southwest the appellation "Anglo" includes also whites of non-English descent, and in isolated cases, even African Americans. Few studies bother to define their use of the term in detail. For my purposes, the most important distinction is that between Hispanic Americans (including Anglos that intermarried) and white newcomers from the East and West. Recent examples of a differentiated use of racial and cultural terms and their connotations in the American Southwest include Jacoby, Shadows at Dawn and Truett, Fugitive Landscapes.

tional-minded Anglo one, connected through the virtual networks of telegraphy and personal acquaintance, from the late 1870s on. It explores how this Anglo elite used both regional and national political connections to gain influence and how it succeeded because of the increasing prevalence of U.S. national institutions, such as the military and the American legal system. Telegraphy played an important role in connecting this group of people to the nerve centers of the American nation. At the same time, the great expense of telegraphy set apart the elite from the non-elite. It also allowed for the creation of a distinct, decidedly more Anglo identity of Arizona, constructed in opposition to that of its sister territory, New Mexico, which had much larger numbers of originally Mexican inhabitants.

Notions of nationalism became more and more important in the Southwest during the period between the Civil War and World War I. Yet, they were also complicated by increasing identification with the (transnational) region and with territorial and local communities. The American Southwest at the turn of the nineteenth and twentieth centuries thus provides insights into processes of nationalization as well as how they coincided with the appearance of new, salient identities on the regional level. These regions were not limited to the nations that they were nominally part of, but could also span across borders. This finding complicates established orthodoxies of national identity as subsuming lower-level forms of identification.[3]

In the years following the 1853/54 Gadsden Purchase, the remoteness of the purchased region created practical and practicable business and personal alliances between the existing Sonoran Spanish-Mexican elite and the few Anglos who had drifted into the area.[4] Mexican norteños, much like Amer-

3 Anderson, *Imagined Communities*, p. 6–7. Stephen Grosby, tasked with writing a concise introduction to the subject of nationalism, argues that end of the spectrum. Grosby, *Nationalism*, p. 1–6. See also Eastman, *Preaching Spanish Nationalism*, p. 49; Herrmann and Brewer, Identities and Institutions, p. 8. Recent work on supranational identity constructions in the European Union has highlighted more complicated "cross-cutting" identities across regional, local, and national levels. Ibid.; Risse, *A Community of Europeans*, p. 24.

4 As with "Anglo," the terminology to describe the non-Anglos who lived at one time in territory that was Mexican is murky, to say the least. While American whites soon took to calling both those that had settled in the Gadsden Purchase area before it was part of the United States, and those who immigrated into the United States from south of the international border "Mexicans," there were differences in their geographic origin, culture, and self-perception. The established elite regarded themselves as Spanish-Americans, harkening back to a Spanish colonial, decidedly European tradition. Like many Anglos, they overwhelmingly saw the Mexican newcomers – many of them not norteños from Sonora, but Guanajuatenses, Jalisences, Michoacanos, or others from southern Mexican sates – as

ican Westerners, were noted for being flexible and pragmatic, and they took to assimilating the newcomers into their circle, using the new connections in trade to personal and community advantage.[5] Anglos readily found a place in an elite that had more in common with its European-minded compeers in Sonoran towns, like the port city of Guaymas, than with central or southern Mexicans. Sonorense elites stressed that they were different than the rest of their country. They not only furnished their homes with European imports, and integrated recent immigrants from Europe into their relations of kin, custom, and deal-making, but stressed their ethnic whiteness, and thus superiority.[6]

Given the remoteness of Southern Arizona in the 1850s and early 1860s, it is hardly surprising that there was such a parallel society, in which Anglo-Sonorenses constituted the elite while central and southern Mexican immigrants made up the bulk of the working class, along with a few poor whites and Native Americans living in Euro-American settlements. Just getting to Southern Arizona was still an achievement in and of itself. Journalist J. Ross Browne, an oft-cited authority on the matter – likely as he was one of the few who made the trip at the time and apparently disliked his experience enough to write about it – found that Southern Arizona was "practically, more distant from San Francisco and New York than either of these cities is from China to Norway. I made the trip from Germany to Iceland and back much more easily, and with much less expense and loss of time, than from San Francisco to Sonora and back."[7]

Browne's singular geographical logic notwithstanding, it was obviously not a place one toured casually. Parish B. Johnson, who made his way to Tucson from California in 1864, the same year as Browne, was less harsh in his judgment, finding a "good road most of the way," at the end of which there was a government post, and "also, a Spanish town." Still, that good road meant a ten-hour trip for a distance of seventeen miles from Point Mountain

lower-class interlopers, and tended to be affronted at being lumped in with this group by the Anglos. Cf. Tinker Salas, In the Shadow of the Eagles, p. 7.

5 Truett, Fugitive Landscapes, 43.

6 This impetus sometimes bordered on the farcical. As late as 1892, "[w]hen Sonora participated in the Chicago Exposition […] the committee in charge of its exhibits also included several photographic portraits of light-skinned Sonoran women" to be shown next to samples of agriculture and industry. Tinker Salas, In the Shadow of the Eagles, p. 27.

7 Truett, Fugitive Landscapes, 44. Browne's chapter on Tucson, scarcely eight pages of his 300-page book, is filled with sarcasm and hyperbole. J. Ross Browne, Adventures in the Apache Country, p. 131–139.

in the west.[8] Shortly after the arrival of the Southern Pacific Railroad in the region during the late 1870s and early 1880s, the picture had changed dramatically. In 1882, Boston Transcript writer Lloyd Vernon Briggs tenderfootedly and foolhardily explored the Southwest with his sister as a traveling companion in the midst of the Apache Wars. Upon his arrival in Tucson, he still saw in the region's foremost town, "a quaint, queer old place with narrow streets and adobe houses of a single story" that was naturally a far cry from "the more progressive cities of California."[9] It had, however, taken Briggs less than a day to get there from Los Angeles after departing on April 20.

Such radical acceleration in travel times between the U.S. American "centers" of politics and trade and its "margins" such as Arizona had palpable consequences. Close personal and even familial ties had bound a smaller Anglo population to a quasi-Sonoran society and hence Mexican territory in the 1850s and 1860s. These ties started to fade as Anglos redirected themselves towards the American nation with its economic power centers on the East and West Coasts, and its political hub in Washington, D.C. Likewise, travel to and from Mexico had sped up considerably, as that country, along with Canada and the United States, became enmeshed in a growing continental rail network.[10] The railroads enabled large numbers of central Mexican immigrants to enter the United States at the same time that large numbers of Anglos came to the Southwest to work for large farms and especially for the growing mining industry.[11] To Anglo eyes, these Mexicans looked even more "foreign" than northern Mexican Hispanos. The two groups competed against one another for jobs, and since Mexicans were willing to work for lower wages, Anglo laborers developed increasing animosity to them.

The changes in the elites' personal lives were also notable. Jacob Mansfield, a German Jew who had settled in Tucson in 1870 and founded a news depot and bookstore, by 1884 already remembered the decade before the arrival of the railroad just past as the "old days."[12] Technology, connected in the minds of many with American "civilization," certainly contributed

8 Huntington Library, San Marino. Parish B. Johnson Papers, 1864–1886.
9 Arizona Historical Society (AHS), Tucson. Lloyd Vernon Briggs Papers. MS 95, "Reminiscences," typescript, p. 1.
10 On the continental reach of nineteenth-century railroads, see White, Railroaded.
11 Gordon, Great Arizona Orphan Abduction, p. 49–50.
12 Sonnichsen, Tucson, p. 96.

to that perception.[13] By the time of Mansfield's writing, his town had – in addition to the railroad – seen the coming of the telephone in 1881 and of gas lights one year later as well as an abortive attempt at establishing an electrical power system, also in 1882.[14] Tucson had reached a momentary peak in population at 7,007 inhabitants listed in the 1880 census, up from 3,215 ten years earlier. Although its population number would decline with the recession of the 1880s, slumping back down to 5,150 in 1890, the town had made up for this decline by the turn of the century, and numbers would only grow from there.[15]

Telegraphy had become more commonplace during the early 1880s to the point where all kinds of news now came across the wires, not just international and national events of varying importance, such as the eruption of Krakatoa in 1883, the outbreak of riots in Brussels in June 1884, or the outbreak of cholera in Southern Europe in the fall of 1884. The Arizona Weekly Citizen had featured a front page "telegraph" column from early 1880 on, which came to feature mostly world news culled from AP reports by the mid-1880s.[16] Beginning in 1875, telegraphic communications had reached Tucson via the Western Union office located on the first floor of a building on the town's main thoroughfare, Congress Street. The building, in which the telegraph office shared space with the Bank of Tucson and Wells Fargo, belonged to photographer Henry Buehman, and was conveniently located next door to the Citizen's press office. Buehman had established his photography studio on the second floor and also sometimes worked as a photographer for the paper.[17] The news that was printed on the Citizen's page one shows an early fascination with not only what was communicated, but with the fact of where it came from, and that it could be reported on quickly after it had occurred because of the telegraph. By 1886, the column was once

---

13 Michael Adas has convincingly argued that both were intrinsically intertwined, and that technological hubris was a chief component in the American idea of "manifest destiny." Adas, Dominance by Design, p. 129–182.

14 Sonnichsen, Tucson, p. 108.

15 Riley Moffatt, Population History of Western U.S. Cities & Towns, 1850–1990, p. 16.

16 Since the AP charged small papers only what it believed they could afford, this was a relatively low-cost way to fill space in the paper and at the same time connect it to the larger discourses prevalent in American newspapers. Tucson was not part of a major news circuit, which is why the "Telegraph" section of the *Citizen* featured chiefly the headlines from the AP's "pony" report, the 1,500–2,500 page smallest common denominator of news sent over the wires to papers all across the U.S. Blondheim, *News Over the Wires*, p. 171–172.

17 Woosley, Early Tucson, p. 78.

again overwhelmingly concerned with U.S. national news, and with regional news concerning Sonora, indicating both an increasing regionalization, and a growing nationalization in the Southwest.[18]

The Citizen's main rival paper, the Arizona Weekly Star, meanwhile, chose to begin its first issue of the year in 1880 with a news item titled "Fatal Accident to Three Drunken Germans Near Napa, Calif,"[19] thus brilliantly combining the era's penchant for ethnic stereotypes and growing fondness for sensationalist reporting. A very different stereotyping of an ethnic group was afoot at the same time. Historian James E. Officer, looking at the Spanish-American experience in late nineteenth century Arizona, found that Spanish surnames disappeared from newspaper accounts that described the comings and goings of local elites in the 1880s.[20] Also, Anglos no longer settled among the established community, choosing instead to live apart from it. In existing settlements, the newcomers tended not to create new neighborhoods. Rather, they increasingly took over downtowns, pushing "Mexicans," as they now derogatorily termed both Spanish-American U.S. citizens and newly arrived migrants from south of the border, to the sidelines.[21] A nationalizing, imperially minded, white male elite now oriented itself to California and the East. Due to their numbers, and in contrast to earlier Anglo arrivals, adapting to the local lifestyle was not a prerequisite. The technological means and the transport system now allowed Anglos to remain in close contact with the national centers of the U.S. so that a regional integration was simply no longer necessary.

Telegraphy, as well as related modes of communication – telegraph news in the local papers, messages relayed during personal meetings scheduled by telegraph and made possible by rail travel – enabled a small number of these late nineteenth century (mostly Anglo) elites in the territory to, for the first time, define themselves as Arizonans. This was a deliberate self-fashioning in order to appear to Washington decision makers as less "Mexican" than their sister territory, New Mexico.

18 Arizona Weekly Citizen, September 15, 1883; June 21, 1884; October 18, 1884. The paper went through several name changes, alternating between Arizona Citizen, Tucson Citizen, and variations thereof.

19 Arizona Weekly Star, January 8, 1880.

20 Officer qtd. in Sonnichsen, p. 88.

21 Sonnichsen, Tucson, p. 107; Sheridan, Tucsonenses, p. 21–23. Tucson's Main Street, center of the former Spanish settlement, at this time became lined with New England style townhouses.

The members of this elite were local, state, and national political figures, newspaper editors, entrepreneurs, and occasionally, newly arrived farmers. Their frequent use of the telegraph, made possible through either political or military privilege (which often allowed them to send telegrams for free), or wealth, allowed them to also dictate the terms of what being Arizonan meant. Lodges, such as the Elks or Odd Fellows, also played a role. Many were national in scope, with membership which was mostly Anglo, though not exclusively, as individual "old stock" Hispanos remained members even though they were losing power as a group. These fraternal orders and other mutual associations provided Anglo elites with a base of less eminent, but still mostly male and white compatriots rooted in their immediate communities.[22] This second-tier elite placed a high social as well as practical value on being connected to their higher-ranking peers.[23]

These men, members in associations and brothers in secret and not-so-secret fraternal societies, formed the core of what constituted the influence of the national in the hidden-in-plain-sight kind of governance typical to "Gilded Age" America. Termed the "associative state" by Brian Balogh and others, this approach to federal authority relied on an assertive legal system, and the organization of people in associations and clubs advocating various ends – out of their own volition and with tacit political support from Washington – rather than an unwanted visible federal presence.[24] The military was the exception to this rule: it was both wanted and visible. The military not only provided some measure of protection against Apache depredations, but it also needed a lot of supplies that local farms could, and did, supply.

Telegraphy had come to Arizona through the military, with a line built in the early 1870s from the West Coast to Yuma (near the Mexican and Californian borders) and on to Tucson as well as the territorial capital at the time, Prescott. Commercial messages traveled across U.S. Signal Service wires until the late nineteenth century. It was not until 1887 that the first

22 While some lodges in Tucson were relatively inclusive, associations in other parts of Arizona, and especially in small mining communities, were organized strictly along racial lines. Huntington Abbott, Beginnings of Freemasonry in Tucson, p. 109, 113; Gordon, Great Arizona Orphan Abduction, p. 188.

23 Relationships to whom could provide them with jobs and favors – sometimes considerable ones.

24 Balogh, A Government Out of Sight, p. 353–378, 380–381. Balogh takes the term from Ellis Hawley's work on the 1920s. Although he strangely does not say it outright, Balogh most likely refers to Hawley, Herbert Hoover.

commercial telegraph company was even established in the territory.[25] This also allowed the military as an institution to become a prolific user of the technology.

A telegraph ledger for Southern Arizona containing records for all the messages sent, received, and forwarded by one unnamed post in 1892 and 1893, shows a clear pattern.[26] Telegrams charged for, and sent by businessmen or other private users of the network were identified on the top of each page as "commercial business." The number sent daily in January 1892, for example, ranged from two on January 31 to a maximum of 18 in two days (for the weekend of January 3 and 4), averaging about twelve a week.[27] "Free business," i.e. military messages, typically were around 25 per week, or twice the number sent on commercial business.[28] The length of the telegrams sent and received is also telling. Commercial telegrams were overwhelmingly under a dozen words long, while military ones were hardly ever less than that, often coming in at word counts in the twenties, thirties, and forties, with some outliers going up to nearly 200.[29] On average, a military telegram contained more than 28 words, while a commercial or governmental one had less than 16.[30]

25 AHS, Tucson. Telegraph Lines in Arizona, 1871–1879. MS 872.

26 AHS, Tucson. Telegraph Ledger. MS 817. Judging from the sent and received lists, the post in question was likely Fort Carlos, although no mention is made on the ledger.

27 Ibid. The exact numbers per day are hard to pin down, as it appears that the keepers of the ledger habitually lumped together weekends, and often forgot to finish their statistics daily, instead tallying up two, three, and in one case, four full days before restarting their count.

28 This included use by other governmental agencies, who received special rates, but still had to pay to use the line, as is evidenced by "government rate" telegrams, such as those sent by the surveyor general. National Archives Western Division, Perris. Surveyor General of Arizona, Letters Sent, 1879–1895. RG 49.

29 AHS, Tucson. Telegraph Ledger. MS 817.

30 For a sample taken for January 1–15, 1892, the average word counts, rounded to the second decimal, are 25.69 and 13.24 respectively for free and commercial telegrams sent and 18.27 and 31.08 for those received. It is possible that these lengths differed in part because codes were used. Ciphers certainly were popular among telegraph users, and telegraph codes in the late nineteenth century U.S. varied greatly in length and complexity, from those that substituted one letter for another one such as the Larrabee Cipher, or those that replaced one word for another one, such as the Pinkerton's Detective Agency cipher, leaving the total number of words essentially unchanged, to those that replaced either whole sentences with one word, or conversely assigned long words to short numbers, such as the code used by W. G. Press & Co., or the ABC Universal Commercial Telegraphic Code. The few telegrams that survive from the period in Arizona are mostly unencrypted.

Commercial messages, then, were not only just about half as frequent as military ones, but they were also only about half as long. They did, however, tend to travel further. Business messages went within the territory, but also as far as Boston, Midland, Columbus, or New Orleans. For a 15-day sample period in 1892, 36 commercial telegrams were either sent to or received from points outside of Arizona, while at the same time, only twelve military communiqués went outside the territories' borders.[31] Military messages went typically between forts within the territory. The picture becomes even more skewed, when one considers that most of the telegrams that crossed the Arizona line constituted contact with other places – commercial centers – in the Southwest. Albuquerque and Los Angeles were the only destinations that showed up with any regularity. Only two of the cablegrams were sent from Southern Arizona to points outside the territory. The recipient in both cases is given as the chargé d'affaires in Washington, D.C. Both times, Washington chose not to reply, at least not by telegram. Most likely, these were status reports, or standard requests, as the messages repeated monthly with almost clockwork regularity. Three more were sent in February, one on March 4 (or 5),[32] and others on April 12, May 5, June 9 (or 10), July 3 (or 4), and so on.

Communication among regional military posts, in contrast to business correspondence, was constant, loquacious, and free-wheeling, with some commanders apparently quite drawn to near-instant messaging.[33] Such no-holds-barred transmissions allowed those involved in military affairs to stay connected to their colleagues in the territory, however far afield. It allowed also for "secondary" communication, such as when officers corresponded on behalf of the private persons they and the military did business with, be it contractors or those called upon for provisions. It is difficult to draw a hard line between military and personal matters here. Expensive commercial traffic, on the other hand, was limited not just in numbers, but also in scope, making elaborate conversations nearly impossible. Yet, it provided an important supplement to the more prevalent writing of letters. Because of

John, Network Nation, p. 29; Larrabee, Larrabee's Cipher; W.G. Press, Telegraphic Cipher; Clauson-Thue, ABC Universal Commercial Electric Telegraphic Code.

31 The same period as above.

32 See footnote 25.

33 Some, like Johnson in Fort Carlos, seem to have made extensive use of their privilege to send free telegrams. On one day, January 16, he managed to send five telegrams to his counterpart, Bean in Fort Apache. These were 21, 50, 172, 127, and 36 words in length respectively. Bean replied equally verbosely that same day, with five messages at 55, 95, 55, 123, and 165 words in length.

its speed as well as the inherent power differential between those who could avail themselves of the service, and those who could not, telegraphy also was an instrument of great power.

The intimate connection of the telegraph to the military – as one of only three at least somewhat visible institutions of the "General Government" that were both functioning and welcomed in the territory at the end of the nineteenth century – shows the immense advantage that those who were in a position to use it had over those who, by station or tradition, did not.[34]

From Mexican Border Elite to Anglo-Federal Arizona

When Arizona's capital moved away from Tucson to Prescott, more than 200 miles to the northwest, in 1879, and was permanently installed in central Arizonan Phoenix in 1889, this marked a symbolic end to the Anglo-Mexican landholding elite's control over the territory.[35] During much of their struggle for recognition as states, parts of both Arizona and New Mexico remained predominantly Mexican both culturally and in population, especially in "Original Arizona," the swath of land bordering Mexico purchased by the United States in the Gadsden Treaty. This meant that the identity, which the new Anglo arrivals who gradually came to dominate Arizona politics carved out for themselves and their territory, was an uneasy amalgam in constant transition. On the one hand, they could not ignore the centuries of Spanish and Mexican rule that had left indelible cultural marks.[36] Furthermore, the sparsely settled territory's frontier status had allowed some leeway

34 The U.S. Federal Government was commonly termed General Government in the nineteenth century. Balogh, p. 13. The other two such institutions being the ubiquitous postal service and, to a lesser degree, the courts. Ibid., p. 220–226, 314–315.

35 Tucson had unceremoniously dropped out of the running four years earlier. A good summation on the machinations behind the move can be found in: Ehrlich, Arizona's Territorial Capital Moves to Phoenix.

36 I focus here on interactions between Anglos and Mexicans. Anglo interactions with Native Americans, while undoubtedly important, resembled patterns seen elsewhere in the Western United States during "Greater Reconstruction," a useful term coined by Elliott West to differentiate Post Civil War developments of attempting to integrate freed slaves in the East from Indian Wars fought to integrate and "civilize" the remaining tribes in the West. West, The Last Indian War, p. xvi–xix, 59, 151, 292, 318–319.

in the construction of racial identities.[37] On the other hand, modernization and civilization in the conception of the age meant increasingly conforming to Victorian models of behavior and thought as they were observed by self-identifying elites in the East and coastal West, and national standards concerning race and gender roles.[38]

Similar creations and fortifications of identity had occurred elsewhere in the United States during much of the nineteenth century, always dependent on economic and settlement patterns. The development of very distinct yet intertwined regional identities in Arizona and New Mexico during their territorial periods is thus tied closely to both territories' racial composition and dominant economy during the late nineteenth and early twentieth centuries. While New Mexico remained comparatively more agricultural, and its Hispanic population – larger in numbers than Arizona's by a factor of, at the very least, two – could therefore retain cultural as well as political influence, Arizona was dominated by the extractive economy of mining, directed by nationally acting capitalists with the institutional or financial wherewithal to make use of telegraphy. [39]

One powerful instrument in curtailing the importance of the old Spanish elite was the American legal system and its labyrinthine bureaucracy, which was fueled and connected by the postal and telegram systems of the nation. Along with communication and transportation, the replacement of one legal

37 Although thoroughly problematic, the term is still useful. In this context, it represents a distinction that the actors in question themselves made, and which is therefore salient to their self-perception. Most startlingly, the region's "frontier status" in some cases allowed a wide range of racial self-fashionings, even to the point where some blacks could consider themselves as whites. See Gordon, Great Arizona Orphan Abduction, p. 104.

38 Although this appears never to have been complete, and as far as race is concerned, less so than would be expected of a former Confederate territory.

39 Although mining dominated in its southwestern parts, close to Arizona, and in the northeast. Lamar, The Far Southwest, p. 165. The data on race is not easy to parse, as the U.S. Census did not distinguish between Hispanics and Anglos at the time, lumping them all in the same "White" category. The earliest somewhat reliable Census data stems from 1940, when a five percent sample was put into either the Hispanic or White categories. In Arizona, the percentage for that year is 20.4, New Mexico's more than double that at 41.7. Raw population numbers, however, allow some conclusions for earlier years as well. While in 1870, Arizona had a total population of 9,658, New Mexico the same year had 91,874. As the railroad arrived in the region only at the end of that decade, most of the white inhabitants must have been from formerly Spanish families. New Mexico's population roughly tripled to 327,301 in 1910, while Arizona's by that year had increased to twenty times the 1870 number at 204,354. Gibson and Jung, Historical Census Statistics; Lamar, The Far Southwest, p. 171; Truett, Fugitive Landscapes, p. 55–77.

system by another was one of the most important changes in the Southwest after the American takeover. According to land grant historian Malcolm Ebright, it was even the most important reason for the resulting inequities. Since the common law system of the United States did not concern itself with certain aspects of the Hispanic system – such as the customary law that had governed the distribution of parcels of land – the validity of land grants could literally get lost in translation from Spanish-Mexican law to Anglo-American law.[40] Frequent moves and changes of venue and responsibility of courts and government agencies on both sides of the border created additional confusion. Mexican frontier recordkeeping was also no friend of Spanish-American claims in the United States. George A. Hilzinger, a special agent of the Surveyor General for Arizona, reported from Hermosillo, Sonora in a letter dated May 28, 1886 that:

> The government offices have been moved into the new building, which is only partly finished, and not much time has been spent on the arrangement of papers. A great many of the old papers have been culled from the heaps that are piled up loose upon the floor, but an industrious man might work constantly for a year sorting and arranging and still find much to do.[41]

In addition, Washington held immense power in reassigning land titles, because Congress had to confirm land grants. Officials at the local and territorial level in the 1880s and 1890s were still connected with the Spanish-speaking landed elite, and thus almost without fail recommended the confirmation of land titles from the Mexican era. In doing so, they found ways to oppose the purposeful misinterpretation of the Gadsden Purchase contract, which guaranteed that Spanish land titles should unequivocally be confirmed.[42]

At the same time, some transparent attempts by absentee Anglo land companies to claim valuable properties were nipped in the bud. John

40 Ebright, Land Grants, p. 1.

41 National Archives Western Division, Perris. Arizona Court of Private Land Claims, Misc. Files, 1789–1907. RG 49.

42 One opinion by George A. Williams, dated May 8, 1878, riles against an attempt to declare a land title illegal because it cannot be found in the Mexican National Archive in Mexico City, quoting the Treaty of Guadalupe Hidalgo and its provisions that such sales had to be located "in the archives of Mexico." The opinion declares that the National Archives were not the proper place for land titles. The plural "archives" was used because this fact was known and the land title could be found in the proper government records center in Sonora. National Archives Western Division, Perris. Arizona Court of Private Land Claims, Misc. Files, 1789–1907. RG 49.

Wasson, Surveyor General for Arizona during the 1880s and until 1892, recommended the confirmation of land grants to several prominent Spanish farming families of the region – among them the Eliases, the Oteros, the Ortizes and the Ainsas – along with grants to Anglo "pioneers."[43] He did, however, reject a claim for "five square leagues, more or less" at Paso de Algodones by the Colorado Commercial and Land Company, concluding in his final report on August 12, 1880 "that the papers are forged and ante-dated, and that the Grant be not confirmed, but rejected." Wasson also noted that "the town of Yuma, is situated on this Grant." Yuma was an important crossing on the Colorado River into California, and, in 1880, a river port connected to the Pacific only beginning its slow but inevitable decline due to the railroads' imminent transportation monopoly; its position was likely a factor in Wasson's skepticism about the Coloradans' intentions and the veracity of their documentation in the first place.[44] Needless to say, the land company from the Centennial State was not alone in its scheme to legalize illegal land claims.

While Spanish-Americans were defrauded of their land quite frequently, at least in Southern Arizona this seems not to have happened on the watch of the local surveyors or the Surveyor General's office. Wasson went to great pains to make sure Spanish speakers understood the importance of providing documentation, often even writing letters to them in Spanish, explaining what was needed and asking them to come to his office in person to explain the circumstances of how parcels of land were inherited or purchased. He did so in the cases of Sabino Otero and of Juan Elías, whom he wrote repeatedly, and for both of whom he recommended confirmation of land titles. On the other hand, Congress did not always feel the need to comply with such recommendations, as men like Albert Beveridge advocated at the same time for an American Empire and against Southwestern statehood.[45] Like much that was done in and to the Southwest in the years between Reconstruction and World War I, the practice of transferring old Mexican land titles to new Anglo owners was national, rather than local.[46] This shows the strong

43 In addition to being a prolific newspaper publisher and anti-Apache agitator.

44 National Archives Western Division, Perris. Surveyor General of Arizona, Letters Sent, 1879–1895. RG 49.

45 Braeman, Albert J. Beveridge.

46 There is, of course, another, even darker chapter related to the defrauding of land claimants. Native Americans, whether allied with the Euro-American newcomers or not, constantly got short shrift in this matter, as they did most elsewhere on the continent. This

influence of the federal government over the region. This influence was not always direct and obvious, but that made it no less effective.

Mining interests were another factor and perhaps the most influential case of national agents meddling with local and regional affairs. Copper doubly hauled the territory into the modern world. It was used in telegraph lines constructed to connect it to the centers of the East and West. But, perhaps even more importantly, copper was an essential material in all telegraph lines and electrical wires. This, in connection with rail transport, made copper mining profitable, and began an upward spiral of exploitation that enabled further extraction.[47]

After the turn of the century, the 1902 creation of the United States Reclamation Service was a first step towards the construction of a plethora of federally funded irrigation and water diversion projects, the most important of which would be, by mid-century, Hoover Dam, and the Arizona Highland Canal.[48] But the Reclamation Service (and later, the Bureau of Reclamation) was not the only visible federal presence in the territory that bound Arizonans to the East. Communication among territorial elites in many places was by necessity always associated with the federal government. As we have seen, the army built the first telegraph line to Arizona. Tucson was connected in 1873, and the station was opened for personal and business use that same year.[49] The "Old Pueblo" thus stood somewhat apart in the region by being wired to the world informationally before it was connected physically, through the Southern Pacific, in 1880. Conventionally, telegraph lines were built alongside rail lines during their construction – and thus, by extension civilian telegraphy also profited from land grants and other tax breaks and rebates given to the railroads by the federal government.

cannot be discussed at length within the confines of this paper, but it has been thoroughly investigated and astutely described in Sheridan, Landscapes of Fraud.

47 See e.g. Gordon, Great Arizona Orphan Abduction, p. 27.

48 For the most recent account of the Reclamation Service's history, especially in the lead-up to Hoover Dam construction, as well as on the federal involvement in water rights negotiation in the Western states, and its role as decision-making arbiter of these rights, see e.g. Hiltzik, Colossus, p. 56, 99, 396.

49 Sonnichsen, Tucson, p. 98. The telegraph line itself was built from San Diego to Yuma and on to Tucson through an Indian reservation, like the railroad line would be in 1877. This required military intervention, yet another reflection on federal policy affecting construction of communication lines in the region. National Archives, Washington, D.C. Fort Yuma Railroad Construction Correspondence. NN 368 52.

Telegrams as a Marker of Social Capital in the Secondary Elite

The telegraph did not always have an obvious influence on Arizona Territory, but that did not render it less pervasive. The case of Joseph Lee Wiley, a professional gambler, Southern Pacific worker for 60 years, and one-time Sergeant of Arms of the State Senate, speaks volumes about the importance that those in the second tier attached to telegrams from well-connected, in any sense of the word, primary elites. Born in Fairmont, West Virginia in either 1862 or '63, Wiley arrived in Tucson in 1882 or '84.[50] He married Irish-born Anna "Annie" O'Sullivan in 1904 and was active in Arizona Democratic (and later Republican) politics. In 1916, he was tied to the killing of one Josephine Welsh Bates, and sentenced to 10–15 years in prison for murder in the second degree. Although there do not appear to be exact records of what transpired, it seems that his political connections saved him from ever serving jail time, however.[51] His wife Annie became a Notary Public by appointment of Governor George W. P. Hunt in 1919, and the couple was even on Hunt's Christmas card list. Wiley died in Tucson in 1947. Although these connections did not put them squarely in the territorial elite, it marked them clearly as members of its secondary elite: they were people who kept contact with the political and economic movers and shakers in the Southwest and who could profit from such associations, given the right circumstances.

Of the four telegrams Wiley kept, two told of the deaths of his mother and father, while two came from important political figures – one from Arizona's longtime U.S. Senator, Marcus Aurelius "Mark" Smith: "Letter received[.] I rely on your judgment and have done what you think best[.] Will speak Tucson night of primary election[.]"[52] Communication from first-tier elites, like politicians navigating national and territorial politics, to those in the second row, hangers-on like the Wileys, often took the tone of favors asked or commands issued. The prohibitively high cost of replying in the negative automatically constrained impulses not to act on the impositions of one's high ranking political friends. This left second-tier elites with only

50 Sources for the area and period are notoriously unreliable for such details.

51 Despite this, Wiley seems not to have been the black sheep of the family. He took out an ad in the Arizona Daily Citizen on January 20, 1900 stating that "I warn the public that I will not be responsible for any debts contracted by Roy Wiley, as he is a drunkard and dead beat." Arizona Daily Citizen, January 20, 1900 / AHS, Tucson. Joseph Lee Wiley Papers. MS 863.

52 AHS, Tucson. Joseph Lee Wiley Papers. MS 863.

the option to follow through on the wishes of senators, congressmen, or the governor, if their favor was not to be lost altogether. The second telegram Wiley kept illustrates this well: "Am short three hundred names to initiate state reclamation bill[.] Will you not send me twenty names Wednesday evening mail or by some one on train leaving Tucson 1230 pm Thursday[.] Please don't fail[.]"[53]

Thus, the primary elite of politicians (who in many cases did not have to pay for telegrams out of their own pocket) and businessmen who could afford such costly communication wielded subtle but real power merely by their ability to communicate. When the recipient was perpetually relegated to secondary elite status with no pretenses to equal standing in the fraternity of white businessmen because of race or gender, the same dynamic was at play, with a further component of inequity added.

Government agencies other than the military paid special government rates for telegrams, but their communications were logged as "commercial business" by the Signal Service, and they seem to have been on a tight budget when it came to telegraphy. The army itself played a negligible role in actively using telegraphy to further nationalizing goals, despite its prolific telegram-sending habit. This seems surprising at first glance. Along with the powerful U.S. postal service, the army was the most visible representative of the federal government in the Southwest. In light of the kind of communication by the military, however, the explanation becomes plain to see: as has been mentioned, officers usually sent messages back and forth between posts. While they might have included other kinds of news in addition to the plain requirements of the troops every now and then, the military was not in the business of disseminating such information to the public.[54] Instead, only a few well-connected members of the elite would have profited, making the military's role even as a secondary communicator less important than it could have been, given its frequency of communication.

As might be expected, the civilian movers and shakers also used such power mostly when they had something immediate to gain as senders of a telegram, particularly in the telegraph's early days in the territory. Whether such communication could be regarded as "immediate," though, was decidedly a matter of one's interpretation of what immediate meant. A message sent via the Signal Service's wires from Yuma in the afternoon of August 27,

53 Ibid.

54 The long telegrams Johnson and Bean sent each other certainly seem to suggest so.

1878, reached Tucson, roughly two hundred miles distant, only the next day at 8:30 p.m. In this case, the telegram was a request for money to be sent by a merchant, L. M. Jacobs, to his business partner P. Drachman, to enable Drachman to pay for a delivery for Jacobs waiting at Yuma. Only circumstances like these, where quick action was necessary, justified the steep expense of telegraphy.[55]

Arizonan Identity

The increased use of the telegraph coincided with the influx of more and more white settlers. Telegraph lines multiplied in Arizona's Anglo-dominated north and center.[56] This made it possible for a small itinerant elite instead of a locally rooted one to supplant the existing Mexican landholding class and their Anglo allies as the powers-that-be in the territory. In this process, they managed to (re)define regional identity as Arizonan and white and in opposition to New Mexican, latinized and mixed-race. This is especially remarkable as the division of Mexican-American New Mexico Territory into the familiar shapes of today's Arizona and New Mexico only dated to 1863. A roughly North-South division of Anglo vs. Mexican thus was reinterpreted as West-East; with Arizona remaining even today the more Anglo state.[57]

As the creation of Arizonan identity coincided not only with a marked increase in communication, but also with the growing influence of the national government in the region, both became inextricably linked, and defined as purposely against the identity of a more Hispanic New Mexico.[58] While traditionally the absence of federal government action and presence has been considered a sign of its weakness on the internal periphery of the United States, like Brian Balogh, I argue that the government, while out of sight, was far from out of mind for the territories' residents.[59] Howard Lamar

55 AHS, Tucson. Harry A. Drachman Papers, 1870–1975. MS 225.

56 AHS, Tucson. Telegraph Lines in Arizona, 1871–1879. MS 872.

57 Culturally, this can be seen in the two states' wildly differing approaches to Mexican immigration, the teaching of Mexican history in public schools, or their official English policies.

58 Within the ranks of the elites, as well as with the centers of population and power in the East.

59 Balogh, A Government Out of Sight, p. 9–17.

even goes so far as to say that "the government... arrived before the people in Arizona."[60] While ultimately attesting to a failure of the federal government to protect Indians living near a military post northeast of Tucson from violent mob action, the Camp Grant Massacre of April 1871 nonetheless showed that Washington might have sent troops to the margins of its influence, but it could not control immediately the actions of Americans – both Anglo and Hispano – living there. Yet, the U.S. government continued to meddle in the affairs of the territories in less direct ways. While an unprecedented trial against the perpetrating vigilante posse predictably went nowhere, it was nonetheless a judicial show of force that federal influence and rules – up until then hidden rather than obvious – were not to be ignored going forward, or at least not completely.[61]

Meanwhile, telegrams remained a terse and expensive mode of correspondence into the 1900s, when market inventions like the day and night letter provided room for more frequent and comprehensive, if not immediately more affordable communication.[62] In 1866, Western Union charged more than $2 for 10 words sent between New York and Chicago. A typical rate for point-to-point communications within the continental U.S. by 1884 was $1.[63] This was still out of the reach of many blue-collar workers, as it was typical day's pay for miners at the turn of the century.[64] By 1905, the price of sending telegrams (notably within Arizona) had decreased further, to significantly under ¢50, while a letter in 1899 cost just ¢2.[65] With a typical white miner's daily wage of $4 at that time, this meant more Anglos could theoretically partake in telegraph exchanges. In practice, the remoteness of quickly founded and abandoned mining towns often still constrained communications. Mexicans earned only $2.50, of which they typically sent home to Mexico two-fifths, leaving a miner only $1.50 to subsist.[66] In this respect,

60 Lamar qtd. in Gordon, Great Arizona Orphan Abduction, p. 28.

61 The incident is best described and assessed in Jacoby, Shadows at Dawn. The trial takes up only eight short pages in that account of close to 300; 181–189. It collapsed under the requirement of jury judgment, as none of their fellow citizens, who in many cases approved of the killers' acts, would commit them.

62 Downey, Telegraph Messenger Boys, p. 88.

63 U.S. Census Bureau, Historical Statistics of the United States, p. 775–810.

64 Gordon, Great Arizona Orphan Abduction, p. 51, 222.

65 AHS, Tucson. Mary Hughston Papers, 1861–1969. MS 1389.

66 The expenses list for a Memorial Day celebration in Bisbee that year lists the price of several telegrams sent to Phoenix in connection with the festivities as ¢40. University of Arizona Special Collections, Tucson, Bisbee Memorial Day Celebration. AZ 74. Another invention that made quick – though not instantaneous – communication more affordable

they experienced both a monetary disadvantage and a cultural outlook that did not include Arizona as a measure of identification. They identified as natives of the borderlands, not of either of the territorial entities that bisected the region.[67]

Considering oneself an Arizonan thus would have been a decidedly Anglo mode of identification. It was never possible to extricate it from being an English-speaking American. This is especially interesting if Arizona is considered in conjunction with its sister territory and later state, New Mexico, where the legislature at the time even paid for translators to accommodate a substantial number of delegates whose primary language was Spanish, and public schools habitually used interpreters.[68]

By the beginning of the twentieth century, it was unthinkable for Arizonans to reunite with their neighbors to the east even if that would have meant statehood, and thus a long-desired move away from the kiddie table of national politics. In 1906, the state legislature opposed a joint statehood bill for the two territories because it "would subject us to the domination of another commonwealth of different traditions, customs and aspirations."[69] Mark Smith, Arizona's tireless man in Washington, contended that Arizonans would rather remain a territory for another fifty years than join the Union as one political unit with New Mexico.[70] This certainly complicates the standard Andersonian model of nationalization, which would explain the Anglo settlement of the American West as an expansion of a preexisting Eastern nationalism.[71] In part, the aversion to New Mexico stemmed from that territory's continuously greater comparable importance and influence. As Arizona's manufacturing and mineral production star began to rise, however, this became less important. In time, calcifying standards of race and

were day or night lettergrams (or day/night letters as Western Union called them). These telegrams were low priority, longer form messages, typically passed on during the least busy hours. Introduced in 1909, they could be as much as 80 percent discounted for a message 50 words or longer. See Downey, Telegraph Messenger Boys, 140. A national telegram by 1918 cost only 30 cents to send, plus a 5 cent war tax. By 1919, the rates further dropped, but the night letters were delivered by U.S. mail from that year on. New York Times, February 9, 1919.

67 Gordon, Great Arizona Orphan Abduction, p. 56, 59.

68 Crawford, Language Loyalties, p. 61. New Mexico, unlike Arizona, is today de jure, in addition to de facto, bilingual.

69 Berman, Arizona Politics & Government, p. 30; Crawford, Language Loyalties, p. 59.

70 Braeman, Albert J. Beveridge, p. 322. Smith's hyperbole was likely also self-serving, as New Mexicans outnumbered Arizonans two to one in 1910.

71 Anderson, Imagined Communities, p. 64.

racism that had made their way inwards from the coasts during the late nineteenth and early twentieth centuries became a convenient way to exclude a now-minority Hispanic population from the Anglo majority's vision for a future white, "civilized" state.

Conclusion

During the formative years of both Arizonan territorial identity and its inclusion into the national fabric of the United States, frequent use of the telegraph set apart the first-tier political elite from everyone else and gave them an undeniable advantage in winning the power of interpretation over what the future state would look like.

During a transitional period, however, a joint Sonoran-Anglo elite still persisted. They continued to operate farms and businesses, but increasingly had to concede the territory's economy to the national corporations that defined its extractive phase. For a timespan of several decades, the members of this group continued to cooperate and often considered themselves "pioneers" of the now settled and civilized region. This showed itself most clearly in the sometimes more, sometimes less successful attempts of Anglo members of this elite to aid their Hispano counterparts, who were less well versed in the new political realities of an American Arizona.

The example of Surveyor General John Wasson explicitly helping his Hispanic peers navigate the U.S. legal system, but ultimately succumbing to the national legislative body with the final say, speaks clearly of a continued mutual respect and kinship between Hispanos and early Anglo settlers. It also shows how the American takeover of the region, initially welcomed by many Sonorans, proved to be a double-edged sword.

On the one hand, the United States, especially after the Civil War, had the means and manpower to control the movement of Southwestern Apache bands through its military, thus providing relief for Hispanic farmers as well as for Anglos. On the other hand, the invisible presence of the American Common Law legal system was a complication that proved detrimental to Hispanos wherever its decisions were not made locally, but nationally. National standards of racial difference, propagated by telegraphic communication, the postal system, and the presence of the railroad after 1880, entered the Southwest at a time when ever higher numbers of Anglo settlers were drowning out the culture and ways of life of the relatively few inhabitants who had called Arizona their home during the Mexican era. This coincided with an economic changing of the guard, also made possible by the combination of modern communication technology and railroad transport, which changed the kinds of industry and businesses that would be profitable from now on. Additionally, the first railroad in the territory, which was built from west to east, altered the direction of trade, which had flowed south–north during the time when Arizona had been part of northern Mexico, and had continued to do so for the first three decades under American control.

For the new purely Anglo elite which made use of this realignment, sending telegrams was not only a matter of practicality in business, which now was conducted across the nation and no longer primarily face-to-face within the region, but also a marker of political influence. As elsewhere around the world, it was a day-to-day means of communication for only a select few. For most, it was tied to telegraph news in local papers and ceremonial messages or bad news. Secondary elites received, but did not send, messages – or instructions – this way. While the military was instrumental in bringing telegraphy to the Southwest, it remained a background actor when it came to the actual use of telegraph lines for nationalizing purposes. Instead, while directing to their constituents the various flows of federal funds that even Gilded Age America provided, an itinerant, telegraphically networked elite of politically active citizens constructed from the ground up a territorial identity as salient, or at times more so, than that of the nation. From them, this identity would spread, albeit mediated, to the bulk of Arizona's Anglo, and through their increasing dominance, even parts of its non-Anglo population. For all, it became a standard to either be followed or foregone, but never ignored.

Bibliography

Archives and Libraries Consulted

Arizona Historical Society (AHS), Tucson. Joseph Lee Wiley Papers. MS 863.
— Lloyd Vernon Briggs Papers. MS 95.
— Samaniego Family Papers. MS 706.
— U.S. Signal Service Telegraph Ledger 1892–93. MS 817.
— Mary Hughston Papers, 1861–1969. MS 1389.
Huntington Library, San Marino. Parish B. Johnson Papers, 1864–1886.

National Archives, Washington, D.C. Fort Yuma Railroad Construction Correspondence. NN 368 52.

National Archives Western Division, Perris. Surveyor General of Arizona, Letters Sent, 1879–1895. RG 49.

— Arizona Court of Private Land Claims, Misc. Files, 1789–1907. RG 49.

University of Arizona Special Collections, Tucson. Bisbee Memorial Day Celebration. AZ 74.

Newspapers

Arizona Weekly Star.

Arizona Citizen.

New York Times.

Published Primary and Secondary Sources

Adas, Michael, Dominance by Design. Technological Imperatives and America's Civilizing Mission, London 2006.

Anderson, Benedict R., Imagined Communities. Reflections on the Origin and Spread of Nationalism, revised and extended edition, London/New York 1991.

Balogh, Brian, A Government Out of Sight. The Mystery of National Authority in Nineteenth Century America, New York/Cambridge 2009.

Berman, David R., Arizona Politics & Government. The Quest for Autonomy, Democracy, and Development, Lincoln, NE 1998.

Blondheim, Menahem, *News Over the Wires: The Telegraph and the Flow of Public Information in America, 1844–1897*, Cambridge/MA 1994.

Braeman, John, "Albert J. Beveridge and Statehood for the Southwest 1902–1912," Arizona and the West, vol. 10, no. 4 (1968), pp. 313–342.

Browne, J. Ross, Adventures in the Apache Country. A Tour Through Arizona and Sonora, 1864, annotated re-edition, Tucson, AZ 1974.

Clauson-Thue, William, The ABC Universal Commercial Electric Telegraphic Code Specially Adapted for the Use of Financiers, Merchants, Shipowners, Brokers, Agents, London 1881.

Crawford, James, Language Loyalties. A Source Book on the Official English Controversy, Chicago, ILL 1992.

Downey, Gregory John, Telegraph Messenger Boys. Labor, Technology, and Geography, 1850–1950, New York 2002.

Eastman, Scott, Preaching Spanish Nationalism Across the Hispanic Atlantic, 1759–1823, Baton Rouge, LA 2012.

Ebright, Malcolm, Land Grants and Lawsuits in Northern New Mexico, third edition, Santa Fe, NM 2008.

Ehrlich, Karen Lynn, "Arizona's Territorial Capital Moves to Phoenix," Arizona and the Southwest, vol. 23, no. 3 (1981), pp. 231–242.

Gibson, Campbell and Kay Jung, "Historical Census Statistics on Population Totals By Race, 1790 to 1990, and By Hispanic Origin, 1970 to 1990, For The United States, Regions, Divisions, and States," U.S. Census Bureau Working Paper 56 (2002), Tables 17 and 46, http://www.census.gov/population/www/documentation/twps0056/twps0056.html.

Gordon, Linda, The Great Arizona Orphan Abduction, Cambridge, MA 1999.

Grosby, Stephen. *Nationalism: A Very Short Introduction*, Oxford 2005.

Hawley, Ellis, "Herbert Hoover, the Commerce Secretariat, and the Vision of an 'Associative State,' 1921–1928," The Journal of American History, vol. 61, no. 1 (1974), pp. 116–140.

Herrmann, Richard K. and Marilyn Brewer, "Identities and Institutions: Becoming European in the EU," *Transnational Identities: Becoming European in the EU*, Lanham, MD, 2004. pp. 1–22.

Hiltzik, Michael A., Colossus. Hoover Dam and the Making of the American Century, New York 2010.

Huntington Abbott, Mary, "Beginnings of Freemasonry in Tucson, 1875–1882," Arizona and the West, vol. 24, no. 2 (1982), pp. 105–118.

Jacoby, Karl, Shadows at Dawn. An Apache Massacre and the Violence of History, New York 2008.

Lamar, Howard R., The Far Southwest, 1846–1912. A Territorial History, revised edition, Albuquerque, NM 2000.

Larrabee, Charles S. Larrabee's Cipher and Secret Letter and Telegraph Code, with Hogg's Improvements, New York 1884.

Lavender, David Sievert, The Great Persuader. The Biography of Collis P. Hunting ton, Boulder, CO 1999.

Risse, Thomas, *A Community of Europeans?: Transnational Identities and Public Spheres*, Ithaca, NY 2010.

Sheridan, Thomas E., Los Tucsonenses. The Mexican Community in Tucson, 1854–1941, Tucson, AZ 1986.

— Arizona. A History, Tucson, AZ 1995.

— Landscapes of Fraud. Mission Tumacácori, the Baca Float, and the Betrayal of the O'odham, Tucson, AZ 2006.

Sonnichsen, C. L., Tucson. The Life and Times of an American City, Norman, OK 1982.

Truett, Samuel, Fugitive Landscapes. The Forgotten History of the U.S.–Mexico Borderlands, New Haven, CT 2006.

U.S. Census Bureau, Historical Statistics of the United States. Bicentennial Edition, Colonial Times to 1970, part I, Washington, D.C. 1970.

W. G. Press & Co., Telegraphic Cipher Compiled by W. G. Press & Co for the Exclusive Use of Themselves and Their Correspondents, Chicago, ILL 1885.
West, Elliott, The Last Indian War. The Nez Perce Story, New York 2009.
Woosley, Anne I., Early Tucson, Charleston, SC 2008.

Between the Ends of a Wire: Electricity, Instantaneity and the Globe of Telegraphy

Florian Sprenger

With the advent of electromagnetic telegraphy in the 1830s, a notion emerging from the history of the science of electricity diffused into popular knowledge: the instantaneous transmission of electric action. Ever since Stephen Gray, more of an amateur than what nowadays would be called a scientist, explored the possibility of electric transmissions through long copper wires in 1730, the speed of electricity was an object of interest and of several investigations. This knowledge fosters the rise of telegraphy a century later and defines the epistemological foundations of the new medium. Experiments with electric transmission during this period revealed that electricity apparently had no speed. It was impossible to measure any temporal difference between the events at the two ends of an electrically charged wire. The notion of instantaneity, which was quickly introduced to explain the strange occurrences, implied that electricity was present at both ends of the wire at the same instant – instantaneous, immediate.

Contemporaries of the nineteenth century experiencing the advent of telegraphic transmissions over distances as long as oceans described the new medium as timeless and without speed. My article follows some of the traces of such transmissions. First, I describe the early experiments conducted by Gray. In the second part, I concentrate on one specific case: Samuel Morse's reference to Charles Wheatstone's experiments on the velocity of electricity. In the third part, I want to highlight some of the consequences of instantaneity for what may be called "cultural techniques of synchronisation."[1] The historical context of the notion of instantaneity were the foundations of physics and of philosophy, because for Newtonian physics and Aristotelian logics nothing can act where it is not or be at two places at the same time. But this seems to be the case with the electric *actio in distans* of Gray's transmissions or telegraphic communications.[2] Thus,

1 I borrow this term from Kassung/Macho, *Kulturtechniken der Synchronisation.*

2 Hesse, *Forces and Fields.*

throughout the eighteenth and nineteenth century and even up to the present day of smart, ubiquitous or mobile media, debates about transmission, communication and their technical time are related to fundamental, but historically grounded questions. By investigating their impacts on the establishment of the electromagnetic telegraph, I intend to explore what may be called a "phantasmatic dimension of technical media."[3] Consequently, scientific truths and mere curiosities have to be treated on the same level.[4]

My goal is to emphasize what Jacques Derrida calls a "coherence in contradiction:"[5] that electricity can be coherently described as instantaneous *and* having speed, as immediate *and* mediate. In the historical discourses I want to explore, these notions can stand shoulder to shoulder without contradiction. This coherence, I want to argue borrowing from Derrida, "expresses the force of a desire,"[6] which accompanies the history of media and communication. Methodically, the discussion of a "coherence in contradiction" should not aim at any solution, as a desire cannot be true or false. It is part of the history of media and in the case of the electric telegraph, it exerts a distinct productivity in the formation of global communication networks.

In telegraphy, much depends on the difference between slow speed and no speed. This difference may seem small, but it means everything to physics and to media. If something has speed, it cannot be instantaneous. For the discourses I wish to investigate, electricity seems to be instantaneous and have no speed, but for the sake of physics, it must exhibit speed. When it has speed, it takes time and is delayed. If it is delayed, there is a gap or an abyss between both ends of the transmission. This abyss is disguised or negated by instantaneity, while it is necessary for communication, media and their materialities.[7] Communication needs at least two communicators and an in-between. There must be a difference between the communicators. They must be two to become one and there must be a third: the medium.

3 In a similar vein this has been explored in Marvin, *When Old Technologies Were New* and Morus, 'The nervous system of Britain'.

4 Bloor, *Knowledge and the Social Imaginary.*

5 Derrida, Structure, Sign and Play, p. 279.

6 Ibid.

7 Gumbrecht/Pfeiffer, *Materialities of Communication.*

Stephen Gray and the Ends of a Wire

To investigate the properties of the force called electricity that was almost unknown at the time, Stephen Gray hung a wire from one end of a garden in the south of England to the other. On a warm summer day in 1730, as he reported in the *Philosophical Transactions of the Royal Society*, electricity was transmitted over a distance for the first time. There was a distance of 666 and later 886 feet between the two ends and there was a distance between cause and effect, which nonetheless seemed to happen at the same time. Gray called this connection a "line of communication."[8] He had no idea about cables and did not know anything about induction, isolation or charges. In the sciences of his time, physical transmissions were named "communication," a term often used by Isaac Newton. It did not mean to send a message to someone else, but to transmit an action to another place, for example between bodies, stars or the ends of a wire (see Fig. 7 in illustration 1). Gray touched his end with a charged tube of glass. At the same instant, with "no perceivable difference,"[9] the electric action was carried to the other end, resulting in the dance of some pieces of brass gold located under the wire. Gray's friend Granvile Wheler, priest and owner of the estate, shouted that something had happened, while Gray was still at the other end of the garden. As both were not able to recognize any temporal difference between the electric occurrences at both ends of the wire, even as the wire was arranged in a circle in following experiments, they were described as instantaneous, as happening at the same time, as having no difference and no mediation.

After Gray, William Watson, Louis Guillaume Le Monnier and Charles Dufay continued these experiments until the 1750s.[10] As no useful application of electricity was in sight, the investigations soon ended. The experimenters' senses were too slow to see what was happening in the interval between the two events, and their instruments did not provide enough exactness to account for any difference. But even when exact measuring instruments were available a century later, electricity was, as I will show, thought to be instantaneous.

8 Gray, A Letter to Cromwell Mortimer, p. 27. See also Heilbron, *Electricity in the 17th and 18th Centuries.*

9 Gray, A Letter to Cromwell Mortimer, p. 28.

10 Watson, A Continuation of a Paper concerning Electricity; Le Monnier, Recherches sur la communication de l'électricité; Dufay, *Versuche und Abhandlungen von der Electricität der Körper.*

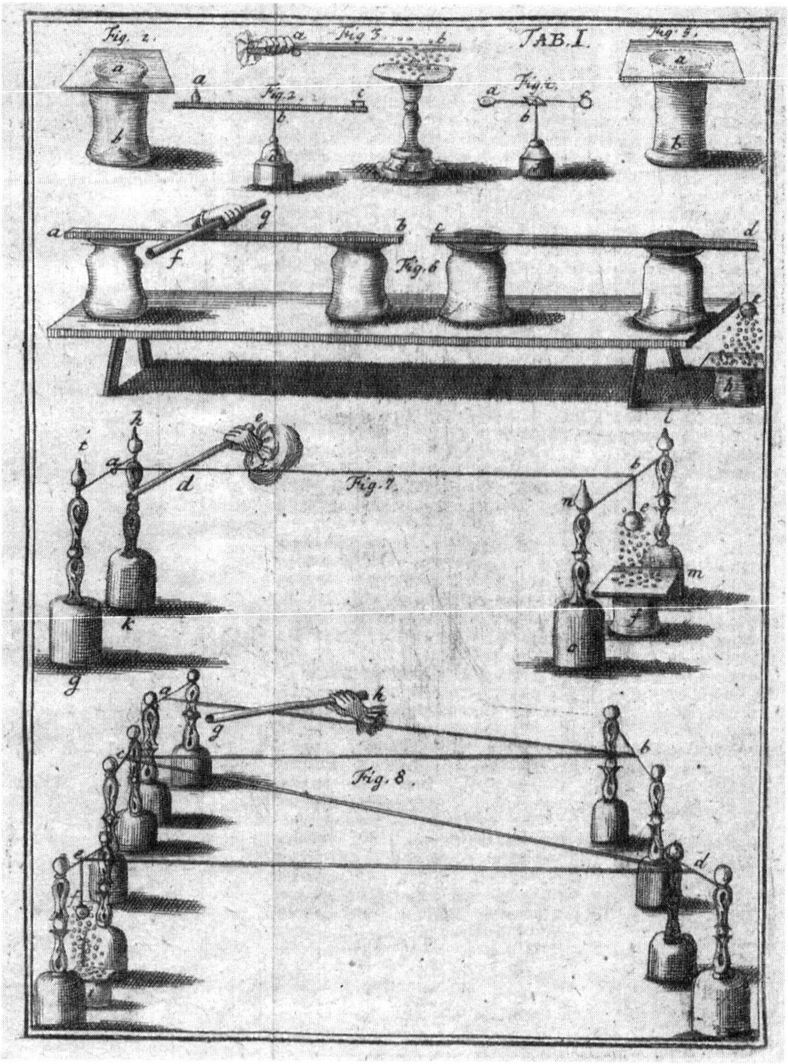

Illustration 1: Doppelmayr, Johann Gabriel: Neu-entdeckte Phaemomena von bewundernswürdigen Würckungen der Natur. Nürnberg, Table II.

Besides the technical aspects of transmission, there is a paradox at hand here. If there is a distance and a communication in the physical sense, that means, if there are two separate elements communicating and trying to bridge the abyss in-between and to cross an empty space like the garden between Gray and Wheler, then there is a mediating transmission between separate elements.[11] And if there is mediation, then there is a contradiction to the notion of instantaneity which results in what may be called phantasms of immediacy. These phantasms, up to today and all the time during the great days of telegraphy in the long nineteenth century, were important to balance the imaginative economy of communication, its relation of here and there or local and global. These were not simply false conceptions of electric transmission that would soon be refuted. The notions were important to deal with the transformations of a new medium and new forms of communication. As a transversal phenomenon, these phantasms occupied several fields from the physical sciences to literature and from engineering to marketing. On the one hand, they show the uncertainty related to each new medium offering formerly unknown forms of connection between people or places and leading to new conceptions of space, time, presence and absence. On the other hand, these phantasms articulate an oscillation historically connected to media upheavals: the oscillation between immediacy and mediation.[12]

However, all electricians of the time, whether physicists or engineers, knew very well that nothing could move with an instantaneous speed because action at a distance is impossible –even though this was a complicated issue for the Newtonian physics of that time and still is for Quantum mechanics. The premise that nothing can act where it is not, at least if there is no medium, dates back to antiquity. Aristotle announced that every cause needs to be in contact with and have proximity to its effect.[13] This contact was disconnected with actions at a distance such as gravitation, magnetism or electricity. The theoretical framework of physical sciences had to be reorganized repeatedly to manage these interruptions of causality and of the nexus of time and space. If there is a causal relation of two separate bodies, then a medium must communicate between them, even if it is immediate, that means, if it is at two places at the same time. This constellation is the matrix for the media of physics: for ethers, vacua and "most subtle spirits."[14] When unknown causalities had to be described, they were introduced as supposed connections of unconnected elements in space in the form of specific

11 Peters, *Speaking into the Air* and Chang, *Deconstructing Communication*.

12 In a similar direction, Jay David Bolter and Richard Grusin have described immediacy and transparency for media interfaces, while I intend to expand on media history. Bolter/Grusin, *Remediation* and Sprenger, *Medien des Immediaten*.

13 Aristotle, *Physics*, p. 267a.

14 Newton, *Mathematical Principles of Natural Philosophy*, p. 574.

actions at a distance. Electricity quickly became part of this group of "argumentative resources."[15] Electric immediacy, as it was discursively established, acts as a medium, though it contradicts mediation. In both the discourses of physicists and in public debates "argumentative resources" provided possible ways to domesticate the new medium of telegraphy and to deal with the uncertainty it introduced. The common denominator of most arguments in this debate was a desire to solve the paradox of the absent presence or present absence an *actio in distans* brought with it. It is still present in the history of global communication networks, their economies and politics.[16]

When dealing with techniques or apparatuses and producing objects like telegraphs in all their different forms, scientists and engineers pondered the immediacy of electricity and what it would offer to humankind. As the commercial success of the telegraph was enabled to some extent by the notion of instantaneity, this idea gained a wider influence in the public sphere too.[17] It can be found in poems, newspaper articles, manuals, patent applications and religious manifestos. Without the idea that this new medium created a form of presence at another place but at the same time, telegraphic communication may not have been asserted so successfully, though it never became a mass medium and served rather specific interests and needs of economy, press and politics. Instantaneity provided economic benefits as well as metaphysical forms of connection. A new conception of communication and of community could emerge: a worldwide connection of all people in a sometimes mediumistic realm of electricity, which was eligible for spiritism.[18] This destiny is obvious in a popular song which served as the motto of the contemporary magazine *The Telegrapher*: "Intellect hath conquered time."[19] By conquering time intellect erased it. Electricity could be everywhere at once and resulted in essentialistic ideas of a worldwide community, anticipating what would later be called a *global village* by Marshall McLuhan, who implicitly relied on these discourses.[20]

15 Cantor, The Theological Significance of Ethers, p. 152.

16 Tomlinson, *The Culture of Speed. The Coming of Immediacy*. For example, the sociological concept of "time-space-convergence" still functions this way in the works of Manuel Castells, Anthony Giddens or Paul Virilio.

17 See also Martin Doll's article in this book.

18 Asendorf, *Batteries of Life*.

19 Anonymous, The Electric Telegraph, quoted in: Shaffner, *Shaffner's Telegraphy Companion*, p. 26.

20 McLuhan, *Understanding Media*.

The new relation of the local and the global introduced by the telegraph, and subsequently applied by the telegraphic networks all around the globe, emerged from the new meaning of time and space. Time and space were supposed to be transcended in no time and from anywhere. In an article dating back to 1857, a member of the American Telegraph Company, launched in 1855 to operate the terrestrial lines connecting to a possible transatlantic telegraph cable, described this "telepresence." This term was used by media theorists in the 1990s with regard to the Internet, as a means to be present at a distant place. Similar constellations can be found in 1857:

> We publish the following novel and interesting account of a meeting of the employees of the American Telegraph Company on the 3rd instant at – what place? That is the question – at no place, or at all places where there were Telegraph offices, within the circuit of seven hundred miles. A large room, that – seven hundred miles in diameter – for a meeting to convene […]. The members together in spirit – in communication, and yet in body seven hundred miles apart![21]

In this sense, the place of presence is everywhere, connected by cables and immediate electricity. Such phrases were repeated throughout the subsequent history of telecommunications.[22] When two or more places are connected in such a way, the space in between is nothing more than a time to bridge. Electricity allowed telegraphers to build this bridge, even though it took time, relays were used at every station and every message had to be decoded in time. The "annihilation of time and space," a popular phrase in the nineteenth century, led to conceptions of a global worldwide unity.[23] These ideas influenced, as I want to show in the final part of my article, the political conceptions of community and shaped the identity of those belonging to the community.

This long and winding history had a starting point in a garden. When Gray for the first time communicated an electric action through a wire, but without any message or code, a space was established between the two

21 "Telegraphic Meeting," Tiffany's Monthly, 3, 1857, p. 142, quoted in: Sconce, *Haunted Media*, p. 21.

22 Müller-Pohl, By Atlantic Telegraph.

23 The idea of an "annihilation of space and time" is not only part of the history of global media, but also of media theory, because it deals with the paradox of how something can be where it is not – one of the main aspects of the rephrasing of the term medium in the nineteenth century. This discourse of near and far and speed is also part of the Hegelian, teleological legacy of media theory (McLuhan, *Understanding Media*; Virilio, *Speed and Politics*).

ends of the garden and the two communicators. Nothing is communicated than communicapility. Through the expansion of a global communication network by means of wired and wireless telegraphy, this space would later become the global space of electric media. Even though there were many scientific developments to come and failures to overcome, the physical structure of communication and its phantasmatic dimension was still the same when Samuel Morse sent his first telegram from Washington to Baltimore in 1844.

Samuel Morse, Charles Wheatstone, Telegraphy and Time

Morse, who can be regarded as the founding father of telegraphy at least in economic respects, also knew very well that electricity had a speed and was not immediate, as British physicist Charles Wheatstone had shown. In 1834, Wheatstone, who later became a competitor together with William Cooke, tried to measure the actual speed of electricity. The results were quite ambiguous and their interpretation widely misunderstood, but nonetheless it was obvious that electricity had a speed.[24] Yet, both Wheatstone and Morse relied upon the gift of instantaneity when talking about the practical uses and the impact of the telegraph on society.

For his experiments, Wheatstone installed a revolving mirror and observed the light of electric sparks in this mirror. By comparing the sparks at different positions in a circuit seen through the mirror and with the help of the number of revolutions of the mirror, he wanted to calculate the velocity of electricity. As Wheatstone's experimental system is extremely sophisticated, I cannot go into deeper detail in this context.[25] Strictly speaking, Wheatstone mistook the light of electric sparks for the speed of electricity. Furthermore, he only stated that his instruments were able to measure a speed of 288,000 miles per second, but never said that this was the actual speed of electricity. Many publications attributed this velocity to electricity, as did Morse. Wheatstone himself never refuted this incorrect interpretation of his experiments. In his paper, he even talked about the "instantaneousness

24 Wheatstone, An Account of Some Experiments. See also Bowers, *Sir Charles Wheatstone FRS, 1802–1875*.

25 For further explanation see Sprenger, *Medien des Immediaten*, p. 288.

of the light of electricity"[26] and reflected on the possibility of telegraphic transmissions.

In his biography, Morse announced that as soon as he had heard about the possibilities of electromagnetism around 1830, he had said: "If the presence of electricity can be made visible in any part of the circuit, I see no reason why intelligence may not be transmitted instantaneously by electricity."[27] Both Morse and Wheatstone built on the benefits of instantaneity while building telegraphs and experimenting with electricity, knowing that media and physics cannot be immediate. Referring to Wheatstone's experiments Morse wrote that:

> [...] no time is consumed that is necessary to write the intelligence to be conveyed, and to convert the words into the telegraphic numbers. The numbers are then transmitted nearly instantaneous, (or if I have been rightly informed in regard to some recent experiments in the velocity of electricity, two hundred thousand miles a second) to any distance, where the numbers are immediately recognized, and reconverted into the words of intelligence.[28]

Many histories of telegraphy hinge on Morse's apparatus and his famous code, which became a worldwide standard by 1865.[29] Even though his actual importance for economic and technological developments can be questioned, Morse is important for my perspective. He extensively wrote about the impacts of the telegraph and reflected on the opportunities of the new medium in contrast to established means of communication. The fact that Morse's method gained such an enormous influence against other competitors – also against Wheatstone's inventions since the 1840s – is equally due to an aggressive patent policy, a narrative deeply embedded in North American culture and a reliable, elementary technical procedure, regarding both the used instruments and coding.[30] Of course, even before the telegraph was established, there were technical and semiotic ways to transmit signs and words and to translate them into signals. Morse's apparatus made it a mass

26 Wheatstone, An Account of Some Experiments, p. 591.

27 Morse, *His Letters and Journals*, p. 6.

28 Morse, Memorandum to the "Secretary of the Treasury," 27. September 1837, quoted in: Aschoff, *Geschichte der Nachrichtentechnik*, p. 192.

29 Czitrom, *Media and the American Mind*; Morus, Telegraphy and the Technology of Display; Silverman, *Lightning Man*.

30 "The steed called Lightning, say the Fates/Was tamed in the United States/'Twas Franklin's hand that caught the horse/'twas harnessed by Professor Morse" (Czitrom, *Media and the American Mind*, p. 14.)

business for the first time. By coding long and short electric impulses and the pauses between them, Morse's system separated communication and transport.[31] But above all, electromagnetic force, which was discovered by Hans-Christian Oersted in 1820, resonated with a new order of knowledge.[32] Morse's telegraph connected the electromagnetic tact and the tact of coding. He optimized the channel's rate of transmission with an effective method of coding. In both cases, electromagnetism and telegraphy, tact means time, because the dots and dashes follow each other and cannot be simultaneous. Against all this knowledge, the notion of immediacy prevailed among physicists and engineers when they pondered about the cultural or epistemological impact of the telegraph. In regard to Morse, I want to show that the electromagnetic transmission of the telegraph and its instantaneity are so profoundly intertwined that transmission could be described as instantaneous against all better knowledge, or even had to be described this way due to the imaginative economy of a new medium.

Let me briefly sum up the historical background. At the end of the 1820s, the academic painter Samuel Finlay Breese Morse visited the art galleries of Europe to find examples for his historical paintings. He became acquainted with the optical telegraph, whose revolutionary effects were the subject of an exhibition at the Louvre. Some years before, Morse had spent his time on the unsolved problems of electricity and done a few experiments. So he was no stranger to this field and even had some friends working on electricity when he perceived the potentials of this force. On his homeward journey from Europe, as he repeatedly remembered later, he had enough time to ponder about this strange discovery: even on the fastest transport medium of the time, a steamer named Sully, the crossing of the ocean took two weeks. At sea, no message could be faster than the messenger who carried it. At that time, no transatlantic communication media was faster than marine transport. During the long evenings on board, the conversations circled around the fashionable topic of electricity.[33] At the time, every newspaper reported on the rapid succession of new discoveries and inventions. Hence, it was possible to refer to electricity a common knowledge, a knowledge which was bewildered by phantasms.

31 This transformation has most prominently been described by Carey, Technology and Ideology.

32 Oersted, *On Electro-magnetism*. See also Caneva, *Ampère, the Etherians, and the Oersted Connection*.

33 Morse, His Letters and Journals, p. 19.

Until 1835, when Morse was appointed Professor of the Literature of Art and Design at University College in New York, he found no time to work out his ideas and to convert from arts to engineering. At the time Wheatstone's experiments became known, he was in the first exploratory stages of his invention and the technical outline was still in an open process of development. It took him almost 15 years from the first small experiments shortly after his voyage to successful fund-raising and the introduction of his system in 1846, followed by wealth and fame.[34] During these years, Wheatstone and Cooke introduced their telegraph in Great Britain and Morse was forced to stage himself as the first inventor of telegraphy by mythologizing the events on board the Sully.

Morse's telegraph can be understood as a combination of different regimes of time. A media archaeological perspective offers the tools to describe the technical characteristics as solutions to temporal differences on several levels.[35] Together with Alfred Vail, Morse developed an apparatus composed of two operators, the transmitter or the Morse key, which changed letters to code, the sender, the writing apparatus and of course the cable. They replaced the group of small batteries formerly used with one huge battery with high voltage to intensify the signal. The first receiver consisted of a pen attached to an electromagnet. Paper was pulled across this pen, which moved depending on the movement of the electromagnet and wrote down a zigzag line. The signal inscribed itself, because it consisted of electric impulses that could be made visible by inscription. Electricity did not transmit any substance of matter, but rather conveyed forces and signals arriving in time. The telegraph operator simply had to take the paper and decode the message manually. Electricity was the invisible hand of this writing. The telegraph only became operational in November 1837, after incorporating the work of Joseph Henry on the use of relays.[36] "Local circuits" were needed to move the pen because when applying such long lines, the electromagnetic force was not sufficient.

Thus, the telegraph can be described as a manifestation of time-critical synchronizations.[37] Similar to the embossed printing that Louis Braille presented in Paris in 1829 during Morse's visit, his apparatus operated with an element of code and its absence. The signal was short or long, separated by

34 Pichler, Die Einführung der Morse-Telegraphie.

35 Huhtamo/Parikka, Media Archaeology.

36 Henry, On the Application of the Galvanic Multiplier to Electro-Magnetic Apparatus.

37 Hirt/Volmar, Kanalarbeit.

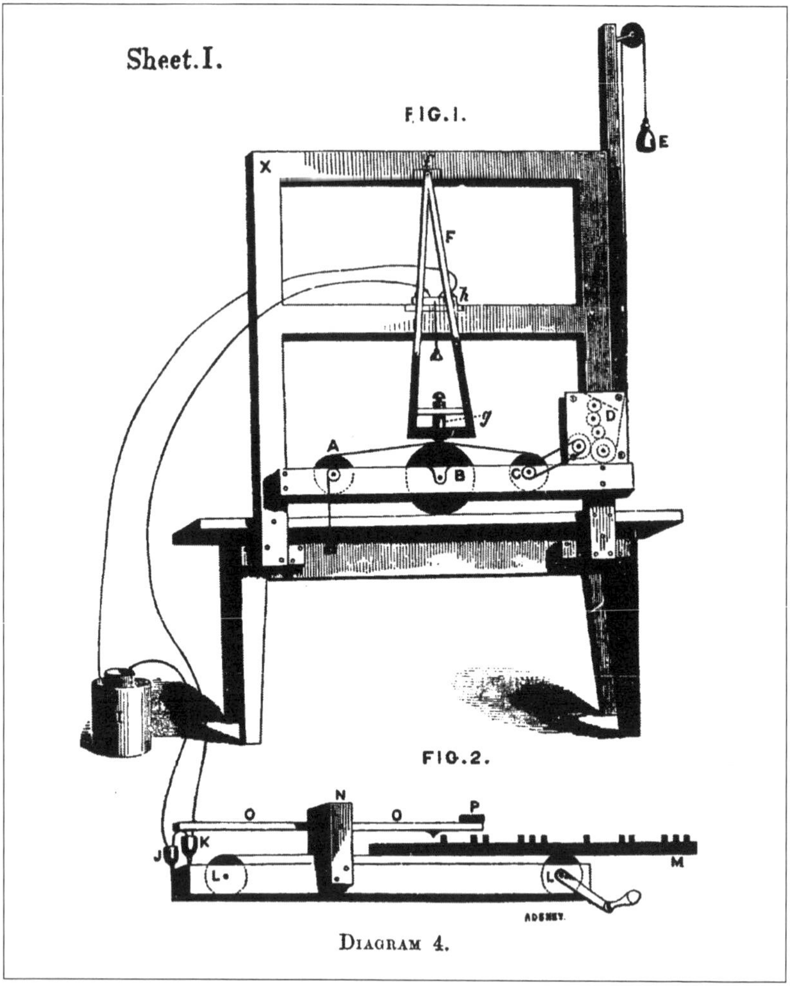

Illustration 2: Morse, Samuel. Modern Telegraphy: Some Errors of Dates and Events and of Statement in the History of Telegraphy. Paris, 1867. S. 20.

brief interruptions for a continuous sign or long breaks for the distinction between signs. There was writing and non-writing: functions or manipulations of time. Transmission was serial, as discrete signals were sent in rapid succession. The signs were interrupted by breaks. Points, lines, usually three after another, and the spaces in between were recorded, though the technical devices changed rapidly. The discrete code of dots and dashes created a tact, because it transformed the continuity of time into a sequence of signals. Each sequence of signals generated tact and tact is a sequence in time, because for the machine signals were only readable in sequence.[38] A perfect, instantaneous signal would switch from one state into the other immediately and write a rectangular line instead of a curve. In practice, things were not quite as easy. The writing apparatus produced irregular curves and jags because in reality, every transmission needs time and signals are never equally strong. For telegraphers, signals were hard to distinguish from each other and from noise, and these problems became more and more urgent with the first attempts of long-distance-transmission.[39] For physicists it became obvious that the problems were the result of a delay in the cable a delay that, like in Wheatstone's experiments, undermined all instantaneity.

By implementing electromagnetism, Morse was able to conjoin transmission and storage to a powerful media network. The introduction of automatic writing put an end to the laborious writing by hand and furthermore offered a more economic operation of emergent global telegraphy. The inscriptions of the telegraph, its ability to translate movements into signs and to store their temporal order, diffused in many directions: for example Hermann von Helmholtz' physiological determination of the velocity of nerve conduction in 1850 – of a similar importance as Wheatstone's experiments – was conducted with telegraphic equipment and presented using telegraphic metaphors.[40] Almost every physiologist of that time used telegraphic apparatuses to translate movement into inscription by recording their duration. As John Durham Peters has shown, the Morse telegraph constitutes a historical intersection of "tele" media such as the telescope, telepathy or later television and "graphic" media such as photography, phonography or seismography.[41] Telegraphic action at a distance is inscription at a distance, and this was applied for different purposes of time-critical recording. With the Morse telegraph, numerous "discourse networks" (*Aufschreibesysteme*) were conceivable because it became possible to record signals of all kinds: muscle movements, blood pressure or earthquakes.[42] The new medium allowed the translation of numerous flows, moves and waves into data by transcribing and storing them in a temporal order.

38 Kittler, Real Time Analysis – Time Axis Manipulation.
39 Hunt, Michael Faraday, Cable Telegraphy and the Rise of Field Theory.
40 Lenoir, Helmholtz and the Materialities of Communication.
41 Peters, Technology and Ideology; Rieger, Schall und Rauch.
42 Kittler, Discourse networks.

Nonetheless, the innate time of the electromagnetic telegraph was occupied by phantasms from electric research. For the practical uses of communication, instantaneity may have been negligible. The signals could be transmitted only one by one. For each telegrapher, it was obvious that his message took time and that even an instantaneous transmission had to be decoded. The sender had to feed signs one by one and the recipient had to decode them one by one. Telegraphy was technically considered clocked economy of time. Any electrical or electromagnetic transmission is a series of differences, a succession of changes. Since it has to be decoded, the time of transmission is shorter than the time of receipt, and thus it is framed by practical margins. The speed of electricity is not the same as the speed of the signal. A signal has to be on-time (*rechtzeitig*) instead of real-time (*echtzeitig*) to assure successful communication.[43] In-time means that transmission has to conform to a given interval of time, a duration in which it has to arrive. These time-critical synchronizations form the technical basis of all telegraphs. But still, the interventions of the telegraph in society's self-descriptions were marked by instantaneity. That does not simply mean that instantaneity, as it was articulated in writings on telegraphy, and its ability to reach distant places, relatives or stock exchanges were merely a dream or a reverie, as Gaston Bachelard called such poetic imaginaries.[44] It also means that the twentieth century's term *global village* and all the ideas of a worldwide connection and community were and still are related to an idea of immediacy. It is historically grounded in the discourses of electric research. Their knowledge had to cross these phantasms.

In 1843, the US Senate decided to fund a 40-mile telegraph line from Washington to Baltimore for 30,000 US dollars. The two cities were connected and their new relation of "distant proximity" provided what appeared as a spiritual unity. The protagonists of the events foresaw a brighter future of "global connectivity." According to Morse, it was no longer "visionary to suppose that it would not be long ere the whole surface of this country would be channeled for those nerves which are to diffuse, with the speed of thought, a knowledge of all that is occurring throughout the land."[45]

On March 21st 1844, after a few teething troubles, the first and for a long time very successful electromagnetic telegraph line started service. The first message ever sent said "What hath God wrought," a quote from the

43 For these terms see Rohrhuber, *Das Rechtzeitige.*

44 Bachelard, On Poetic Imagination and Reverie.

45 Morse, *His Letters and Journals*, p. 85.

Bible (Book of Numbers, 23–23).[46] God's words, carefully chosen by Morse, were sent back to Washington to confirm the success of transmission. In this sense, telegraphy verified itself as a Will of God. Morse initially charged one cent for the transfer of four letters. The revenue from the first days was one cent. The first commercial message significantly asked "4," which means "what time is it?" and the response from Baltimore said "1." The impact of telegraphy was documented by the fact that it synchronized time between two places. Baltimore said "1" when it was one o'clock and this message reached Washington when it was one o'clock and not when it was past one o'clock, provided their clocks were synchronized. As the interest declined after a few days, another offer was made: for half a cent, the customers could carry their name to the other end of the cable, and for a full cent, they even got it back. A few years later, a similar epistemological confusion took place in England: "We have heard of things being done 'in less than no time,' and always looked on the phrase as a figure of speech signifying great dispatch. The paradox seems, however, to have been actually realised in the case of Wheatstone's Great Western Telegraph, a message having been sent in the year 1845, and received in the year 1844." On New Year's Eve, just one second after midnight, the station supervisor at Paddington Station sent a message to his brother in Slough. But in Slough, another time was used. The answer came directly, "suggesting that the wish was premature, as the New Year had not yet arrived at Slough."[47]

These examples of the curiosities and epistemological irregularities of temporal difference and instantaneous transmission give an impression of the new medium's potential to transform common conceptions of time between instantaneity, synchronization and mediation. In this respect, the telegraph transformed local times by embedding both the phantasms of instantaneity and the delay of transmission. The use of local times became inevitably obsolete, because times had to be synchronized for the railway. The railway demanded a temporal uniformity, finally leading to the estab-

46 Morse, His Letters and Journals, p. 187; see also Siegert, *Von der Unmöglichkeit, Mediengeschichte zu schreiben.*

47 Anonymous, Time and the Electric Telegraph, 1845a, p. 416. There are several instances for this quote: Anonymous, Facts connected with the Telegraph; and Anonymous, Time and the Electric Telegraph, 1845b. This newspaper article spread from England to the US and to New Zealand in a month by use of the telegraph. At that time it was common to publish articles in different newspapers. Thanks to Roland Wenzelhuemer/Heidelberg for this note.

lishment of worldwide standard time by Sandford Fleming.[48] The nineteenth century is famous for its technologies of synchronization, which allowed to keep technical media, different time schemes and cultural practices in balance to achieve simultaneity. As James Beniger has shown in detail in *The Control Revolution*, the impacts of these developments shaped global interactions.[49] Different forms of synchronization helped to deal with the complexity of working processes: the timetables of the railway, the production of energy in steam engines, photography or telegraphy. From Greek "synchronos," (together and time), synchronization means to produce simultaneity and to adjust differences. There should be no place for instantaneity in this technical world. Another far-reaching and exemplary case comes from astronomy. In 1854, the observatories in Brussels and Greenwich were connected telegraphically for astronomical research.[50] This research aimed at accuracy to the nearest second in order to determine the degrees of longitude of the two cities. More than 3000 signals were interchanged to synchronize clocks. A master clock was used to coordinate several slave clocks. The transmission took approximately 0.1 second to cover 270 miles. This delay was factored in by simply letting the slave clocks run ahead by 0.1 second. For an exact measurement, the signal would have had to be at two places at the same time. Consequently, a metaphysical immediacy was part of this synchronization: "The immediate result of such a connexion [sic] of different Observatories is the power of bringing into combination the astronomical observations made at different Observatories as if they were made at one."[51] The time of observation was subtracted from the time of transmission to achieve synchronicity. Synchronization was needed where simultaneity was impossible but essential. As Peter Galison has shown, a similar epistemological problem led Albert Einstein to the formulation of the theory of relativity, which finally excluded all instantaneity from the fields of physics.[52]

After several days of successful telegraphing through Morse's first line, cheques were validated and rumors of deceased relatives in the neighboring city were clarified. News was sent from New York and appraised in Baltimore with much curiosity. But still, the Senate refused to buy Morse's patent for the required 100,000 US dollars. Instead, Morse redeemed the line from the

48 Blaise, *Time Lord.*

49 Beniger, *The Control Revolution.*

50 Schmidgen, *Telegraphie, Zeit und Lärm.*

51 Times, The Year-Book of Facts, p. 206.

52 Galison, *Einstein's Clocks, Poincaré's Maps.*

state. He made a better deal: in 1854, there were 36,972 miles of telegraph lines across America and Morse was the head of a cartel.

After some years, the newly founded Telegraph Company recognized that the telegraph operators could read and write in code as code. They could even think in code. The Morse Code could be used like any alphabet if the operator had the necessary feeling for tact. Time had become money for the new medium. The circuit could be closed without any intermediate step, while the receiving apparatus could remain automatic because the telegraph operator was not able to write faster than the arrival of signals. Experienced receiving operators left this work to their ears and translated the clicking sounds of the apparatus into plain language without looking at the dots and dashes. As late as 1955, when the demise of the telegraph as a communication tool was obvious, its signal was still command and connected the operator's nervous system with the wire: "What the ether or the wire commands, is an irrevocable command for every Morsist."[53] Every second counted, and if the transmission was instantaneous or nearly instantaneous, decoding should not take too long.

Under the influence of Wheatstone's attempts to measure the velocity of electricity, which quickly became known in England and North America, Morse went a step further and stated: "It was perceived by the philosophers of a past age that Electricity was so rapid in its passage through a conductor that, in the attempt to measure the speed through many miles, it was pronounced to be instantaneous." This assumption of instantaneity, Morse continued, had a practical plausibility and was the basis of his own inventions. But even if, as Wheatstone had shown, electricity was not instantaneous, this fact would not interfere with the foundations of telegraphy: "It is, nevertheless, evident that while the discovery may be of practical use for other purposes, it does not affect nor modify the invention of a telegraph based on the formerly received hypothesis of the instantaneity of electrical transmission."[54] Instantaneity means that there is no delay and consequently no interval between two events. Wheatstone's experiments supposedly determined this interval. But still, the velocity of electricity seemed irrelevant to Morse because the desired speeds for a successful business were easily achieved. Even if an electrical signal travelled at the assumed speed of 288,000 miles per second, this was ten times longer than any practical cable on the planet.

53 "Was der Äther oder der Draht befiehlt, ist ein unumstößliches Gebot für jeden Morsisten." Sonnleithner, *Morsen*, p. 12.

54 Morse, *Modern Telegraphy*, p. 37.

Thus, even if the telegraph was not instantaneous, as Morse had initially presupposed, it would be wrong to turn this deficit into an accusation. At that time, no one knew that the speed of an electric signal differs in relation to the length and material of the cable, as Michael Faraday and William Thomson were to demonstrate a decade later.[55] Morse's attempt to justify the instantaneity of telegraphy while at the same time insisting on its velocity seemed plausible from a pragmatic point of view. At such speeds, delay is not crucial (at least not for his intentions of telegraphing; it is of utmost importance for all technical media ever since). But this was not evident. On the contrary, this evidence was produced by grounding the telegraph in the phantasmatic dimension of telegraphy and scientific research.

The subsequent developments are well documented and have been described from multiple perspective. The prevalence of telegraphy began in the 1850s. At that time, 43 North American cities were connected. In 1846, there were only Morse's 40 miles between Washington and Baltimore. Two years later, there were 2000 and four years later 12,000 miles. From 1850 on, many telegraph companies all around the world were founded and connected the lines to networks. In the 1860s, the first long-distance lines from London to Calcutta (in operation until the 1930s) or from San Francisco to New York, and a direct sea route from London to New York were established.[56] Companies such as the Western Union Telegraph Company, Eastern and Associated Telegraph Companies and the American Union Telegraph Company rose as communication empires, without which no traditional empire could exist.[57]

With the transmission of information made possible by the success not only of Morse's telegraph, the desire for immediacy found new articulations as a desire for the latest business news or gossip. News agencies like Reuters, Havas or Wolffs began to fill the newspapers, which started to use "Telegraph" in their name.[58] The proximity induced by the telegraph required a redefinition of diplomacy, war and economy, a reorganization of news and journalism, love and the world. During the nineteenth century, telegraphy and other media became elements of economic planning and objects of investment.[59] However: "It tells us almost on the instant what is said in Paris,

55 Hunt, Michael Faraday, Cable Telegraphy and the Rise of Field Theory.

56 Aschoff, *Geschichte der Nachrichtentechnik.*

57 Boyce, *Submarine Cables as a Factor in Britain's Ascendancy as a World Power.*

58 See Amelia Bonea's article in this book.

59 Winseck/Pike, *Communication and Empire.*

Berlin, Vienna, or Trieste, but the story is not always worth telling."[60] All this was related to the discussion of the new status of near and far which was related to the telegraph. In a concluding step, I wish to highlight some of the consequences of this for the self-descriptions and negotiations of identity in the emerging global networks.

Vital Cords of Each with All

During the nineteenth century, the economic and technological arrangements of information retrieval changed rapidly.[61] The need to control no longer referred to the accessible environment or to face-to-face arrangements between factory directors and merchants, but to relations of spaces that were distant and therefore required time-critical media. Control, understood as the ability to exercise power at other places, had to be fast and increasingly depended on the speed of the railroad or the express boat. At the same time, factories began to ship by rail and to produce on the fly. Raw material was processed immediately after delivery, because storage became expensive and transportation available. To ensure the time-sensitive distribution of these resources, communication was needed that was faster than physical transport: telegraphy. The "crisis of control" that characterized the middle of the century was soothed and sometimes also intensified by telegraphic transmission. It established the means to exercise power in other places at the same time. Telegraph and railways as well as modern bridges, channels and cable networks allowed a management of contingency. But these achievements were reinforced by instantaneity. According to the London *Times* of August 6th, 1858:

> We may, now that this the most difficult problem of all has been solved, be justified in anticipating that there is no portion of the earth's surface which may not be placed in immediate communication with us. We now know that we have in our hands the means of a practical ubiquity. Distance as a ground of uncertainty will be eliminated from the calculation of the statesman and the merchant.[62]

60 Anonymous, *The Electric Storyteller*, p. 2.

61 Beniger, *The Control Revolution*.

62 The London Times, August 6, 1858. Holtorf, *Das erste transatlantische Telegraphenkabel von 1858*, p. 24.

Instantaneity was a common antidote to all forms of uncertainty. During the long nineteenth century, these reveries of electricity operated with a whole repertoire of metaphors and narrations which connected different sources of knowledge. They contributed to of the deparadoxization of the new relation from near and far, local and global. In this respect, they were of utmost importance for the cohesion of new global communities. "Community" is used here in the sense of recent philosophical debates mainly from France,[63] in which community is described as always already imaginative. Community always is a product of imaginations about the shape and connections inside of it. Only as such does it become a political actor. Thus, its self-descriptions play a crucial role for the definition of their agency. With the rise of the telegraph, a new global structure of space emerged in which the categorical status of near and far had to be negotiated in different ways. A new idea of community emerged which relied on the media of its connection. The cable did not connect one with absentees, but distributed a precarious form of presence on its network.

The "intimate connection between nations, with race and race"[64] which had already been expected in 1858 with the advent of transatlantic telegraphy, was supposed to expose the outdated nationalistic politics of exclusion as "stagnation and death,"[65] as Charles Briggs and August Maverick said in their influential and exemplary book *The Story of the Telegraph* in 1858. In the middle of the nineteenth century, the cable was described as a "vital cord" of "free and unobstructed interchange of each with all."[66] I will highlight two dimensions of these discourses.

First, the connection was often described as animate. As Laura Otis has shown, comparisons of telegraphic cables and nerves were common and of importance for both engineering and physiology.[67] For example, Carl August von Steinheil, a pioneer of telegraphy in Germany, called it "the most powerful bond of living creation."[68] It was typical for such comparisons to expand the metaphoric realm: cables were not only like nerves, they were nerves, as Ernst Kapp underlined in his *Grundlinien einer Philosophie der Technik* in

63 See for example Nancy, *Being Singular Plural.*

64 Briggs/Maverick, *The Story of the Telegraph*, p. 21.

65 Ibid.

66 Ibid.

67 Otis, *Networking.*

68 "Das gewaltsamste Band der lebendigen Schöpfung [...]Steinheil, *Ueber Telegraphie,* p. 3.

1877.[69] Thomas Soemmering described his telegraph in the same manner. If we take these metaphors or their expansion seriously, connection is alive when it is immediate – as a "bond of living creation" – and dead when it embodies division and difference. Instantaneous telegraphy, Steinheil and Kapp implied, provided a presence without interruption or difference, because electricity was supposed to be everywhere at the same time. This presence came alive as the opposite of absence and death. In this sense, difference was emptiness, vacuum, forlornness and loneliness. The separate elements, the reveries suggested, came to life only as a unity. The cable brought its related elements into a proximity of living touch, which has been connected to life traditionally.[70] The presence created by immediate connections is a presence of touch and of life while a mediate connection including difference can only be dead from the start.

Symptomatic for the imaginative economy was, secondly, the connection of everyone with everyone else, as Briggs and Maverick suggested when they wrote that the telegraph provided a "free and unobstructed interchange of each with all."[71] This kind of connection is specific for the new communities of electricity, which described themselves as constituted by instantaneous communications and media over all distances. Connection meant not only that each individual was connected with other individuals, but with all at the same time. The connection existed not only between isolated people, but as a network of everybody with all. The telegraph, said Steinheil in almost the same words, "reproduces in each what is given to all."[72] If everyone was connected with all others and all with each through a "vital cord," then there can be no other and no outside.[73]

This idea is an idea of unity occupied by metaphysical and religious thought. Difference cannot be a part of this unity. And a unity without difference may be excluding, even in political ways. It leads to an essentialistic and politically explosive conception of community, which is still in use today when, for example, Manuel Castells writes about the "annihilation of space and time by electronic means."[74] His idea of a network society is the con-

---

69 Kapp, *Grundlinien einer Philosophie der Technik*; Soemmering, Ueber Sömmerings Electrischen Telegraphen.

70 Derrida: *On Touching*.

71 Briggs/Maverick, *The Story of the Telegraph*, p. 21.

72 Steinheil, *Ueber Telegraphie*, p. 3.

73 Ibid.

74 Castells, *The Information Age*, p. 379.

ceptual result of an extensive analysis of social practices of synchronisation. Everyone in Castells' "network society" is connected simultaneously by the power of instantaneous electricity. The notion that this society exists "without reference to either past or future" by electronic means is analogous to McLuhan's "electric nowness."[75] Both have their genealogical origins in the instantaneity of telegraphy.

The vitality of cable-connections and the "each with all" were accompanied by an ignorance of the related mechanisms of exclusion. Electric presence had no difference. The white spots on the map, the places which were not part of the network, or the villages without radio or television, were excluded. Such a space necessarily has a blind spot: the outside that is not part of the figure. But this outside is needed to determine what is inside. The cohesion of the community of instantaneous electricity, as it has been commonly described at least since Morse's telegraph, differs from other communities because the essence of this community and its similarity to other such communities derives from the media of their connection.

As a result of these new modes of community, the world became a project and the projection of bodies, organs and senses, similar to a figure resembling Shakespeare's Puck aka Robin Goodfellow.[76] The frontispiece of Alexander Jones' *Historical Sketch of the Electric Telegraph*, which appeared more than ten years before the laying of the transatlantic cable in New York, showed the imagination of what would occur some years later.[77] The actual success only fulfilled what had been imagined. This figure holds the two ends of a world-connecting cable in her hands and pulls them on one end. Instantly, the other end moves. The figure is receiver and transmitter at the same time. Significantly, this cable is not routed across the poles, but across the equator. The cable not only connected the individual spaces with each other but formed a connection that ended where it began, as the phrase quoted from Shakespeare's *Midsummer Night's Dream* notes: "I'll put a girdle round about the earth in forty minutes." Electricity, these reveries say, conducts its power instantly around the world like a vital cord to each with all.

The presence of electricity was also used against telegraphy: perhaps electricity only produces apparitions of presence and threatens the social community. Telegraphy may violate the laws of nature. The tamed flash could

75 Ibid., p. 386; McLuhan/Nevitt, *The Argument*, p. 2.

76 For the world as a project see Krajewski, Restlosigkeit.

77 Jones, *Historical Sketch of the Electric Telegraph*.

Illustration 3: Jones, Alexander. Historical Sketch of the Electric Telegraph. *New York, 1852. Frontispiece.*

become wild again, as Presbyterian Reverend Cortlandt van Rensselaer preached in 1858:

> It is not, indeed, to be disguised that the telegraph may also be employed for purposes of evil. If Satan transformed himself into an angel of light, it is no marvel if he still uses the agency of light in strengthening his influence and dominion. But, for the purposes of the wicked, light is the most hazardous and self-destructive of all weapons. The devil, in his attempts to quote Scripture, was overwhelmed by the replies of the Son of Man. All assaults upon the cause of truth and liberty through the telegraph, will be repelled by the avenging power of right, in the Providence of the Most High.[78]

These sentences were part of a sermon to celebrate the first transatlantic cable at September 1st, 1858 in Burlington/New Jersey. The event was accompanied by an hour of ringing the church bells. To praise the new global connection, the event featured not only *The Star-Spangled Banner* and *God Save the Queen*, but also the *Marseillaise*. At the end of this festivity, an "Introductory,

78 Van Rensselaer, *Signals from the Atlantic Cable*, p. 19.

written by request, by a young Lady," was read. Electricity, it implies, is necessarily, contrary to better knowledge and despite all satanic fears, "Immediate!" because otherwise it could not maintain its promises:

Time is, yet is not! Distance stands,
Thought, flying on electric chain, across
The deep, dark Ocean, now can find response
Immediate! God, in mercy, deigns,
To give to man another wondrous power.[79]

The electric victory over the ocean accomplished a few weeks earlier was supposed to become stasis in instantaneity: a presence that has waited at the end of acceleration and will eat up all distances, space and time in an "electric allatonceness"[80] or an "electric union"[81] of a *global village*. To follow such notions of an "annihilation of space and time by electronic means" in understanding media means to risk losing sight of the phantasmatic dimension of technical media, their actual *modi operandi* and their micro-temporal politics.

Bibliography

Anonymous, "Time and the Electric Telegraph," *Journal of the Franklin Institute*, vol. 3, no. 10 (1845).

Anonymous, "Time and the Electric Telegraph," *The Mechanic's Magazine*, vol. 42, no. 1 (1845), p. 416.

Anonymous, "Facts connected with the Telegraph," *New Zealand Spectator and Cook's Strait Guardian*, 26th August (1848).

Anonymous, "The Electric Storyteller," *The Moreton Bay Courier*, October 14 (1854).

Aristotle, *Physics*, Oxford 2008.

Aschoff, Volker, *Geschichte der Nachrichtentechnik*, Berlin 1995.

Asendorf, Christoph, *Batteries of Life. On the History of Things and their Perception in Modernity*, Berkeley, CA 1993.

Bachelard, Gaston, *On Poetic Imagination and Reverie*, Dallas, TX 1987.

Beniger, James R., *The Control Revolution: Technological and Economic Origins of the Information Society*, Cambridge, MA 1986.

79 Ibid., p. 3.

80 McLuhan, *The Gutenberg Galaxy*, p. 63.

81 Van Rensselaer, *Signals from the Atlantic Cable*, p. 5.

Blaise, Clark, *Time Lord. Sir Sandford Fleming and the Creation of Standard Time*, New York 2000.

Bloor, David, *Knowledge and the Social Imaginary*, London 1976.

Boyce, Robert, "Submarine Cables as a Factor in Britain's Ascendancy as a World Power, 1850–1914," in: Michael North (ed.), *Kommunikationsrevolutionen. Die neuen Medien des 16. und 19. Jahrhunderts*, Wien 2001, pp. 81–100.

Briggs, Charles/Maverick, August, *The Story of the Telegraph*, New York 1858.

Caneva, Kenneth L., "Ampère, the Etherians, and the Oersted Connection," *British Journal for the History of Science*, vol. 23 (1980), pp. 121–138.

Carey, James W., "Technology and Ideology: The Case of the Telegraph," in James W. Carey, *Communication as Culture: Essays on Media and Society*, New York 2009, pp. 155–177.

Castells, Manuel, *The Information Age: Economy, Society, and Culture: End of Millennium*, Malden, MA 1998.

Chang, Briankle, *Deconstructing Communication*, Minneapolis, MN 1996.

Czitrom, Daniel J., *Media and the American Mind: From Morse to McLuhan*, Chapel Hill, NC 1982.

Derrida, Jacques, "Structure, Sign, and Play in the Discourse of the Human Sciences," in: Jacques

— (ed.), *Writing and Difference*, London 1967, pp. 278–294.

—, *On Touching. On Jean-Luc Nancy*, Stanford, CA 2005.

Dufay, Charles, *Versuche und Abhandlungen von der Electricität der Körper*, Erfurt 1745.

Galison, Peter, *Einstein's Clocks, Poincaré's Maps. Empires of Time*, New York 2003.

Gray, Stephen, "A Letter to Cromwell Mortimer, M. D. Secr. R. S. containing several Experiments concerning Electricity," *Philosophical Transactions of the Royal Society of London*, vol. 37 (1731), pp. 18–44.

Gumbrecht, Hans Ulrich/Pfeiffer, Karl L. (eds.), *Materialities of Communication*, Stanford, CA 1994.

Heilbron, John L., *Electricity in the 17th and 18th Centuries: A Study in Early Modern Physics*, Berkeley, CA 1979.

Henry, Joseph, "On the Application of the Galvanic Multiplier to Electro-Magnetic Apparatus," *American Journal of Science*, vol. 19 (1831), pp. 400–408.

Hesse, Mary B., *Forces and Fields: The Concept of Action at a Distance in the History of Physics*, London 1961.

Hirt, Kilian/Volmar, Axel, "Kanalarbeit: Das Übertragungsproblem in der Geschichte der Kommunikationstechnik und die Entstehung der Signalverarbeitung," in: Axel Volmar, *Zeitkritische Medien*, Berlin 2009, pp. 213–238.

Holtorf, Christian, *Das Erste Transatlantische Telegraphenkabel von 1858 und seine Auswirkungen auf die Vorstellungen von Raum und Zeit*, Dissertation, Berlin 2009.

Hunt, Bruce, "Michael Faraday, Cable Telegraphy and the Rise of Field Theory." in: *History of Technology*, vol. 13, pp. 1–19.

Jones, Alexander, *Historical Sketch of the Electric Telegraph: Including Its Rise and Progress in the United States*, New York 1852.

Kapp, Ernst, *Grundlinien einer Philosophie der Technik*, Braunschweig 1877.

Kassung, Christian/Macho, Thomas (eds.), *Kulturtechniken der Synchronisation*, München 2012.

Kittler, Friedrich, "Real Time Analysis Time Axis Manipulation" in Georg Christoph Tholen (ed.), *Zeit-Zeichen: Aufschübe und Interferenzen zwischen Endzeit und Echtzeit*, Weinheim 1990, pp. 362–377.

Kittler, Friedrich A., *Discourse Networks, 1800–1900*, Stanford, CA 2005.

Krajewski, Markus, *Restlosigkeit. Weltprojekte um 1900*, Frankfurt a. M. 2006.

Le Monnier, Louis Guillaume, "Recherches sur la communication de l'électricité," *Histoire de l'Académie Royale des Sciences*, vol. 1 (1746), pp. 447–464.

Lenoir, Timothy, "Helmholtz and the Materialities of Communication," *Osiris*, vol. 9, no. 1 (1994), pp. 185–207.

Marvin, Carolyn, *When Old Technologies Were New. Thinking About Communications in the Late Nineteenth Century*, Oxford 1988.

McLuhan, Marshall, *The Gutenberg Galaxy*, Toronto 1962.

McLuhan, Marshall, *Understanding Media: The Extensions of Man*, New York 1964.

McLuhan, Marshall/Nevitt, Barrington, "The Argument: Causality in the Electric World," *Technology and Culture*, vol. 14, no. 1 (1973), pp. 1–18.

Morse, Samuel, *Modern Telegraphy: Some Errors of Dates and Events and of Statement in the History of Telegraphy*, Paris 1867.

—, *His Letters and Journals*, Charleston, SC 2007.

Morus, Iwan R., "Telegraphy and the Technology of Display: The Electricians and Samuel Morse," *History of Technology*, vol. 13, no. 1 (1991), pp. 20–40.

—, "'The Nervous System of Britain.' Space, Time and the Electric Telegraph in the Victorian Age," *British Journal for the History of Science*, vol. 33, no. 4 (2000), pp. 455–475.

Müller-Pohl, Simone, "'By Atlantic Telegraph.' A Study on Weltcommunication in the 19th Century," *medien & zeit*, vol. 25, no. 4 (2010), pp. 40–55.

Nancy, Jean-Luc, *Being Singular Plural*, Stanford, CA 2000.

Newton, Isaac, *Mathematical Principles of Natural Philosophy*, edited by Florian Cajori, Berkeley, CA 1687/1934.

Oersted, Hans C., "On Electro-magnetism," *Annals of Philosophy*, November (1821), pp. 321–337.

Otis, Laura, *Networking. Communicating with Bodies and Machines in the Nineteenth Century*, Ann Arbor, MN 2001.

Peters, John Durham, *Speaking Into the Air: A History of the Idea of Communication*, Chicago, ILL 2000.

—, "Technology and Ideology: The Case of the Telegraph revisited," in:

Packer, Jeremy/Robertson, Craig (eds.), *Thinking with James Carey: Essays on Communications, Transportation, History*, New York 2006, pp. 137–56.

Pichler, Franz, "Die Einführung der Morse-Telegraphie in Deutschland und Österreich," *Elektrotechnik & Informationstechnik*, vol. 123, no. 9 (2006), pp. 402–408.

Rieger, Stefan, *Schall und Rauch: Eine Mediengeschichte der Kurve*, Frankfurt a.M. 2009.

Rohrhuber, Julian, "Das Rechtzeitige: Doppelte Extension und Formales Experiment," in: Axel Volmar (ed.), *Zeitkritische Medien*, Berlin 2009, pp. 195–212.

Sconce, Jeffrey, *Haunted Media: Electronic Presence from Telegraphy to Television*, Durham, NC 2005.

Shaffner, Tal P. (ed.), *Shaffner's Telegraphy Companion*, New York 1854.

Siegert, Bernhard, "Von der Unmöglichkeit, Mediengeschichte zu schreiben," in: Philipp von Hilgers/Ana Ofak (eds.), *Rekursionen*, München 2009, pp. 157–175.

Silverman, Kenneth, *Lightning Man: The Accursed Life of Samuel F.B. Morse*, New York 2003.

Soemmering, Thomas Samuel, "Ueber Sömmering's Electrischen Telegraphen," *Journal für Chemie und Physik*, vol. 39, no. 2 (1811), pp. 217–230.

Sonnleithner, Karl, *Morsen: Morseschulung, Morsepraxis, Morsewirklichkeit*, Wien 1955.

Sprenger, Florian, *Medien des Immediaten. Elektrizität, Telegraphie, McLuhan*, Berlin 2012.

Steinheil, Carl A. von, *Ueber Telegraphie, Insbesonders Durch Galvanische Kräfte*, München 1838.

Times, John, *The Year-Book of Facts*, London 1854.

van Rensselaer, Cortlandt, *Signals from the Atlantic Cable: An Address Delivered at the Telegraphic Celebration*, Philadelphia, PA 1858.

Virilio, Paul, *Speed and Politics*, Los Angeles, CA 2006.

Watson, William, "A Continuation of a Paper concerning Electricity," *Philosophical Transactions of the Royal Society of London*, vol. 44 (1746), pp. 695–749.

Wheatstone, Charles, "An Account of Some Experiments to Measure the Velocity of Electricity and the Duration of Electric Light," *Philosophical Transactions of the Royal Society of London*, vol. 124 (1834), pp. 583–591.

Winseck, Dwayne/Pike, Robert M., *Communication and Empire. Media, Markets and Globalization, 1860–1930*, Durham, NC 2007.

Contributors

M. Michaela Hampf is Professor of North American History at the John F. Kennedy Institute at Freie Universität Berlin. She received her Ph.D. from the University of Bern, Switzerland in 2006. After having taught at the Universities of Cologne and Bonn she directed a DFG-funded research project on transatlantic telegraphy, cultural networks and epistemic communities in maritime space. Most recently she has been a Visiting Scholar at the University of California at Berkeley. Her research focuses on cultural and social history of the United States, transnational media history, cultural history of violence and the military as weil as the history of the body. She is the author and co-editor of two volumes on media history (Freies Radio in den USA: Die Pacifica Foundation 1946–1965, 2000 and Radio Welten: Politische, sozial und kulturelle Aspekte atlantischer Mediengeschichte, 2006).

Simone Müller-Pohl is Assistant Professor for North American History at the University of Freiburg. She received her Ph.D. from Freie Universität Berlin with a study on the social and cultural constructions of global telegraph networks during the first age of globalization. She has been a research fellow at the German Historical Institute as well as the Smithsonian Institutions in Washington D.C. and has published widely on the actor networks driving global communication. In 2013 she co-organized a workshop on the Intellectual Foundations of Global Commerce and Communications at Harvard. Her research focuses on global history, maritime history, the history of science and technology, media studies and environmental history.

Dwayne Winseck is Professor at the School of Journalism and Communication, Carleton University, Ottawa, Canada. His research on the political economies of communication, media history, new media, surveillance and national security, and theories of democracy and globalization has been published widely in leading scholarly journals and his latest book (co-authored with Robert M. Pike), *Communication and Empire: Media, Markets and Glo-*

balization, 1860–1930 (Duke University, 2007), won the Canadian Communication Association's G.G. Robinson Award for book-of-the-year in 2008.

Léonard Laborie has been a Research Fellow at the Centre National de la Recherche Scientifique (CNRS), Paris, since 2010. He holds a Ph.D. in contemporary history from Paris-Sorbonne University with a study on international cooperation in the field of post and telecommunication from the 1850s to the 1950s and its articulation with the idea and project of a united Europe (Peter Lang, 2010). Laborie serves as scientific secretary of the Committee for the History of Electricity and is editor of the *Annales historiques de l'électricité.* Recent publications include an essay from his teaching on the history of environmental communication, an edited volume on the landscapes of electricity, and an article on the Cold War governance of telecommunication.

Martin Doll is Research Associate at the University of Luxembourg within the ATTRACT research project "Aesthetical Figurations of the Political". He has received his Ph.D. from the University of Gießen with a study on the critical effects of forgeries and hoaxes. From 2008 to 2010 he was Fellow at Institute for Cultural Inquiry in Berlin. He has published widely on the discursive construction of fake and deceit as well as, most recently, on the interrelation of media systems and Arab Spring.

Daqing Yang is an Associate Professor of History and International Affairs at George Washington University, where he teaches modern Japanese and East Asian history. A native of Nanjing, China, he received his Ph.D. from Harvard University with a study on the interrelation of telecommunications networks and prewar Japanese expansion (*Technology of Empire*, Harvard University Press 2012). He has also co-edited a book on the history of submarine telegraphy (*Communications under the Sea*, MIT Press 2009). Yang has been a Visiting Associate Professor at Harvard University as well as the University of Tokyo in Japan. His interests include Japan's colonial and diplomatic history, historical reconciliation, as well as the history of technology.

Lars Bluma is a Senior Research Associate at the German Mining Museum Bochum while concurrently serving as an Assistant professor of History at Ruhr University Bochum. He received his Ph.D. in 2004 from Ruhr University Bochum with a study on cybernetics. Bluma served as member and chair of ICOHTEC's program committee from 2007 to 2009 and is also Chairman of the Committee for the History of Technology at the Association of German Engineers (VDI). His fields of interest include mining history, pictorial turn, cybernetics and information technologies, history of industrialization and workplace studies. He has published widely in these fields.

Wendy Gagen completed her Ph.D. at the University of Essex, funded by the Economic and Social Research Council. Focusing on the relationship between gender and corporeality, her thesis considered masculinity and physical disability during the First World War in Britain. Most recently she is part of the Arts and Humanity Research Council funded project, 'Connecting Cornwall', at the University of Exeter (teamed with Porthcurno Telegraph Museum) that aimed to uncover the hidden histories surrounding the rise of submarine telegraphy. More specifically, she looked at body movement, masculinity and technology. As a historian, her teaching and research has also focused on the relationship between medicine, disability, war, gender, the body, and mental health in the nineteenth and twentieth centuries.

Michael Mann is Professor for History and Society of South Asia at the Institute for Asian and African Studies, Humboldt-Universität zu Berlin. He completed his Ph.D. at the South Asia Institute, Heidelberg University and worked as Assistant Professor at the Fern Universität in Hagen. He was awarded two individual research grants by the Deutsche Forschungsgemeinschaft (German Research Council). He works and publishes widely on the history of South Asia, (Geschichte Indiens, Schöningh, 2005), the complex of colonialism, slavery and Europe's civilizing mission (Sahibs, Sklaven und Soldaten, Wissenschaftliche Buchgesellschaft, 2012; together with Carey A. Watt: Civilzing Mission in Colonial and Postcolonial South Asia, Anthem Press, 2011; together with Harald Fischer-Tiné: Colonialism as Civilizing Mission, Anthem Press, 2004) and media history of British-India.

Amelia Bonea received her Ph.D. in Modern History from the University of Heidelberg and is currently a postdoctoral researcher at the Wellcome Unit for the History of Medicine, University of Oxford. Her main research interests are media history, with a focus on journalism and the development of global communications, and the history of science, technology and medicine, with a regional focus on South Asia and the British Empire. She has published on war reporting and telegraphy in colonial India and on missionary journalism and indentured labor in colonial Fiji. During her doctoral studies at the University of Heidelberg, Amelia acted as co-organizer of two international workshops: "Telecommunication and Globalization: Information Flows in the Nineteenth and Early Twentieth Century" (2009) and "Pre-colonial and Colonial Perspectives on Science and Technology in South Asia" (2010).

Volker Barth is an Assistant Professor of modern history at the University of Cologne. In 2004 he received a co–tutelle Ph.D. from the Ludwig–Maximilians–University (LMU), Munich, and the École des Hautes Études en Sciences Sociales (EHESS), Paris with a study on the world exhibition in Paris 1867. Further teaching and research positions include the Université Paris VII–Denis Diderot, the German Academic Exchange Service (DAAD) and the Bureau International des Expositions (BIE), Paris. His current research project focuses on the cooperation of globally operating news agencies between the mid–nineteenth century and the interwar period.

Gordon Winder is a geographer with interests in the historical experiences of industrialization, urbanization, environmental transformation, globalization, and imperialism. Born in New Zealand, he gained his Ph.D. at the University of Toronto, and worked at the University of Auckland before he joined LMU as a visiting research fellow from 2008 to 2010. In 2010 Dr. Winder was named an honorary research fellow at the University of Auckland's School of Environment, and in 2011 he became a LMU Fellow at the Carson Center. Winder is a member of the editorial board of *The Journal of Historical Geography*. His recent research is on historical geographies of the news and news agencies. The central focus of this research is on the ways narratives of distant disaster have changed in North American newspapers over the course of the twentieth century.

Jonas Harvard is an Associate Professor at Mid Sweden University and Adjunct professor in the History of Political Discourse and Communication at Jyväskylä University. His research focuses on the history of public opinion formation and communication technologies. He is currently finishing the four-year research project "Distant News and Local Opinion: How the Telegraph Affected Spatial and Temporal Horizons in Northern Scandinavia, 1850–1880" and has been a visiting scholar at the Department of the History of Science and Technology at the Royal School of Technology in 2009 and at the Department of History and Ethnology at University of Jyväskylä. Between 2010 and 2012 he coordinated the international research program Nordic Spaces at Södertörn University and was the series editor of the book series *The Nordic Experience* (Ashgate 2012–2013). He has also published widely on media history (with Peter Stadius, 'Communicating *the North, Media structures and Images in the Making of the Nordic Region*' Asghate 2013; *'1800-talets mediesystem' [The Media System of the 19th Century]*, with Patrik Lundell, Royal library, 2010).

Torsten Kathke is Postdoctoral Fellow for North American History at the German Historical Institute in Washington D.C. He studied American Cultural History, American Literary History, and Drama at Ludwig-Maximilians-University in Munich, graduating with an M.A. in 2006. From 2007 to 2012, he was a Research Assistant and Lecturer at the LMU's Amerika-Institut, primarily teaching classes on 19th century American history and the American West. He finished his Ph.D. in American Cultural History 2013 with a dissertation on competing regional identities and the construction of a democratic nation through communication in the American Southwest during the late 19th and early 20th centuries. His publications include essays on US environmental issues, as well as on questions of democracy and media. In his current position he is pursuing a book project on the crises of the 1970s represented in popular non-fiction in Germany and the United States.

Florian Sprenger is Postdoctoral Researcher at the Graduate Centre for the Study of Culture at the Justus-Liebig-University of Gießen. Before, he was Junior Fellow at the International Research Centre for Cultural Studies at Vienna and Visiting Scholar at Stanford University. His research interests cover the history of media and of media theory, history of the sciences of media, mediality of monologues, media of media, histories of the future, flashlight and photography. His book *Medien des Immediaten - Elektrizität, Telegraphie, McLuhan* was published in 2012. Currently, he is working on a project on the infrastructures of domestic electricity around 1900 and their architectural impacts.